Studies in Logic
Volume 51

Metalogical Contributions to the Nonmonotonic Theory of Abstract Argumentation

Volume 40
The Lambda Calculus. Its Syntax and Semantics
Henk P. Barendregt

Volume 41
Symbolic Logic from Leibniz to Husserl
Abel Lassalle Casanave, ed.

Volume 42
Meta-argumentation. An Approach to Logic and Argumentation Theory
Maurice A. Finocchiaro

Volume 43
Logic, Truth and Inquiry
Mark Weinstein

Volume 44
Meta-logical Investigations in Argumentation Networks
Dov M. Gabbay

Volume 45
Errors of Reasoning. Naturalizing the Logic of Inference
John Woods

Volume 46
Questions, Inferences, and Scenarios
Andrzej Wiśniewski

Volume 47
Logic Across the University: Foundations and Applications. Proceedings of the Tsinghua Logic Conference, Beijing, 2013
Johan van Benthem and Fenrong Liu, eds.

Volume 48
Trends in Belief Revision and Argumentation Dynamics
Eduardo L. Fermé, Dov M. Gabbay, and Guillermo R. Simari

Volume 49
Introduction to Propositional Satisfiability
Victor Marek

Volume 50
Intuitionistic Set Theory
John L. Bell

Volume 51
Metalogical Contributions to the Nonmonotonic Theory of Abstract Argumentation
Ringo Baumann

Studies in Logic Series Editor
Dov Gabbay   dov.gabbay@kcl.ac.uk

Metalogical Contributions to the Nonmonotonic Theory of Abstract Argumentation

Ringo Baumann

© Individual author and College Publications 2014.
All rights reserved.

ISBN 978-1-84890-143-8

College Publications
Scientific Director: Dov Gabbay
Managing Director: Jane Spurr

http://www.collegepublications.co.uk

Original cover design by Orchid Creative   www.orchidcreative.co.uk
Printed by Lightning Source, Milton Keynes, UK

All rights reserved. No part of this publication may be reproduced, stored in a retrieval system or transmitted in any form, or by any means, electronic, mechanical, photocopying, recording or otherwise without prior permission, in writing, from the publisher.

Abstract

The study of nonmonotonic logics is one mayor field of Artificial Intelligence (AI). The reason why such kind of formalisms are so attractive to model *human reasoning* is that they allow to withdraw former conclusion. At the end of the 1980s the novel idea of using *argumentation* to model nonmonotonic reasoning emerged in AI. Nowadays argumentation theory is a vibrant research area in AI, covering aspects of knowledge representation, multi-agent systems, and also philosophical questions.

Phan Minh Dung's abstract argumentation frameworks (AFs) [Dung, 1995] play a dominant role in the field of argumentation. In AFs arguments and attacks between them are treated as primitives, i.e. the internal structure of arguments is not considered. The major focus is on resolving conflicts. To this end a variety of semantics have been defined, each of them specifying acceptable sets of arguments, so-called extensions, in a particular way. Although, Dung-style AFs are among the simplest argumentation systems one can think of, this approach is still powerful. It can be seen as a general theory capturing several nonmonotonic formalisms as well as a tool for solving well-known problems as the stable-marriage problem.

This book is mainly concerned with the investigation of *metalogical properties* of Dung's abstract theory. In particular, we provide cardinality, monotonicity and splitting results as well as characterization theorems for equivalence notions. The established results have theoretical and practical gains. On the one hand, they yield deeper theoretical insights into how this nonmonotonic theory works, and on the other the obtained results can be used to refine existing algorithms or even give rise to new computational procedures. A further main part is the study of problems regarding dynamic aspects of abstract argumentation. Most noteworthy we solve the so-called *enforcing* and the more general *minimal change problem* for a huge number of semantics.

Acknowledgements

This book is based on my PhD thesis, which I completed at Leipzig University in January 2014. I am grateful to College Publications for publishing this book and making it accessible to wider audience.

I would like to thank my supervisor Gerhard Brewka. In 2009 he gave me the opportunity to start in his team as a PhD student as well as research assistant although I was a *computer science rookie*. During the years I learned a lot from him in thematically as well as non-thematically appropriate talks. Especially, I want to thank him for giving me space and time to pursue my own ideas which is, unfortunately, no longer taken for granted during a PhD.

Thanks go also to my colleagues Frank Loebe and Hannes Straß. It is a great pleasure to work with you and I am looking forward to future collaborations as well as joyful and highly interesting discussions in front of our whiteboard.

I was lucky to meet and get in contact with many people from all over the world which have influenced my research interest and/or contributed to the successful completion of my thesis. In particular, I would like to mention Stefan Woltran and Heinrich Herre.

Finally, I do thank my wife, Frances, for being there and my daughter, Freya-Ruthlinda, for having arrived.

Contents

1	**Introduction**	**1**
	1.1 Argumentation in Artificial Intelligence	1
	1.1.1 The Overall Instantiation Process	2
	1.1.2 *Why* Metalogical Analysis?	3
	1.2 Structure of the Book and Main Contributions	5
	1.3 Publications	8
2	**Background of Abstract Argumentation**	**11**
	2.1 Abstract Argumentation Frameworks	11
	2.1.1 Semantics	13
	2.1.2 Abstract Principles	18
	2.1.3 Reasoning	20
	2.1.4 Dynamics and Expansions	21
	2.1.5 Equivalence	22
3	**Monotonic Aspects**	**27**
	3.1 Monotonicity Results	27
	3.2 Computation of Justification States	29
	3.3 Application: Reasoning about Action	31
	3.3.1 Background: Action Theories	31
	3.3.2 Specifying Action Domains	32
	3.3.3 Encoding	35
	3.3.4 Encoded Priorities	38
	3.3.5 Stratification and Computing Extensions	41
	3.4 Conclusions and Related Work	44
4	**Splitting Results**	**47**
	4.1 Classical Splitting Results	47
	4.1.1 Classical Splitting, Reduct, Undefined Set and Modification	48
	4.1.2 Splitting Theorem	51
	4.1.3 Splittings, Directionality and Monotonicity	56
	4.1.4 Dynamic Scenario: Reusing Extensions	58
	4.1.5 Static Scenario: Computing Extensions	59
	4.2 Generalized Splitting Results	62
	4.2.1 k-Splitting, Conditional Extension and Match	63

		4.2.2 Generalized Splitting Theorem	67

    4.2.2   Generalized Splitting Theorem 67
  4.3 Splittings for Logic Programs 68
      4.3.1   Background: Logic Programs 69
      4.3.2   k-Splitting, Conditional Answer Set and Match 70
      4.3.3   Generalized Splitting Theorem 72
  4.4 Conclusions and Related Work 73

5 Notions of Equivalence and Replacement   75

  5.1 Recapitulation: Expansion Equivalence 76
      5.1.1 σ-Kernel 77
      5.1.2   Characterization Theorems 77
  5.2 Characterizing Strong Expansion Equivalence 78
      5.2.1   Splitting Results: A Tool for Simplifying Proofs 79
      5.2.2   Strong Expansion Equivalence for Stable and Stage Semantics 81
      5.2.3   Strong Expansion Equivalence for Semi-Stable and Eager Semantics 82
      5.2.4   Strong Expansion Equivalence for Admissible, Preferred and Ideal Semantics 85
      5.2.5   Strong Expansion Equivalence for Grounded Semantics 92
      5.2.6   Strong Expansion Equivalence for Complete Semantics 98
      5.2.7   Strong Expansion Equivalence for Naive Semantics . . . 104
  5.3 Characterizing Normal Expansion Equivalence 105
      5.3.1   Normal Expansion Equivalence for Stable and Stage Semantics 106
      5.3.2   Normal Expansion Equivalence for Semi-Stable, Eager, Admissible, Preferred and Ideal Semantics 106
      5.3.3   Normal Expansion Equivalence for Grounded Semantics 107
      5.3.4   Normal Expansion Equivalence for Complete Semantics 109
      5.3.5   Normal Expansion Equivalence for Naive Semantics . . 109
  5.4 Characterizing Local and Weak Expansion Equivalence 110
      5.4.1   Weak Expansion Equivalence for Stable Semantics . . . 111
      5.4.2   Weak Expansion Equivalence for Preferred Semantics . 112
      5.4.3   Local Expansion Equivalence for Stage Semantics 113
  5.5 Summary of Results and Implications 118
      5.5.1   Overview: "Strength" of Kernels 118
      5.5.2   Relations Between Different Notions of Equivalence . . 120
      5.5.3   The Role of Self-loop-free AFs 123
  5.6 Conclusions and Related Work 124

6 Cardinality Results   127

  6.1 Background 128
  6.2 Analytical Results 130
      6.2.1   Maximal Number of Stable Extensions 130
      6.2.2   Average Number of Stable Extensions 133
  6.3 Empirical Results 136
      6.3.1   Experimental Setup 137

		6.3.2 Average Number of Stable Extensions 137

- 6.3.2 Average Number of Stable Extensions 137
- 6.3.3 Number of AFs with at most one Stable Extension . . . 137
- 6.4 Conclusions and Related Work 139

7 Enforcing and Minimal Change 143
- 7.1 Enforcing Problem 146
  - 7.1.1 Conservative and Liberal Enforcements 146
  - 7.1.2 Impossibility Results 147
  - 7.1.3 Possibility Result 149
- 7.2 Minimal Change Problem 151
  - 7.2.1 Characteristics - A Formal Representation 152
  - 7.2.2 Value Functions - Local Criteria 155
  - 7.2.3 Summary of Results 164
- 7.3 Spectrum Problem 165
  - 7.3.1 Formal Properties of Spectra 165
  - 7.3.2 The (stb,ss,pr,Φ)-Spectrum ($\Phi \in \{E,N,S\}$) 166
  - 7.3.3 Properties of the (stb,ss,pr,W)-Spectrum 170
  - 7.3.4 A Note on the (stb,ss,pr,U)-Spectrum 173
- 7.4 Minimal Change Equivalence 174
  - 7.4.1 Formal Definitions and General Results 174
  - 7.4.2 Stable and Preferred Semantics: Analyzing the Equivalence Zoo 178
- 7.5 Conclusions and Related Work 189

8 Conclusion 193
- 8.1 Summary 193
- 8.2 Open Problems and Future Research 196

List of Figures

1.1	The Entire Instantiation Process	4
2.1	Running Example $\mathcal{F}$	12
2.2	Evaluation Table of $\mathcal{F}$	15
2.3	Relations between Semantics	16
2.4	Evaluation Criteria: A Complete Picture	19
2.5	Complexity of $Cred_\sigma$ and $Scept_\sigma$	21
2.6	Notions of Expansions	22
2.7	Standard Equivalence	23
2.8	Non-replaceability	23
2.9	Relations between Equivalence Notions	25
3.1	The Revised AF $\mathcal{F}_{rev}$	28
3.2	Weak Expansion Chains	30
3.3	Computation of Justification States	31
4.1	Classical Splitting	48
4.2	The Reduct of $\mathcal{F}_2$ with respect to E_1 and R_3	49
4.3	Counterexample Reduct ($\sigma \in \{pr, co, gr\}$)	49
4.4	The Modification of $\mathcal{F}_2$ with respect to E_1 and R_3	50
4.5	Counterexample for $\sigma \in \{ss, stg, eg\}$	55
4.6	Counterexample for $\sigma = id$	56
4.7	Modification of $\mathcal{F}_2$ with respect to $\{a_1, a_2\}$ and R_3	56
4.8	Reusing Extensions - The AF $\mathcal{G}$	58
4.9	Reusing Extensions - Different Cases	59
4.10	SCCs and Splittings	60
4.11	A 4-splitting of $\mathcal{F}$	63
4.12	Conditional Extensions of $\mathcal{F}_1^S$	64
4.13	(E, S)-match of $\mathcal{F}$	67
5.1	Stable and Grounded Kernel of $\mathcal{F}$	77
5.2	Partially Known AFs	80
5.3	Strong Expansion and Classical Splitting	80
5.4	Uniquely Determined Reducts	80
5.5	Non-Expansion Equivalent AFs	85
5.6	A Possible Scenario	85

5.7	Counter-example for Semi-stable and Eager Semantics	92
5.8	Grounded-*- vs. Admissible-*-Kernel	94
5.9	Strong Expansion Equivalent AFs	98
5.10	Non-coincidence of Expansion and Strong Expansion Equivalence	98
5.11	Grounded-*- vs. Complete-*-Kernel	99
5.12	Non-coincidence of Expansion and Strong Expansion Equivalence	104
5.13	Unchanged Attack-relation	106
5.14	Non-coincidence of Stable and Stage Semantics	110
5.15	Local Expansion Equivalent AFs	110
5.16	Weak Expansion Equivalent AFs	112
5.17	A *Huge* Equivalence Class	117
5.18	The Whole Landscape of Characterizations	119
5.19	Relations for Stage Semantics	121
5.20	Relations for Stable Semantics	121
5.21	Relations for Semi-stable and Eager Semantics	121
5.22	Relations for Admissible, Preferred and Ideal Semantics	122
5.23	Relations for Grounded and Complete Semantics	122
5.24	Relations for Naive Semantics	122
6.1	Generating vs. Constraining	128
6.2	Different Operators on Graphs	129
6.3	The AF $\mathcal{F}_{3\cdot 3+1}$	133
6.4	Average Number of Stable Extension ($n = 20$)	138
6.5	Average Number of Stable Extension ($n = 50$)	139
6.6	Absolute Frequencies of 0-AFs and 1-AFs	141
7.1	Snapshot of a Dialogue	143
7.2	Minimal Change Equivalence	144
7.3	Liberal Enforcement	147
7.4	Conservative Enforcement	147
7.5	Strict- vs. Non-strict Enforcements	149
7.6	Standard Construction	151
7.7	Alternative Construction	152
7.8	Distances	153
7.9	Inbetween Zero and Infinity	156
7.10	Inadequateness of Stable and Admissible Value	160
7.11	Inadequateness of Stable and Admissible Value	164
7.12	State of the Art: Characterizing Characteristics	164
7.13	The AF $\mathcal{F}_{3,2,4}$	168
7.14	Non-coherence of the Weak Spectrum	170
7.15	The AF $\mathcal{F}_{\infty,3,0}$	172
7.16	Minimal Change Equivalence and Minimal-$\{a_1,a_2\}$-Equivalence	176
7.17	Necessity of I-maximality	178
7.18	Stable Semantics (Arbitrary AFs)	179
7.19	Stable Semantics (Same Arguments)	181

7.20 Stable Semantics (Self-loop-free AFs) 183
7.21 Preferred Semantics (Arbitrary AFs) 184
7.22 Preferred Semantics (Same Arguments) 187
7.23 Preferred Semantics (Self-loop-free AFs) 188

Chapter 1

Introduction

1.1 Argumentation in Artificial Intelligence

Classical logics, such as first-order logic, can be seen as approaches which are mainly concerned with a formalization of universal truths. In such a logic, whenever a formula ϕ is a logical consequence of a set of axioms Σ, then it remains true for all time and without exception even if we add new axioms to Σ. Without doubt such a kind of *monotonic reasoning* perfectly fits together with the purpose or self-image of mathematics but it is inadequate to model commonsense reasoning. In daily life we are often faced with incomplete knowledge. Nevertheless, we want or have to draw conclusions. In such a situation we typically assume that the world behaves as expected. Consequently, we conclude what is normally true as long as there is no evidence to the contrary. If we later learn that the drawn conclusion is not justified since the normality assumption is invalidated we have to withdraw it. The realization that classical logic is not suitable to model such defeasible reasoning was the main reason for the increasing interest in *nonmonotonic logics* within the Artificial Intelligence (AI) community in the late 1970s. Among the pioneers of the field were John McCarthy, Raymond Reiter, Drew V. McDermott and Jon Doyle (cf. [McCarthy, 1990; Reiter, 1980; McDermott and Doyle, 1980] or [Brewka, 1991] for excellent overview).

At the end of the 1980s the novel idea of using *argumentation* to model nonmonotonic reasoning emerged in AI (see [Loui, 1987; Lin and Shoham, 1989; Pollock, 1987] and [Prakken and Vreeswijk, 2002] for a comprehensive overview). The new way to model defeasible inference can be summarized as follows: building arguments based on the existence of proofs in a certain underlying logic, identifying conflicts between them and then determining acceptable sets of arguments which finally justifies a certain decision or conclusion. One piece of work which was highly influential in turning argumentation theory into the popular and vibrant research area it is today was the the landmark paper of Phan Minh Dung [Dung, 1995]. Dung's abstract argumentation frameworks (AFs) treat arguments and attacks between them as undefined primitives, i.e. the internal structure of arguments is not considered. The major focus is on resolving conflicts. To this end a variety of *semantics* have been defined, each of which captures different intuitions about how to reason about conflicting knowledge. Although, Dung-style AFs are among the simplest argumentation systems one can think of, this approach is still powerful. For example, it has been shown how to reconstruct some mainstream nonmonotonic formalisms as special forms, or more precisely as certain instances of this novel theory. Dung himself provided such a correspondence for default logic [Reiter, 1980], defeasible logic [Pollock, 1987] and logic programming under stable and well-founded semantics [Gelfond and Lifschitz, 1988; Gelder et al., 1988].

In the subsequent years further correspondences between existing nonmonotonic formalisms and abstract argumentation theory were shown, e.g. reinterpretations of Nute's defeasible logic [Governatori et al., 2000], a later version of Pollock's system for defeasible reasoning [Pollock, 1994; Pollock, 1995] and logic programming under 3-valued stable model semantics [Wu et al., 2009]. Other research lines in abstract argumentation are the introduction of new semantics as well as defining suitable extensions of Dung's theory. The motivations of the newly invented semantics range from the desired treatment of specific examples to fulfilling a number of abstract principles (c.f. [Verheij, 1996; Baroni et al., 2005; Dvořák and Gaggl, 2012]). The further developments of Dung's theory encompass attacks on attacks, collective attacks as well as the addition of preferences or values to arguments in order to judge the success of an attack (e.g. [Baroni et al., 2009; Nielsen and Parsons, 2006; Amgoud and Cayrol, 2002; Bench-Capon, 2003]).

The investigation of *metalogical properties* like cardinality results [Baumann and Strass, 2013], monotonicity results [Cayrol et al., 2008; Baumann and Brewka, 2010], splitting results [Baumann, 2011; Baumann et al., 2011; Baumann et al., 2012] and replacement theorems [Oikarinen and Woltran, 2011; Gaggl and Woltran, 2011; Baumann, 2012a; Baumann, 2012b; Baumann and Woltran, 2014; Baumann and Brewka, 2013a] has begun only recently and is still at the beginning. This research field is the main subject of this book.

1.1.1 The Overall Instantiation Process

Dung's theory of abstract argumentation can be seen as one major component in a multi-stage reasoning process (compare [Caminada and Amgoud, 2007]). The following steps can be distinguished:

1. building a knowledge base
2. argument construction
3. conflict identification
4. abstraction through instantiating in a Dung-style AF
5. applying a semantics
6. drawing conclusions

One starts with building a knowledge base containing certain and uncertain knowledge about a particular domain of discourse. Such a knowledge is usually represented as a theory in classical or defeasible logic. The next two steps are concerned with the creation of arguments based on a certain construction method as well as the determination of their strength by applying a reasonably defined notion of attack. Arguments typically consist of two parts, namely a *support* grounded in the knowledge base and a *claim* which is derivable from it. A frequently used criterion for two arguments being in conflict is that they support contradictory conclusions. It is important to realize that *argument construction* and *conflict identification* are monotonic in the following sense: adding a new piece of information to a given knowledge base cannot rule out an existing argument and its corresponding attacks given that the construction method and the underlying notion of attack remain the same. Consequently, only new arguments which may interact with the previous ones arise.

The fourth step is the proper instantiation process. The derived arguments are conveyed to an AF by abstracting away from the internal structure of arguments. This means, a structured argument consisting of a support and a claim becomes an abstract argument without any content, i.e. it is just regarded as a node in a directed graph. Analogously, the former identified conflicts are represented as vertices in the corresponding AF. Thus, the reason why an argument attacks another one is not reflected on the abstract layer. After that we have to resolve the given conflicts. This can be done by applying a suitable semantics singling out sets of arguments representing reasonable positions in the given abstract scenario. It is this evaluation step where the nonmonotonicity comes into play. Adding new arguments and their associated attacks may change the outcome of an AF in a nontrivial and nonmonotonic fashion. More precisely, sets of arguments regarded as acceptable earlier may become unaccepted and vice versa. Finally, having the reasonable positions at hand we have to draw *justified* conclusions. Such a conclusion is not in terms of abstract arguments, but rather an expression in the underlying logical language. Drawing a conclusion can be done by considering the particular content of the accepted arguments. In doing so, different levels of scepticism can be applied.

For illustration purpose we sketch an instantiation process (see Figure 1.1) where the underlying knowledge base consists of strict and defeasible rules (indicated by "→" and "⇒"). Such a kind of rule-based argumentation systems are considered in [Caminada and Amgoud, 2007; Prakken, 2009]. An analysis of instantiating abstract argumentation with classical logic can be found in [Amgoud and Besnard, 2010; Gorogiannis and Hunter, 2011].

1.1.2 *Why* Metalogical Analysis?

A metalogical investigation of an abstract formalism has theoretical and practical gains. On the one hand, it yields deeper theoretical insights into how a certain approach works, and on the other the obtained results can be used to refine existing algorithms or may give rise to new computational proced-

1.1. Argumentation in Artificial Intelligence

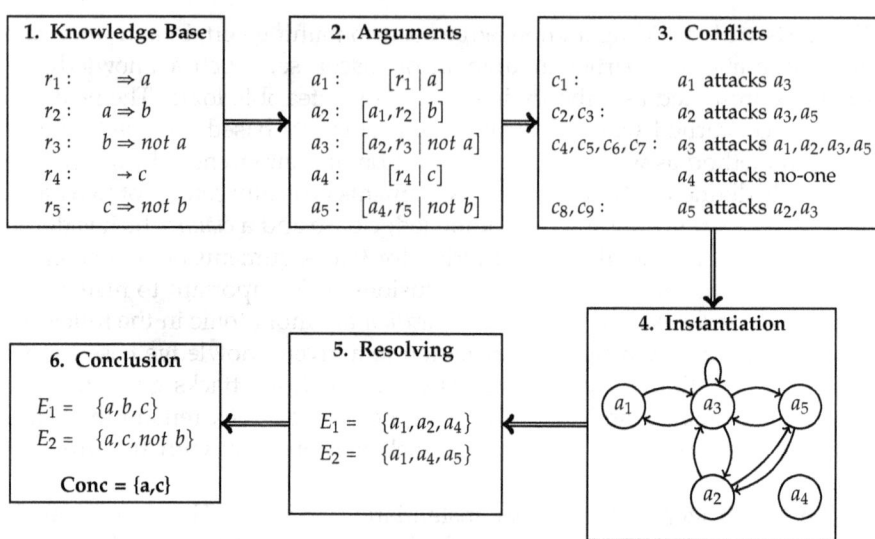

Figure 1.1: The Entire Instantiation Process

ures. In particular, obtained results for abstract argumentation even yield new insights in the instantiation-based context, as long as they are correctly interpreted in the big picture.

In order to illustrate what kind of meta-properties we are interested in we briefly review well-known properties of classical propositional logic. The operator $\text{Mod}(S)$ denotes the classical model operator which returns the set of models for a given propositional theory S. As usual $\sigma(S)$ denotes the signature of S, i.e. $\sigma(S)$ contains the atomic variables occurring in S.

1. anti-monotonicity: $S \subseteq T \Rightarrow \text{Mod}(T) \subseteq \text{Mod}(S)$
2. splitting: $\text{Mod}(S \cup T) = \text{Mod}(S) \cap \text{Mod}(T)$
3. replacement: $\text{Mod}(S) = \text{Mod}(T) \Leftrightarrow \text{Mod}(S \cup U) = \text{Mod}(T \cup U)$ for any U
4. cardinality: $\sigma(S) = \{A_1, ..., A_n\} \Rightarrow |\text{Mod}(S)| \leq 2^n$

The first property provides one with the possibility to *reuse* already computed models. Such a property is of high interest for any logical formalism because computation of semantics is usually much harder than verification. The second characteristic states that it is possible to *divide* a theory in subtheories such that the formal semantics of the entire theory can be obtained by constructing the semantics of the subtheories. Also this property is interesting from a computational point of view, especially in case of large inputs (compare [Gelfond and Przymusinska, 1992; Lifschitz and Turner, 1994; Turner, 1996] for splitting results in major nonmonotonic formalisms). In general, equivalence tells us whether two syntactically different objects represent the same information - which is relevant, for instance, for simplification issues. It is a special attribute of classical logic that standard or classical equivalence, i.e. sharing the same models even guarantees inter-substitutability in any logical context without lost of information. Such a strong notion of equivalence is of great importance for dynamically evolving scenarios since it allows to *simplify* parts of a theory without looking at the rest (see [Maher, 1986; Lifschitz et al., 2001; Turner, 2004; Truszczynski, 2006] for strong equivalence notions in nonmonotonic theories). A priori *bounds* of the number of models for a given input are very useful for algorithms computing semantics and may improve existing complexity results. The last property states a quite naive bound for propositional formulae over n variables. Unfortunately, without any further knowledge about the considered formula it is impossible to find better bounds.

1.2 Structure of the Book and Main Contributions

The book is organized in eight chapters, the first one being this introduction. The rest of the work is structured as follows:

- Chapter 2 reviews the necessary background and definitions at work in abstract argumentation. In particular, we present ten prominent argumentation semantics and survey their behaviour with respect to certain abstract principles. For the sake of completeness we briefly introduce the two most important reasoning problems together with their computational complexity. Furthermore, we discuss several kinds of expansion as well as their corresponding equivalence relations. We additionally provide some preliminary relations between these different notions of equivalence.

- The subsequent Chapter 3 studies the impact on the set of extensions if finitely or even infinitely many new arguments are added. One main result is that the class of weak expansions and semantics satisfying the directionality principle guarantee a monotonic behaviour with respect to subset relations and cardinality (Theorem 3.2). We then show how to benefit from the obtained results. In particular, we show how to simplify the computation of justification states (Proposition 3.6).

   In the last section of this chapter we turn to *reasoning about actions* which is a subfield of artificial intelligence that is concerned with representing and reasoning about dynamic domains. We propose to employ abstract argumentation for this purpose. The theoretical results proven before will play a key role in showing that our approach can be efficiently implemented.

- Chapter 4 is devoted to the fundamental principle of *splitting*. The main questions regarding splitting is whether it is possible to divide a formal theory $\mathcal{T}$ in disjoint subtheories $\mathcal{S}_1, ..., \mathcal{S}_n$ such that the formal semantics of $\mathcal{T}$ can be obtained by constructing the semantics of $\mathcal{S}_1, ..., \mathcal{S}_n$. In the first section we consider partitions into two parts such that the remaining attacks are restricted to a single direction, so-called *classical splittings*. We prove the main splitting theorem and establish a general re-

1.2. Structure of the Book and Main Contributions

lation between abstract principles and the splitting property (Theorems 4.10, 4.13).

In the second part we show how the conditions under which splitting is possible can be relaxed. In fact, in case of stable semantics we prove a generalized splitting result allowing for *arbitrary* splits (Theorem 4.28). Finally, we convey our idea of a non-classical splitting to logic programs (Theorem 4.34) generalizing former results proven in [Lifschitz and Turner, 1994].

- In Chapter 5 we address the issue of characterizing equivalence notions lying in-between standard and expansion equivalence. This means, rather than considering arbitrary expansions we focus here on restricted classes like normal and strong expansions.

  First, we briefly summarize already existing results regarding expansion equivalence. Then, in Section 5.2 we present characterization theorems with respect to strong expansion equivalence for all semantics considered in this book. Two unexpected as well as remarkable results are that strong expansion equivalence with respect to stable and stage semantics coincide (Theorem 5.10) and furthermore, stage, stable, semi-stable, eager and naive semantics do not *distinguish* between expansion and strong expansion equivalence (cf. Figure 5.18). In Section 5.3 we proceed with normal expansion equivalence. The main result is that two AFs are expansion equivalent if and only if they are normal expansion equivalent (see Figure 5.18).

  Additionally, in Section 5.4 we characterize weak expansion equivalence for stable and preferred semantics (Theorems 5.38 and 5.40). Interestingly, the obtained characterization theorems are not purely syntactical in contrast to the other results established in this chapter. Furthermore, we consider local equivalence with respect to stage semantics (Theorem 5.44). Finally, in Section 5.5, we summarize the obtained results, provide relations between different notions of equivalence and discuss the role of self-attacking arguments. One main result is that any equivalence notion on the set of AFs characterizable through kernels presented in this book collapses to identity if we restrict ourselves to self-loop-free AFs (Theorem 5.47).

- In Chapter 6 we provide an analytical and empirical study of the maximal and average numbers of stable extensions in abstract argumentation frameworks. As one of the analytical main results, we prove a tight upper bound on the maximal number of stable extensions that depends only on the number of arguments n in the framework, namely $3^{\frac{n}{3}}$ (Theorem 6.5). More interestingly, our empirical results indicate that the distribution of stable extensions as a function of the number of attacks in the framework seems to follow a universal pattern that is independent of the number of arguments (cf. Figures 6.4 and 6.5).

- Chapter 7 is dedicated to *dynamic* aspects of abstract argumentation. The first question we study is the so-called *enforcing problem*. This is, in brief, the question whether it is possible to modify a given AF in such a way that a desired set of arguments becomes an extension or at least a subset of an extension. We show both impossibility and possibility results (Theorems 7.5, 7.6 and 7.7). In particular, enforcing is always possible given that the desired set of arguments is conflict-free.

In Section 7.2 we go an important step further. We are not only interested whether enforcements are *possible*, but also in the *effort needed* to enforce a set of arguments. In a nutshell, the *minimal change problem* considered in this section corresponds to the mathematical problem of determining the minimal number of modifications (additions or removals of attacks) needed to enforce a certain set. We present characterization theorems for weak, strong, normal, arbitrary expansions as well as arbitrary modifications for the stable, preferred, complete and admissible semantics. The most remarkable result is that the *characteristic* (representing the minimal change problem formally) does not change if we switch simultaneously between strong, normal or arbitrary expansions and preferred, complete or admissible semantics (Theorem 7.29).

In the subsequent Section 7.3 we proceed with a formal analysis of characteristics. In particular, given a collection of semantics and a modification type, what are the corresponding tuples of characteristics one may obtain for an arbitrary argumentation framework and set of arguments. In other words, we want to determine the set of all tuples of natural numbers which may occur as characteristics simultaneously, the so-called *spectrum*. In case of stable, semi-stable and preferred semantics we present a complete characterization with respect to strong, normal and arbitrary expansions (Theorem 7.43). One important consequence is that it may be arbitrarily more difficult to enforce arguments using stable rather than semi-stable semantics, and also using semi-stable rather than preferred semantics. Additionally, we provide some first results for the spectra with respect to weak expansions as well as arbitrary modifications.

Section 7.4 studies novel dynamic notions of equivalence which guarantee equal minimal efforts needed to enforce certain subsets, namely *minimal-D-equivalence* and the more general *minimal change equivalence*. We present characterization theorems for stable, preferred, complete and admissible semantics (Theorems 7.51 and 7.52). Furthermore, we study the relation between minimal change equivalence and expansion or standard equivalence in general, i.e. rather than considering specific semantics we provide our results for a whole range of semantics satisfying certain abstract principles (Theorems 7.54 and 7.55). Furthermore, we provide the complete picture for all equivalence relations considered in this book (so-called *equivalence zoo*) for the most important semantics. In particular, in case of stable semantics the different forms of minimal change equivalence are shown to be intermediate forms between strong expansion and weak expansion equivalence (see Figure 7.18).

- Finally, in Chapter 8 we conclude this work. We summarize and discuss the results achieved in this work and give pointers to possible future research directions.

1.3 Publications

Most of the results presented in this book are already published at international conferences, refereed workshops, journals papers and in a book in honour of Vladimir Lifschitz. In the following we list the involved publications together with the chapters or sections where they are mainly used.

- journal articles

    1. [Baumann, 2012a] *Normal and Strong Expansion Equivalence for Argumentation Frameworks*, Journal of Artificial Intelligence
    (Chapter 5)

    2. [Baumann and Woltran, 2014] *The Role of Self-attacking Arguments in Characterizations of Equivalence Notions*, accepted for Journal of Logic and Computation (Chapter 5)

- conference papers

    3. [Baumann and Brewka, 2010] *Expanding Argumentation Frameworks: Enforcing and Monotonicity Results*, Conference on Computational Models of Argument (Sections 3.1, 3.2 and 7.1)

    4. [Baumann, 2011] *Splitting an Argumentation Framework*, International Conference on Logic Programming and Non-Monotonic Reasoning (Section 4.1)

    5. [Baumann, 2012b] *What Does it Take to Enforce an Argument? Minimal Change in abstract Argumentation*, European Conference on Artificial Intelligence (Sections 7.2, 7.4)

    6. [Baumann and Strass, 2012] *Default Reasoning about Actions via Abstract Argumentation*, Conference on Computational Models of an Argument (Section 3.3)

    7. [Baumann and Brewka, 2013b] *Spectra in Abstract Argumentation: An Analysis of Minimal Change*, International Conference on Logic Programming and Non-Monotonic Reasoning (Section 7.3)

- workshop papers

    8. [Baumann et al., 2011] *Splitting Argumentation Frameworks: An Empirical Evaluation*, Workshop on Theorie and Applications of Formal Argumentation (Section 4.1)

    9. [Baumann and Brewka, 2013a] *Analyzing the Equivalence Zoo in Abstract Argumentation*, International Workshop on Computational Logic in Multi-Agent Systems (Section 7.4)

10. [Baumann and Strass, 2013] *On the Maximal and Average Numbers of Stable Extensions*, Workshop on Theorie and Applications of Formal Argumentation (Chapter 6)

- book contribution

    11. [Baumann et al., 2012] *Parameterized Splitting: A Simple Modification-Based Approach*, Correct Reasoning - Essays on Logic-Based AI in Honour of Vladimir Lifschitz (Sections 4.2, 4.3)

Chapter 2

Background of Abstract Argumentation

In this chapter we introduce the necessary definitions at work in abstract argumentation. As pointed out in the introductory part argumentation theory as considered in the AI-community can be divided into *deductive* and *abstract argumentation* (see [Besnard and Hunter, 2008] for a comprehensive overview). In the latter area Dung's abstract argumentation frameworks (AFs) [Dung, 1995] play a dominant role. Here, arguments and attacks between them are treated as primitives, i.e. the internal structure of arguments is not considered. The major focus is on resolving conflicts and therefore, a variety of semantics have been defined, each of them specifying acceptable sets. The motivations of these semantics range from the desired treatment of specific examples to fulfilling a number of abstract principles. We present ten different semantics and survey their behaviour with respect to certain evaluation criteria proposed by [Baroni and Giacomin, 2007]. We then introduce the two most important reasoning problems together with a survey of their complexity for all semantics under consideration.

The last two subsections are devoted to the inherently dynamic nature of argumentation. In a dispute, for instance, new arguments are put forward in response to former arguments with the objective to convince the participants of a certain opinion. In order to analyze such argumentation scenarios we introduce several kinds of expansions capturing the specific dynamics of typical argumentation processes. Finally, we introduce and motivate equivalence relations which allow for replacements in such typical argumentation scenarios without loss of information.

2.1 Abstract Argumentation Frameworks

We start with the definition of an argumentation framework (AF) given by [Dung, 1995].

2.1. Abstract Argumentation Frameworks

Definition 2.1. An *argumentation framework (AF)* is a pair $\mathcal{F} = (A, R)$, where A is a set whose elements are called *arguments* and $R \subseteq A \times A$ a binary relation, called the *attack relation*.

In the following we introduce some notations and technical terms we will use throughout the book. If $(a, b) \in R$ holds we say that *a attacks b*, or *b is defeated* by *a* in $\mathcal{F}$. Furthermore, we will slightly abuse notations, and write $(A, b) \in R$ for $\exists a \in A : (a, b) \in R$. In this case we say (the set) A attacks b. Likewise we use the abbreviations $(b, A) \in R$ and $(A, A') \in R$. An argument $a \in A$ is *defended* by a set $A' \subseteq A$ in $\mathcal{F}$ if for each $b \in A$ with $(b, a) \in R$, $(A', b) \in R$. For a set $E \subseteq A$ we use $R^+_{\mathcal{F}}(E)$, or simply R^+_E, for $E \cup \{b \mid (a, b) \in R, a \in E\}$. This set is called the *range* of E in $\mathcal{F}$. For an AF $\mathcal{F} = (B, S)$ we use $A(\mathcal{F})$ to refer to B and $R(\mathcal{F})$ to refer to S. We introduce the union for two AFs $\mathcal{F}$ and $\mathcal{G}$ as expected, namely $\mathcal{F} \cup \mathcal{G} = (A(\mathcal{F}) \cup A(\mathcal{G}), R(\mathcal{F}) \cup R(\mathcal{G}))$. An AF $\mathcal{F} = (A, R)$ is said to be *finite* iff it contains finitely many arguments only, i.e. $|A| = n$ for some $n \in \mathbb{N}$. Furthermore, $\mathcal{F}$ is called *finitary* iff each argument $a \in A$ possesses at most finitely many attackers, i.e. it exists $n \in \mathbb{N}$, such that $|\{b \in A \mid (b, a) \in R\}| = n$.

If not specified otherwise we restrict ourselves to finite AFs (including the empty framework) as done in most work on abstract argumentation during the last 20 years. However, this convention does not mean that non-finite AFs are of no interest. To the contrary, in recent times some work showing practical applications and theoretical properties of non-finite AFs has been presented in the literature [Modgil, 2009; Baroni et al., 2011b; Weydert, 2011]. For this reason we will give pointers to the non-restrictive case or sometimes we even prove results for infinite AFs.

As one can easily notice AFs are set-theoretically just directed graphs where nodes are interpreted as arguments and edges represent conflicts between them. The following AF $\mathcal{F}$ will be used throughout the whole book. We will take this running example to exemplify the above introduced definitions and vocabulary.

Example 2.2. Given $\mathcal{F} = (A, R)$ where

- $A = \{a, b, c, d, e, f\}$ and
- $R = \{(a, b), (a, d), (b, c), (c, a), (d, d), (e, d), (e, f), (f, e)\}$.

The graphical representation of $\mathcal{F}$ is given as follows.

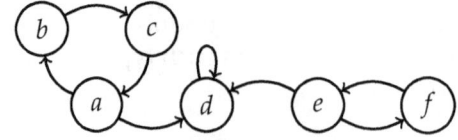

Figure 2.1: Running Example $\mathcal{F}$

The following statements apply to $\mathcal{F}$:

- *c* attacks *a* and is in turn defeated by *b*,

- d is defended by $\{a,c,f\}$ and $\{a,c,f\}$ attacks d,
- the range of $\{a,b,f\}$ is the whole framework, i.e. $R_{\mathcal{F}}^{+}(\{a,b,f\}) = A$.
- $\mathcal{F}$ is finite and thus, finitary.

Finally, we introduce the union of two AFs as well as the restriction of an argumentation framework to a subset of its arguments as expected. These set operations will be frequently used in the book.

Definition 2.3. Given two AFs $\mathcal{F} = (A, R)$, $\mathcal{G} = (B, S)$ and a set $A' \subseteq A$.

- The *union* of $\mathcal{F}$ and $\mathcal{G}$ is given by: $\mathcal{F} \cup \mathcal{G} = (A \cup B, R \cup S)$.
- The *restriction* of $\mathcal{F}$ to A' is defined as: $\mathcal{F}|_{A'} = (A', R \cap (A' \times A'))$.

We say that $\mathcal{F}|_{A'}$ is the by A' induced subframework of $\mathcal{F}$.

2.1.1 Semantics

A semantics σ is a function which assigns to any AF $\mathcal{F}$ a set of sets of arguments denoted by $\mathcal{E}_\sigma(\mathcal{F})$. Each one of them, a so-called σ-extension, is considered to be acceptable with respect to $\mathcal{F}$. For two semantics σ and τ we use $\sigma \subseteq \tau$ to indicate that for any AF $\mathcal{F}$, $\mathcal{E}_\sigma(\mathcal{F}) \subseteq \mathcal{E}_\tau(\mathcal{F})$. Numerous semantics are available for abstract argumentation frameworks among those already defined by [Dung, 1995], namely stable, admissible, preferred, complete and grounded semantics. To overcome the problem of non-existence of stable extensions, semi-stable and stage semantics [Caminada et al., 2012; Verheij, 1996] were suggested. Another approach for semantics is driven by the motivation that semantics which always prescribe exactly one extension to any AF are important but the grounded semantics (the only unique-statues semantics originally proposed) is too cautious. Ideal and eager semantics [Caminada, 2007; Dung et al., 2006] follow this idea. Finally, we consider here also the so-called naive semantics.

Definition 2.4. Let $\mathcal{F} = (A, R)$ be an AF and $E \subseteq A$. E is a

1. *conflict-free set* ($E \in cf(\mathcal{F})$) iff there is no $a \in E$, such that $(E, a) \in R$,
2. *stable extension* ($E \in \mathcal{E}_{stb}(\mathcal{F})$) iff $E \in cf(\mathcal{F})$ and for each $a \in A \setminus E$, $(E, a) \in R$ holds,
3. *admissible extension*[1] ($E \in \mathcal{E}_{ad}(\mathcal{F})$) iff $E \in cf(\mathcal{F})$ and each $a \in E$ is defended by E in $\mathcal{F}$,
4. *preferred extension* ($E \in \mathcal{E}_{pr}(\mathcal{F})$) iff $E \in \mathcal{E}_{ad}(\mathcal{F})$ and for each $E' \in \mathcal{E}_{ad}(\mathcal{F})$, $E \not\subset E'$ holds,
5. *complete extension* ($E \in \mathcal{E}_{co}(\mathcal{F})$) iff $E \in \mathcal{E}_{ad}(\mathcal{F})$ and for any $a \in A$ defended by E in $\mathcal{F}$, $a \in E$,

---

[1] Note that it is more common to speak about admissible sets instead of the admissible extensions. For reasons of unified notation we used the uncommon version.

2.1. Abstract Argumentation Frameworks

6. *grounded extension* ($E \in \mathcal{E}_{gr}(\mathcal{F})$) iff $E \in \mathcal{E}_{co}(\mathcal{F})$ and for each $E' \in \mathcal{E}_{co}(\mathcal{F})$, $E' \not\subset E$ holds,

7. *semi-stable extension* ($E \in \mathcal{E}_{ss}(\mathcal{F})$) iff $E \in \mathcal{E}_{ad}(\mathcal{F})$ and for each $E' \in \mathcal{E}_{ad}(\mathcal{F})$, $R_{\mathcal{F}}^{+}(E) \not\subset R_{\mathcal{F}}^{+}(E')$,

8. *stage extension* ($E \in \mathcal{E}_{stg}(\mathcal{F})$) iff $E \in cf(\mathcal{F})$ and for each $E' \in cf(\mathcal{F})$, $R_{\mathcal{F}}^{+}(E) \not\subset R_{\mathcal{F}}^{+}(E')$ holds,

9. *ideal extension* of $\mathcal{F}$ ($E \in \mathcal{E}_{id}(\mathcal{F})$) iff $E \in \mathcal{E}_{ad}(\mathcal{F})$, $E \subseteq \bigcap_{P \in \mathcal{E}_{pr}(\mathcal{F})} P$ and for each $A \in \mathcal{E}_{ad}(\mathcal{F})$ satisfying $A \subseteq \bigcap_{P \in \mathcal{E}_{pr}(\mathcal{F})} P$ we have $E \not\subset A$,

10. *eager extension* of $\mathcal{F}$ ($E \in \mathcal{E}_{eg}(\mathcal{F})$) iff $E \in \mathcal{E}_{ad}(\mathcal{F})$, $E \subseteq \bigcap_{P \in \mathcal{E}_{ss}(\mathcal{F})} P$ and for each $A \in \mathcal{E}_{ad}(\mathcal{F})$ satisfying $A \subseteq \bigcap_{P \in \mathcal{E}_{ss}(\mathcal{F})} P$ we have $E \not\subset A$ and

11. *naive extension* ($E \in \mathcal{E}_{na}(\mathcal{F})$) iff $E \in cf(\mathcal{F})$ and for each $E' \in cf(\mathcal{F})$, $E \not\subset E'$ holds.

To get a first impression of the different behavior of the introduced semantics we consider the running example.

Example 2.5. Consider the evaluation table of the AF $\mathcal{F}$ depicted in Figure 2.2. The listed sets in the first column represent all conflict-free sets of $\mathcal{F}$, i.e. $cf(\mathcal{F}) = \{\emptyset, \{e\}, \{f\}, \{a,e\}, \{b,e\}, \{c,e\}, \{a,f\}, \{b,f\}, \{c,f\}\}$. The entry "×" in line C column σ indicates that the conflict-free set C is a σ-extension of $\mathcal{F}$, i.e. $C \in \mathcal{E}_{\sigma}(\mathcal{F})$.

We want to mention that complete extensions can be equivalently defined as conflict-free fixpoints of the so-called *characteristic function*. This monotonic function was originally introduced in [Dung, 1995] and is defined as follows.

Definition 2.6. Given an AF $\mathcal{F} = (A, R)$. The characteristic function $\Gamma_{\mathcal{F}} : 2^A \to 2^A$ is defined as

$$\Gamma_{\mathcal{F}}(S) = \{a \in A \mid a \text{ is defended by } S \text{ in } \mathcal{F}\}.$$

To familiarize the reader with the equivalent definition of complete extensions we give two short examples: First, $\Gamma_{\mathcal{F}}(\{e\}) = \{e\}$ since e defends itself against f and no other argument is defended by e. Consequently, $\{e\}$ is a complete extension, in contrast to $\{a, f\}$ where $\Gamma_{\mathcal{F}}(\{a, f\}) = \{c, f\} \neq \{a, f\}$.

Observe that a grounded extension is defined as a minimal (with respect to subset relation) complete extension. Thus, it corresponds to the least conflict-free fixpoint of $\Gamma_{\mathcal{F}}$. The main advantage of this alternative characterization is that the least fixpoint can be achieved by applying iteratively $\Gamma_{\mathcal{F}}$ on the empty set which allows for an efficient computation in case of finite AFs.

Now let us proceed with the analysis of basic properties of the introduced semantics. Considering the table above a first important question that comes to mind is the question of comparability with respect to subset relation of the different semantics. The table indicates that such relations do not exist for any pair of semantics σ and τ. For instance, in case of semi-stable and stage

Chapter 2. Background of Abstract Argumentation

	stb	ss	stg	pr	ad	co	gr	id	eg	na
∅					×	×	×	×		
{e}		×		×	×	×			×	
{f}				×	×	×				
{a,e}			×							×
{b,e}			×							×
{c,e}			×							×
{a,f}			×							×
{b,f}										×
{c,f}										×

Figure 2.2: Evaluation Table of $\mathcal{F}$

2.1. Abstract Argumentation Frameworks

semantics neither $ss \subseteq stg$ nor $stg \subseteq ss$. The following proposition clarifies the question for the considered semantics. In the interest of readability we present the relations graphically instead of providing a long list of propositions.

Proposition 2.7. *For semantics σ and τ, $\sigma \subseteq \tau$ iff there is a link between σ and τ in Figure 2.3.*

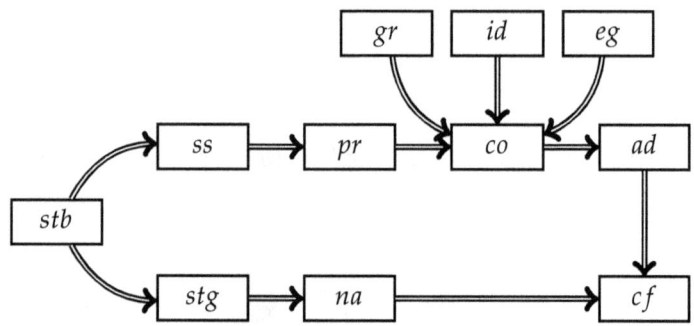

Figure 2.3: Relations between Semantics

Proof. First, consider the lower branch. Obviously, $na \subseteq cf$. Let $E \in \mathcal{E}_{stg}(\mathcal{F}) \smallsetminus \mathcal{E}_{na}(\mathcal{F})$. Consequently there is a set E', such that $E' \in cf(\mathcal{F})$ and $E \subset E'$. We have $R_{\mathcal{F}}^+(E) \subseteq R_{\mathcal{F}}^+(E')$. Note that $R_{\mathcal{F}}^+(E') \subseteq R_{\mathcal{F}}^+(E)$ requires $(E, E' \smallsetminus E) \in R(\mathcal{F})$. This would contradict the conflict-freeness of E' in $\mathcal{F}$. Thus, $stg \subseteq na$. Furthermore, if $E \in \mathcal{E}_{stb}(\mathcal{F})$, then $R_{\mathcal{F}}^+(E) = A(\mathcal{F})$ and consequently, $E \in \mathcal{E}_{stg}(\mathcal{F})$. Hence, $stb \subseteq stg$.

Second, consider the upper branch. By definition we have $co \subseteq ad \subseteq cf$. Let $E \in \mathcal{E}_{pr}(\mathcal{F}) \smallsetminus \mathcal{E}_{co}(\mathcal{F})$. Thus, there is an argument $e \in A(\mathcal{F}) \smallsetminus E$ which is defended by E. Since E is admissible and therefore conflict-free we deduce $(E, e) \notin R(\mathcal{F})$. Thus, $(e, e) \notin R(\mathcal{F})$. This means, $E \cup \{e\} \in \mathcal{E}_{ad}(\mathcal{F})$ contradicting the subset maximality of E. Hence, $pr \subseteq co$. Let $E \in \mathcal{E}_{ss}(\mathcal{F}) \smallsetminus \mathcal{E}_{pr}(\mathcal{F})$. Consequently there is a set E', such that $E' \in \mathcal{E}_{ad}(\mathcal{F})$ and $E \subset E'$. We deduce $R_{\mathcal{F}}^+(E) \subseteq R_{\mathcal{F}}^+(E')$. Observe that equal ranges of E and E' imply $(E, E' \smallsetminus E) \in R(\mathcal{F})$. This in turn would contradict the conflict-freeness of E' in $\mathcal{F}$. Thus, $ss \subseteq pr$ is shown. If $E \in \mathcal{E}_{stb}(\mathcal{F})$, then $R_{\mathcal{F}}^+(E) = A(\mathcal{F})$. Consequently, E defends all its elements. This means E fulfills all criteria of a semi-stable extension. Altogether, $stb \subseteq ss \subseteq pr \subseteq co$ is shown.

Finally, we consider $gr \subseteq co$, $id \subseteq co$ and $eg \subseteq co$. The first proposition holds by definition. Consider $E \in \mathcal{E}_{id}(\mathcal{F}) \smallsetminus \mathcal{E}_{co}(\mathcal{F})$. Then there is an argument $e \in A(\mathcal{F}) \smallsetminus E$ which is defended by E. Let $P \in \mathcal{E}_{pr}(\mathcal{F})$. Since e is defended by E, $E \subseteq P$ and $P \in cf(\mathcal{F})$ we have e is defended by P. Since $pr \subseteq co$ is already shown we deduce $e \in P$, for any preferred extension P. Furthermore, since E is admissible and therefore conflict-free in $\mathcal{F}$ we deduce $E \cup \{e\} \in \mathcal{E}_{ad}(\mathcal{F})$ contradicting the subset maximality of E. In a similar way one may show $eg \subseteq co$. □

The evaluation table of the running example $\mathcal{F}$ (Figure 2.2) shows that stable extensions do not necessarily exist. We call a semantics σ *universally defined* iff for any AF $\mathcal{F}$, $|\mathcal{E}_\sigma(\mathcal{F})| \geq 1$. We mention that the other semantics considered in this book always possess at least one extension in case of finite AFs. This can be seen as follows: First, the empty set is always admissible and conflict-free. Furthermore, the definiens of the semantics are looking for sets maximal in range or maximal/minimal with respect to subset relation. Finally, only finitely many subsets have to be considered since we are dealing with finite AFs. It is an important observation that universally definedness of an semantics σ in case of finite AFs does not necessarily carry over to the infinite case. This property holds for semi-stable semantics. Remarkably, it warrants the existence of extensions for the class of finitary AFs [Weydert, 2011].

A further important property of semantics is the following strengthening of universally definedness, namely guaranteeing at least one and at most one extension for every argumentation framework. If a semantics σ satisfies this property we say that σ follows the *unique status* approach. The following proposition states that grounded, ideal and eager semantics yield exactly one unique extension. Furthermore, ideal semantics accepts more arguments than grounded semantics and eager semantics is even more credulous than ideal semantics.

Proposition 2.8. *For any AF $\mathcal{F}$, $|\mathcal{E}_{gr}(\mathcal{F})| = |\mathcal{E}_{id}(\mathcal{F})| = |\mathcal{E}_{eg}(\mathcal{F})| = 1$. Furthermore, if $G \in \mathcal{E}_{gr}(\mathcal{F})$, $I \in \mathcal{E}_{id}(\mathcal{F})$ and $E \in \mathcal{E}_{eg}(\mathcal{F})$, then $G \subseteq I \subseteq E$.*

Proof. Since we already observed that the considered semantics are universally defined it suffices to prove uniqueness. For a contradiction we assume that $|\mathcal{E}_\sigma(\mathcal{F})| > 1$ where $\sigma \in \{gr, id, eg\}$. Consider ideal semantics. Let $I_1, I_2 \in \mathcal{E}_{id}(\mathcal{F})$. Consequently, I_1 and I_2 are maximal admissible sets, such that $I_1, I_2 \subseteq \bigcap_{P \in \mathcal{E}_{pr}(\mathcal{F})} P$. Therefore, $I_1 \not\subseteq I_2$ and $I_2 \not\subseteq I_1$ Since preferred extensions are conflict-free we have $I_1, I_2 \in cf(\mathcal{F})$. Together with the already assumed admissibility of I_1 and I_2 we have $I_1 \cup I_2 = I \in \mathcal{E}_{ad}(\mathcal{F})$. Furthermore, $I \subseteq \bigcap_{P \in \mathcal{E}_{pr}(\mathcal{F})} P$ is implied contradicting the maximality I_1 and I_2. A proof for eager semantics can be obtained by replacing preferred with semi-stable. In case of grounded semantics we refer to [Dung, 1995, Theorem 25] where it was shown that the set of all complete extensions forms a complete semi-lattice. Thus, there is a least complete extension being a subset of any other complete extension.

In consideration of the last sentence we observe that the unique grounded extension can be equivalently defined as the maximal complete (admissible) set G that is subset of each complete extension. By Proposition 2.7 we have $ss \subseteq pr \subseteq co$ and thus $\bigcap_{C \in \mathcal{E}_{co}(\mathcal{F})} C \subseteq \bigcap_{P \in \mathcal{E}_{pr}(\mathcal{F})} P \subseteq \bigcap_{S \in \mathcal{E}_{ss}(\mathcal{F})} S$. Consequently, if G, I, E represents the unique grounded, ideal and eager extension, then $G \subseteq I \subseteq E$ for any AF $\mathcal{F}$. □

The following two propositions provide us some sufficient conditions for the agreement of certain semantics. The first one states that stable, stage and semi-stable semantics coincide if the considered AF possesses at least one stable extension. The second proposition claims the agreement of preferred, semi-stable and ideal semantics for AFs possessing a unique preferred extension. Note that the prerequisites of both propositions are essential (compare Figure 2.1).

Proposition 2.9. *For any AF $\mathcal{F}$, if $\mathcal{E}_{stb}(\mathcal{F}) \neq \emptyset$, then $\mathcal{E}_{stb}(\mathcal{F}) = \mathcal{E}_{stg}(\mathcal{F}) = \mathcal{E}_{ss}(\mathcal{F})$.*

Proof. In consideration of Proposition 2.7 it suffices to prove that $\mathcal{E}_{stg}(\mathcal{F}) \subseteq \mathcal{E}_{stb}(\mathcal{F})$ and $\mathcal{E}_{ss}(\mathcal{F}) \subseteq \mathcal{E}_{stb}(\mathcal{F})$ in case of $\mathcal{E}_{stb}(\mathcal{F}) \neq \emptyset$. Assume $E \in \mathcal{E}_{stb}(\mathcal{F})$. Since $E \in \mathcal{E}_{stg}(\mathcal{F})$, $E \in \mathcal{E}_{ss}(\mathcal{F})$ and $R^+_{\mathcal{F}}(E) = A(\mathcal{F})$ is implied we deduce $R^+_{\mathcal{F}}(S) = A(\mathcal{F})$ for any $S \in \mathcal{E}_\sigma(\mathcal{F})$ ($\sigma \in \{stg, ss\}$). This means, $S \in \mathcal{E}_{stb}(\mathcal{F})$ concluding the proof. □

Proposition 2.10. *For any AF $\mathcal{F}$, if $|\mathcal{E}_{pr}(\mathcal{F})| = 1$, then $\mathcal{E}_{pr}(\mathcal{F}) = \mathcal{E}_{ss}(\mathcal{F}) = \mathcal{E}_{id}(\mathcal{F}) = \mathcal{E}_{eg}(\mathcal{F})$.*

Proof. Assume $|\mathcal{E}_{pr}(\mathcal{F})| = 1$. Since semi-stable semantics is universally defined and $ss \subseteq pr$ is already shown we obtain $\mathcal{E}_{pr}(\mathcal{F}) = \mathcal{E}_{ss}(\mathcal{F})$. In consideration of the definition of ideal and semi-stable semantics $\mathcal{E}_{pr}(\mathcal{F}) = \mathcal{E}_{id}(\mathcal{F})$ and $\mathcal{E}_{ss}(\mathcal{F}) = \mathcal{E}_{eg}(\mathcal{F})$ follow immediately. □

2.1.2 Abstract Principles

In [Baroni and Giacomin, 2007] several general criteria for comparing and evaluating semantics were introduced. This paper was an important step to classify semantics because until its publication comparisons between semantics were almost exclusively example driven. One main motivation for considering abstract principles instead of dealing with concrete semantics (and AFs) is that possible results using abstract properties are general enough to cover semantics which may be defined in the future. In the following we introduce some abstract properties and provide a full picture for the considered semantics.

Definition 2.11. A semantics σ satisfies

1. *admissibility (AD)*, if for each AF $\mathcal{F}$ and for each σ-extension E we have: $E \in \mathcal{E}_{ad}(\mathcal{F})$,

2. *reinstatement (RE)*, if for each AF $\mathcal{F}$ and for each σ-extension E we have: any argument $a \in A(\mathcal{F})$ defended by E in $\mathcal{F}$, $a \in E$,

3. *conflict-freeness (CF)*, if for each AF $\mathcal{F}$ and for each σ-extension E we have: $E \in cf(\mathcal{F})$,

4. *directionality (DI)*, if for each AF $\mathcal{F}$ and any $U \in US(\mathcal{F}) = \{U \mid U \in A(\mathcal{F}), (A(\mathcal{F}) \setminus U, U) \notin R(\mathcal{F})\}$ we have: $\mathcal{E}_\sigma(\mathcal{F}|_U) = \{E \cap U \mid E \in \mathcal{E}_\sigma(\mathcal{F})\}$ and

5. *I-maximality (IM)*, if for each AF $\mathcal{F}$ and for each two σ-extensions E_1, E_2 we have: if $E_1 \subseteq E_2$, then $E_1 = E_2$.

Chapter 2. Background of Abstract Argumentation

The following table provides an overview of the considered semantics (compare [Baroni and Giacomin, 2007; Baroni et al., 2011a; Gaggl, 2013]). In case of admissible and eager semantics we use red-highlighted entries. This reflects the situation that (to the best of our knowledge) these semantics have not been explicitly studied in the literature. We discuss the obtained results below the table.

	stb	ss	stg	pr	ad	co	gr	id	eg	na
$\mathcal{AD}$	Yes	Yes	No	Yes	Yes	Yes	Yes	Yes	Yes	No
$\mathcal{RE}$	Yes	Yes	No	Yes	No	Yes	Yes	Yes	Yes	No
$\mathcal{CF}$	Yes	Yes	Yes	Yes	Yes	Yes	Yes	Yes	Yes	Yes
$\mathcal{DI}$	No	No	No	Yes	Yes	Yes	Yes	Yes	No	No
$\mathcal{IM}$	Yes	Yes	Yes	Yes	No	No	Yes	Yes	Yes	Yes

Figure 2.4: Evaluation Criteria: A Complete Picture

The presented results with regard to the first three abstract principles are not very surprising. Obviously, semantics which always return admissible, complete or conflict-free extensions satisfy admissibility, reinstatement or the conflict-freeness principle. Roughly speaking, these principles qualify basic requirements concerning single extensions, namely: the arguments of a certain extension E are able to defend E collectively, consists of all arguments defended by E and are compatible with each other.

The directionality and I-maximality criteria are requirements involving the whole set of extensions and even extensions of certain subframeworks. The latter principle is satisfied if no extension may be a strict subset of another one. Consequently, semantics following the unique status approach possess this property. In particular, eager semantics does. The simple AF $\mathcal{F} = (\{a\}, \emptyset)$ proves that the admissible semantics does not satisfy I-maximality. The directionality criterion claims that the evaluation of a certain argument should only be affected by its attackers and the attackers of its attackers and so on. This means, an unattacked set should not be affected by the remaining part of the AF. The proof that admissible semantics satisfies directionality is implicitly given by combining [Baroni and Giacomin, 2007, Proposition 55] and [Baroni and Giacomin, 2004, Proposition 2].[2] The AF $\mathcal{F} = (\{a, b, c\}, \{(a, b), (b, a), (b, c), (c, c)\})$ taken from [Baroni and Giacomin, 2007, Figure 4] shows that eager semantics does not satisfy directionality. Observe that $\mathcal{E}_{eg}(\mathcal{F}|_{\{a,b\}}) = \{\emptyset\} \neq \{\{b\}\} = \{E \cap \{a, b\} \mid E \in \mathcal{E}_{eg}(\mathcal{F})\}$.

---

[2] In Section 4.1.3 we we will present an alternative proof (including infinite AFs).

2.1. Abstract Argumentation Frameworks

A note on infinite AFs. Evidently, if an abstract criterion is not even satisfied in case of finite AFs, then the same applies to the overall class of AFs. If we have a closer look at the proof of Proposition 2.7 we observe that finiteness is not used. Consequently, stable, semi-stable, preferred, grounded, ideal, eager and complete semantics satisfy admissibility, reinstatement and the conflict-freeness principle even in case of arbitrary AFs since they always return complete extensions. In general it is nontrivial to determine whether the fulfillment of a certain criterion carries over to infinite AFs given that the considered criterion focuses on properties of the whole set of extensions.

2.1.3 Reasoning

A certain argumentation semantics σ assigns a set of σ-extensions to a given AF $\mathcal{F}$. We said that any single σ-extension is considered to be acceptable with regard to $\mathcal{F}$. What about the acceptance status of a certain argument a? There are some obvious ways to assign a certain status to an argument. First, an argument is either *justified* or *rejected*. The latter holds if the semantics do not return an extension containing the considered argument. In case of justified arguments we may consider at least two different levels of scepticism, namely *credulous* or *sceptical acceptance* (see [Baroni and Giacomin, 2009] for further classifications). The first acceptance status is assigned if the considered argument a is contained in at least one extension. A strengthening of this status is sceptical acceptance. Here a has to be contained in any extension. Here are the formal definitions.

Definition 2.12. Given an AF $\mathcal{F}$, a semantics σ and an argument $a \in A(\mathcal{F})$. The argument a is

1. *rejected* (with respect to σ and $\mathcal{F}$), if $a \notin \bigcup_{E \in \mathcal{E}_\sigma(\mathcal{F})} E$,
2. *credulously accepted* (with respect to σ and $\mathcal{F}$), if $a \in \bigcup_{E \in \mathcal{E}_\sigma(\mathcal{F})} E$ and
3. *sceptically accepted* (with respect to σ and $\mathcal{F}$), if $a \in \bigcap_{E \in \mathcal{E}_\sigma(\mathcal{F})} E$.

It may be observed that sceptical acceptance implies credulous acceptance. The converse direction holds in case of a unique status approach. Furthermore, an argument a is credulously accepted if and only if it is not rejected. A decision whether a certain argument a is credulously or sceptically accepted with respect to a semantics σ and AF $\mathcal{F}$ (also known as $Cred_\sigma$ or $Scept_\sigma$ respectively) can be easily given if we have access to all σ-extensions. Of course, this approach is computationally costly, especially if there are exponentially many different extensions (in the size of the inputframework $\mathcal{F}$).[3] Hence, there is a great interest in results or better algorithms providing a simplified method for checking the aforementioned decision problems. In Section 3.2 we will show that in some cases the acceptability of a certain argument can be decided by regarding a subframework only.

---
[3]Such an example is given by $\mathcal{F}_n = (\{a_i, b_i \mid 1 \le i \le n\}, \{(a_i, b_i), (b_i, a_i) \mid 1 \le i \le n\})$. Here, $|\mathcal{E}_{stb}(\mathcal{F})| = 2^n$.

For the sake of completeness we provide a complexity classifications of the sceptical and credulous reasoning problems. We want to emphasize that there are further decision problems of interest, i.e. verifying whether a specific set of arguments is an extension (Ver_σ) or simply the question whether there is an σ-extensions for a given AF ($Exists_\sigma$). For a very good introduction into the computational complexity of abstract argumentation we refer the reader to [Dunne and Wooldridge, 2009]. A detailed analysis is given in [Dvořák, 2012].

	stb	ss	stg	pr	ad	co	gr	id	eg	na
$Cred_\sigma$	NP-c	Σ_2^P-c	Σ_2^P-c	NP-c	NP-c	NP-c	P-c	in Θ_2^P	Π_2^P-c	in L
$Scept_\sigma$	coNP-c	Π_2^P-c	Π_2^P-c	Π_2^P-c	triv	P-c	P-c	in Θ_2^P	Π_2^P-c	in L

Figure 2.5: Complexity of $Cred_\sigma$ and $Scept_\sigma$

2.1.4 Dynamics and Expansions

Dung's argumentation frameworks are static: they specify sets of acceptable arguments given a fixed set of arguments and attacks among them. Since argumentation is inherently dynamic, it is natural to investigate the dynamic behavior of AFs. What are typical dynamic scenarios or how does argumentation usually take place? Consider therefore the following citation [Besnard and Hunter, 2009]:

> Argumentation starts when an initial argument is put forward, making some claim. An objection is raised, in the form of a counterargument. The latter is addressed in turn, eventually giving rise to a counter-counterargument, if any. And so on.

Furthermore, with regard to the overall instantiation process it turns out that in almost all deductive argumentation systems older arguments and their corresponding attacks survive and only new arguments which may interact with the previous ones arise given that a new piece of information is added to the underlying knowledge base (see [Besnard and Hunter, 2001]). This kind of dynamics corresponds to the concept of normal expansions firstly introduced in [Baumann and Brewka, 2010]. This class of expansions is further divided into *strong* and *weak expansions*. These are normal expansions which add strong or weak arguments only, i.e. the added arguments never are attacked by former arguments or attack the previous ones. In contrast to strong expansions the study of weak expansions may seem to be more of an academic exercise than a task with practical relevance. Being aware of this fact, we emphasize that there are formalisms, like Value Based AFs [Bench-Capon, 2003] where weak expansions naturally occur. Former arguments may be arguments which advance higher values than the further arguments. Consequently, the new arguments cannot attack the former (compare the idea of "attack-succeed" in [Bench-Capon, 2003]). The last kind of dynamic scenarios we consider are so-called *local expansions*. This is an somehow orthogonal concept to normal expansions where no new arguments are raised, but the attack relation can be augmented. Such an expansion might occur if the underlying notion of attack is changed and the current framework has to be reinstantiated (compare [Gorogiannis and Hunter, 2011] for a detailed discussion).

Definition 2.13. An AF $\mathcal{F}^*$ is an *expansion* of AF $\mathcal{F} = (A, R)$ (for short, $\mathcal{F} \leq_E \mathcal{F}^*$) iff $\mathcal{F}^* = (A \cup A^*, R \cup R^*)$ where $A^* \cap A = R^* \cap R = \emptyset$. An expansion is called

1. *normal* ($\mathcal{F} \leq_N \mathcal{F}^*$) iff $\forall ab\ ((a,b) \in R^* \rightarrow a \in A^* \vee b \in A^*)$,
2. *strong* ($\mathcal{F} \leq_S \mathcal{F}^*$) iff $\mathcal{F} \leq_N \mathcal{F}^*$ and $\forall ab\ ((a,b) \in R^* \rightarrow \neg(a \in A \wedge b \in A^*))$,
3. *weak* ($\mathcal{F} \leq_W \mathcal{F}^*$) iff $\mathcal{F} \leq_N \mathcal{F}^*$ and $\forall ab\ ((a,b) \in R^* \rightarrow \neg(a \in A^* \wedge b \in A))$,
4. *local* ($\mathcal{F} \leq_L \mathcal{F}^*$) iff $A^* = \emptyset$.

For short, normal expansions add new arguments and possibly new attacks which involve at least one of the fresh arguments. Strong (weak) expansions are normal and only add arguments which are never attacked by (attack) former arguments. Finally, local expansions do not introduce any new arguments but possibly new attacks among the old arguments. For the purpose of illustration we present the following simple example.

Example 2.14. The AF $\mathcal{F}$ is the initial framework. Weak and strong expansions of $\mathcal{F}$ are given by $\mathcal{F}_W$ or $\mathcal{F}_S$, respectively. Furthermore, the AFs $\mathcal{F}_E$, $\mathcal{F}_N$ and $\mathcal{F}_L$ show an arbitrary, normal and local expansion of $\mathcal{F}$.

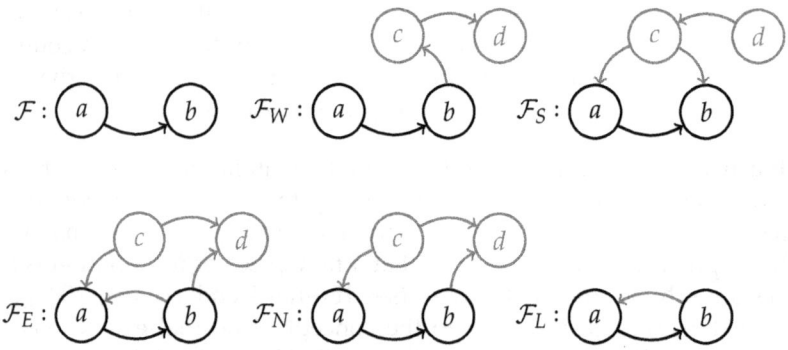

Figure 2.6: Notions of Expansions

2.1.5 Equivalence

In general, equivalence tells us whether two syntactically different object represent the same information - which is relevant, for instance, for simplification issues. In monotonic formalisms, we usually have only one standard, model based notion of equivalence. Consider logical equivalence in propositional or first order logic. It is well-known that this standard notion of equivalence is even a congruence relation with respect to the logical connectives. This property is the main reason for the validity of the so-called *replacement theorem* which states that if two formulae ϕ_1 and ϕ_2 are logically equivalent then no change in the set of models of any formula Φ occurs if we replace one of them with the other (compare Theorem 4.1 in [Rautenberg, 1996]). For short, possessing the same models guarantees inter-substitutability in any logical context. Generally, an analogous statement in case of nonmonotonic formalisms does not hold, as it was shown in [Gelfond and Lifschitz, 1988] for logic programs and [Oikarinen and Woltran, 2011] for abstract argumentation frameworks. Consequently, for the purpose of replacement a stronger notion of equivalence is required. Consider the following example.

Example 2.15. The AFs $\mathcal{F}_1$ and $\mathcal{F}_2$ possess the unique semi-stable extension $\{e\}$. That means, $\mathcal{F}_1$ and $\mathcal{F}_2$ are standard equivalent.

Figure 2.7: Standard Equivalence

The AF $\mathcal{G}$ syntactically results by replacing the subframework $\mathcal{F}_1$ of $\mathcal{F}$ with $\mathcal{F}_2$. One may easily verify that $\mathcal{E}_{ss}(\mathcal{F}) = \{\{e\}\} \neq \{\{e\},\{f\}\} = \mathcal{E}_{ss}(\mathcal{G})$ proving that $\mathcal{F}$ and $\mathcal{G}$ are not standard equivalent.

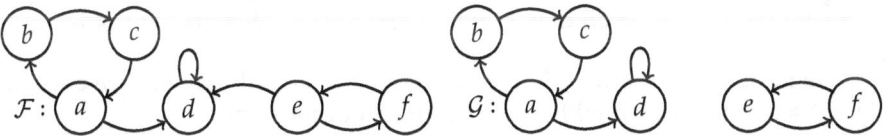

Figure 2.8: Non-replaceability

As shown above, *standard equivalence* of two AFs is not sufficient for their mutual replaceability without loss of information. Nevertheless, possessing the same extensions guarantees that all queries with respect to credulous or skeptical accepted arguments are answered identically. In this sense, standard equivalence is appropriate for non-dynamical, static argumentation scenarios since it allows replacements without changing the semantics. *Strong equivalence*, in contrast, guarantees inter-substitutability in any dynamical scenario. In other words, two AFs $\mathcal{F}$ and $\mathcal{G}$ are strongly equivalent if they share the same acceptable sets of arguments when conjoined with any further framework $\mathcal{H}$.

In Section 2.1.4 we introduced and discussed several kinds of expansion which are compatible with the very nature of argumentation. The observation that argumentation is not static and furthermore, a typical dynamic scenario or re-instantiation process corresponds to a certain subclass of expansions suggests to study the middle ground between the two extremes of standard and strong equivalence, namely so-called *normal expansion, strong expansion, weak expansion* and *local expansion equivalence*.

Now we formally define the aforementioned notions of equivalence. For the sake of clarity and comprehensibility we use *expansion equivalence* instead of *strong equivalence* since any kind of expansions are allowed.

Definition 2.16. Given a semantics σ. Two AFs $\mathcal{F}$ and $\mathcal{G}$ are

1. *standard equivalent* with respect to σ ($\mathcal{F} \equiv^\sigma \mathcal{G}$) iff they possess the same extensions under σ, i.e. $\mathcal{E}_\sigma(\mathcal{F}) = \mathcal{E}_\sigma(\mathcal{G})$ holds,

2. *expansion equivalent* with respect to σ ($\mathcal{F} \equiv^\sigma_E \mathcal{G}$) iff for each AF $\mathcal{H}$, $\mathcal{F} \cup \mathcal{H} \equiv^\sigma \mathcal{G} \cup \mathcal{H}$ holds,

3. *normal expansion equivalent* with respect to σ ($\mathcal{F} \equiv^\sigma_N \mathcal{G}$) iff for each AF $\mathcal{H}$, such that $\mathcal{F} \leq_N \mathcal{F} \cup \mathcal{H}$ and $\mathcal{G} \leq_N \mathcal{G} \cup \mathcal{H}$, $\mathcal{F} \cup \mathcal{H} \equiv^\sigma \mathcal{G} \cup \mathcal{H}$ holds,

4. *strong expansion equivalent* with respect to σ ($\mathcal{F} \equiv^\sigma_S \mathcal{G}$) iff for each AF $\mathcal{H}$, such that $\mathcal{F} \leq_S \mathcal{F} \cup \mathcal{H}$ and $\mathcal{G} \leq_S \mathcal{G} \cup \mathcal{H}$, $\mathcal{F} \cup \mathcal{H} \equiv^\sigma \mathcal{G} \cup \mathcal{H}$ holds,

5. *weak expansion equivalent* with respect to σ ($\mathcal{F} \equiv^\sigma_W \mathcal{G}$) iff for each AF $\mathcal{H}$, such that $\mathcal{F} \leq_W \mathcal{F} \cup \mathcal{H}$ and $\mathcal{G} \leq_W \mathcal{G} \cup \mathcal{H}$, $\mathcal{F} \cup \mathcal{H} \equiv^\sigma \mathcal{G} \cup \mathcal{H}$ holds and

6. *local expansion equivalent* with respect to σ ($\mathcal{F} \equiv^\sigma_L \mathcal{G}$) iff for each AF $\mathcal{H}$, such that $A(\mathcal{H}) \subseteq A(\mathcal{F} \cup \mathcal{G})$, $\mathcal{F} \cup \mathcal{H} \equiv^\sigma \mathcal{G} \cup \mathcal{H}$ holds.

We recall some useful relations concerning standard equivalence and different semantics which will be used throughout the paper.

Proposition 2.17. *For any AFs $\mathcal{F}$ and $\mathcal{G}$,*

1. $\mathcal{F} \equiv^{ad} \mathcal{G} \Rightarrow \mathcal{F} \equiv^\sigma \mathcal{G}$, $\sigma \in \{pr, id\}$,

2. $\mathcal{F} \equiv^{co} \mathcal{G} \Rightarrow \mathcal{F} \equiv^\sigma \mathcal{G}$, $\sigma \in \{pr, gr, id\}$.

Proof. The only non-trivial statement is $\mathcal{F} \equiv^{co} \mathcal{G} \Rightarrow \mathcal{F} \equiv^{id} \mathcal{G}$. The other relations follow immediately by Definition 2.4 and Proposition 2.7. Assume $\mathcal{F} \equiv^{co} \mathcal{G}$. Hence, $\bigcap_{P \in \mathcal{E}_{pr}(\mathcal{F})} P = \bigcap_{P \in \mathcal{E}_{pr}(\mathcal{G})} P$ since $\mathcal{F} \equiv^{pr} \mathcal{G}$ is guaranteed. Without loss of generality let E be ideal in $\mathcal{F}$ but not in $\mathcal{G}$. We have that ideal extensions are complete (Proposition 2.7). Consequently, since both AFs share the same complete extensions we have that E is also complete, thus admissible in $\mathcal{G}$ and furthermore, a subset of $\bigcap_{P \in \mathcal{E}_{pr}(\mathcal{G})} P$. Consequently, the ideal extension E' of $\mathcal{G}$ is a proper superset of E. Note that E' is also complete, thus admissible in $\mathcal{F}$ contradicting the assumption that E is ideal in $\mathcal{F}$. □

Chapter 2. Background of Abstract Argumentation

Now we present some preliminary relations between the mentioned notions of equivalence. The presented implications follow directly from Definition 2.16. Figure 2.9 summarizes all results in a compact way.

Proposition 2.18. *For any AFs $\mathcal{F}, \mathcal{G}$, and **any** (possible) semantics σ the following holds:*

1. $\mathcal{F} \equiv_E^\sigma \mathcal{G} \Rightarrow \mathcal{F} \equiv_N^\sigma \mathcal{G} \Rightarrow \mathcal{F} \equiv_S^\sigma \mathcal{G} \Rightarrow \mathcal{F} \equiv^\sigma \mathcal{G}$
2. $\mathcal{F} \equiv_E^\sigma \mathcal{G} \Rightarrow \mathcal{F} \equiv_N^\sigma \mathcal{G} \Rightarrow \mathcal{F} \equiv_W^\sigma \mathcal{G} \Rightarrow \mathcal{F} \equiv^\sigma \mathcal{G}$
3. $\mathcal{F} \equiv_E^\sigma \mathcal{G} \Rightarrow \mathcal{F} \equiv_L^\sigma \mathcal{G} \Rightarrow \mathcal{F} \equiv^\sigma \mathcal{G}$

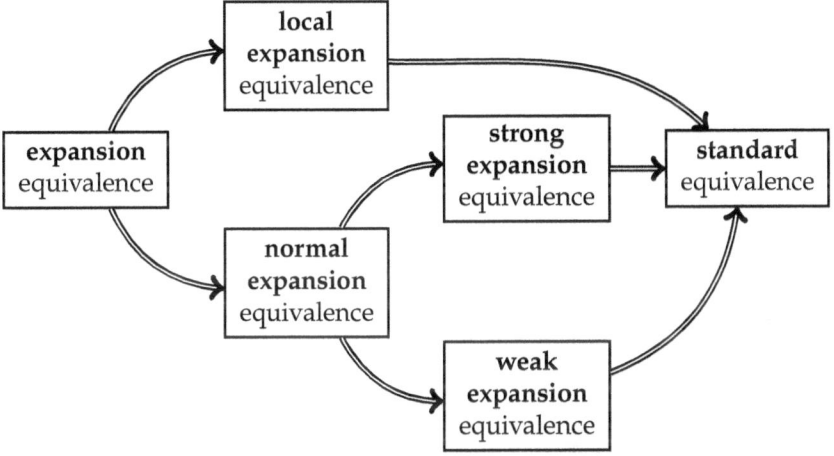

Figure 2.9: Relations between Equivalence Notions

In general the converse directions do not necessarily hold. Since any arbitrary expansion can be split into a normal and local part one could conjecture that normal and local expansion equivalence jointly imply expansion equivalence. In Chapter 5 we will devote particular attention to the question of how to decide the described notions of equivalence. We will see that certain non-redundant frameworks, so-called *kernels*, play a major role in deciding equivalence. Using the results presented in this chapter we will not only verify the addressed conjecture but even the significantly stronger result that normal expansion equivalence and expansion equivalence coincide for all considered semantics (see Section 5.3).

Chapter 3

Monotonic Aspects

This chapter addresses the problem of revising a Dung-style argumentation framework by adding finitely or infinitely many *new* arguments which may interact with *old* ones. We study the relations between the semantics of the initial and the revised argumentation frameworks. In particular, we are interested in subset relations as well as cardinality statements. We will show how to use the obtained results to simplify the computation of justification states. In fact, in certain scenarios, so-called *weak expansion chains*, it suffices to consider a subframework only. Moreover, in some cases the decisive subframework may be finite even if the underlying argumentation framework is infinite. In the third section we turn to *reasoning about actions* which is a subfield of artificial intelligence that is concerned with representing and reasoning about dynamic domains. We propose to employ abstract argumentation for this purpose. The theoretical results proven before will play a key role in showing that our approach can be efficiently implemented.

3.1 Monotonicity Results

Adding new arguments and their associated interactions obviously may change the outcome of an AF in a nonmonotonic way: arguments accepted earlier may become unaccepted, others become accepted; the number of extensions may shrink or increase, depending on the new arguments. For instance, it is easy to verify that we obtain a total collapse of stable extensions if we revise an AF by adding a self-defeating argument. The following example presents a more realistic dynamical evolvement of the running example $\mathcal{F}$ (see Figure 2.1).

Example 3.1. The arguments and attacks marked with red represent the added information. We use $\mathcal{F}_{rev}$ to denote the revised version of the initial framework $\mathcal{F}$.

3.1. Monotonicity Results

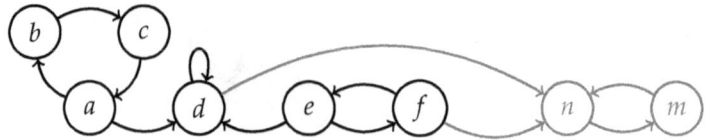

Figure 3.1: The Revised AF $\mathcal{F}_{rev}$

We already observed that $\mathcal{E}_{pr}(\mathcal{F}) = \{\{e\}, \{f\}\} = \{E_1, E_2\}$ (Figure 2.2). Furthermore, we have $\mathcal{E}_{pr}(\mathcal{F}_{rev}) = \{E_1 \cup \{n\}, E_1 \cup \{m\}, E_2 \cup \{m\}\}$. Consequently, the following interrelations hold:

1. the number of extensions increased
2. every old belief set is contained in a new one
3. every new belief set is the union of an old one and a new argument

The previous example contrasts with the general observation that adding new arguments and attacks may change the outcome of an AF in a non-monotonic fashion. Such a behaviour allows for reusing already computed extensions and is therefore worth studying. What are sufficient conditions for monotonic evolvements? The following theorem proves that the class of weak expansions and semantics satisfying the directionality principle guarantee the desired behaviour.

Theorem 3.2. *Given an AF $\mathcal{F} = (A, R)$ and a semantics σ satisfying directionality, then for all AFs $\mathcal{G} = (B, S)$, such that $\mathcal{F} \leq_W \mathcal{G}$ we have:*

1. $|\mathcal{E}_\sigma(\mathcal{F})| \leq |\mathcal{E}_\sigma(\mathcal{G})|$, *(cardinality)*
2. $\forall E \in \mathcal{E}_\sigma(\mathcal{F}) \ \exists E' \in \mathcal{E}_\sigma(\mathcal{G}) \ \exists C \subseteq B \setminus A: \ E' = E \cup C$ *and* *(subset)*
3. $\forall E' \in \mathcal{E}_\sigma(\mathcal{G}) \ \exists E \in \mathcal{E}_\sigma(\mathcal{F}) \ \exists C \subseteq B \setminus A: \ E' = E \cup C$ *(representation)*

Proof. Without loss of generality we may assume that $B = A \cup A'$ and $S = R \cup R'$, such that $A \cap A' = R \cap R' = \emptyset$ (compare Definition 2.13). Since $\mathcal{G}$ is assumed to be a weak expansion of $\mathcal{F}$, $A \in \mathcal{US}(\mathcal{G})$. Furthermore, the directionality of σ implies $\mathcal{E}_\sigma(\mathcal{G}|_A) = \{E' \cap A \mid E' \in \mathcal{E}_\sigma(\mathcal{G})\}$ (Definition 2.11). Since $\mathcal{G}|_A = \mathcal{F}$ we obtain

$$\mathcal{E}_\sigma(\mathcal{F}) = \mathcal{E}_\sigma(\mathcal{G}|_A) = \{E' \cap A \mid E' \in \mathcal{E}_\sigma(\mathcal{G})\} \quad (3.1)$$

which will be frequently used throughout the proof.
1. Obviously, $|\mathcal{E}_\sigma(\mathcal{F})| = |\{E \cap A \mid E \in \mathcal{E}_\sigma(\mathcal{G})\}|$. Furthermore, $|\{E \cap A \mid E \in \mathcal{E}_\sigma(\mathcal{G})\}| \leq |\mathcal{E}_\sigma(\mathcal{G})|$ concluding the proof.
2. (reduction to the absurd) Assume $\exists E \in \mathcal{E}_\sigma(\mathcal{F})$, such that $\forall E' \in \mathcal{E}_\sigma(\mathcal{G}) \ \forall C \subseteq B \setminus A: \ E' \neq E \cup C$. Consequently, for any $E' \in \mathcal{E}_\sigma(\mathcal{G})$, $E' \cap A \neq E$. Assuming $E' \cap A = E$ would imply $E' = E \cup C$ where $C = E' \setminus A$. This means $\mathcal{E}_\sigma(\mathcal{F}) \nsubseteq \{E' \cap A \mid E' \in \mathcal{E}_\sigma(\mathcal{G})\}$ in contrast to equation 3.1.
3. (reduction to the absurd) Assume $\exists E' \in \mathcal{E}_\sigma(\mathcal{G})$, such that $\forall E \in \mathcal{E}_\sigma(\mathcal{F}) \ \forall C \subseteq$

$B \setminus A$: $E' \neq E \cup C$. Now, for any $E \in \mathcal{E}_\sigma(\mathcal{F})$, $E' \cap A \neq E$ because assuming the existence of an σ-extension E with the property $E' \cap A = E$ implies $E' = E \cup C$ with $C = E' \setminus A$. Consequently, $\{E' \cap A \mid E' \in \mathcal{E}_\sigma(\mathcal{G})\} \nsubseteq \mathcal{E}_\sigma(\mathcal{F})$ in contrast to equation 3.1. □

The theorem above proves that class of weak expansions behaves monotonically with respect to the cardinality of extensions given that the considered semantics satisfies directionality. Furthermore, every old belief set is contained in a new one and every new belief set can be represented as the union of an old one and a (possibly empty) set of new arguments. Having a closer look at the proof of Theorem 3.2 reveals that the same assertion applies to the infinite case provided that the considered semantics still satisfies directionality (see Section 4.1.3).

3.2 Computation of Justification States

We already observed in Section 2.1.3 that a decision whether a certain argument a is credulously or sceptically accepted with respect to a semantics σ and AF $\mathcal{F}$ can be easily given if we have access to all σ-extensions. Such an approach is computationally costly, especially if there is a huge number of extensions. In this section we will show how to benefit from the monotonicity results.

Proposition 3.3. *Given an AF $\mathcal{F}$ and a semantics σ satisfying directionality, then for all AFs $\mathcal{G}$, such that $\mathcal{F} \leq_W \mathcal{G}$ we have:*

1. $\bigcup_{E \in \mathcal{E}_\sigma(\mathcal{F})} E \subseteq \bigcup_{E' \in \mathcal{E}_\sigma(\mathcal{G})} E'$ *(credulously justified arguments persist),*

2. $\bigcap_{E \in \mathcal{E}_\sigma(\mathcal{F})} E \subseteq \bigcap_{E' \in \mathcal{E}_\sigma(\mathcal{G})} E'$ *(sceptically justified arguments persist).*

Proof. 1. Given $a \in \bigcup_{E \in \mathcal{E}_\sigma(\mathcal{F})} E$. Thus, it exists a σ-extension E, such that $a \in E$. Using statement 2 of Theorem 3.2 justifies $a \in E'$ for some $E' \in \mathcal{E}_\sigma(\mathcal{G})$. Consequently, $a \in \bigcup_{E' \in \mathcal{E}_\sigma(\mathcal{G})} E'$.
2. Given $a \in \bigcap_{E \in \mathcal{E}_\sigma(\mathcal{F})} E$. This means, for any $E \in \mathcal{E}_\sigma(\mathcal{F})$ we have $a \in E$. Statement 1 of Theorem 3.2 guarantees that there is at least one σ-extension. Furthermore, by statement 3 we have that any $E' \in \mathcal{E}_\sigma(\mathcal{F})$ has a representation as $E' = E \cup C$ with $E \in \mathcal{E}_\sigma(\mathcal{F})$. Consequently, $a \in E'$ for any $E' \in \mathcal{E}_\sigma(\mathcal{F})$ and thus, $a \in \bigcap_{E' \in \mathcal{E}_\sigma(\mathcal{G})}$ concluding the proof. □

That means, a sceptically (credulously) justified argument in $\mathcal{F}$ is sceptically (credulously) justified in $\mathcal{G}$. Remember that the preferred, admissible, complete, grounded and ideal satisfy the directionality principle (compare Figure 2.4). We proceed with the introduction of a new concept, a so-called *expansion chain*.

Definition 3.4. *Let $C = \langle \mathcal{F}_1, ..., \mathcal{F}_n \rangle$ be a sequence of AFs, $\mathcal{F}$ an AF. C is called expansion chain of $\mathcal{F}$ iff*

3.2. Computation of Justification States

1. $\mathcal{F} = \mathcal{F}_n$ and
2. $\mathcal{F}_i \leq_N \mathcal{F}_{i+1}$ ($\mathcal{F}_{i+1}$ is a normal expansion of $\mathcal{F}_i$) for all i: $0 \leq i \leq n-1$.

C is called *weak* (resp. *strong*) if all expansions in the chain are weak (resp. strong).

The following example illustrates the new concept.

Example 3.5. The sequence $C_1 = \langle \mathcal{F}_1, \mathcal{F}_2 \rangle$ is a weak expansion chain of the running example $\mathcal{F}$ (compare Figure 2.1). Furthermore, $C_2 = \langle \mathcal{F}_1, \mathcal{F}_2, \mathcal{F}_3 \rangle$ is a weak expansion chain of $\mathcal{F}_{rev}$ depicted in Figure 3.1. The AFs are defined as follows:

- $\mathcal{F}_1 = (\{a,b,c\}, \{(a,b),(b,c),(c,a)\})$,
- $\mathcal{F}_2 = \mathcal{F}_1 \cup (\{d,e,f\}, \{(a,d),(d,d),(e,d),(e,f),(f,e)\})$ and
- $\mathcal{F}_3 = \mathcal{F}_2 \cup (\{n,m\}, \{(d,n),(f,n),(n,m),(m,n)\})$.

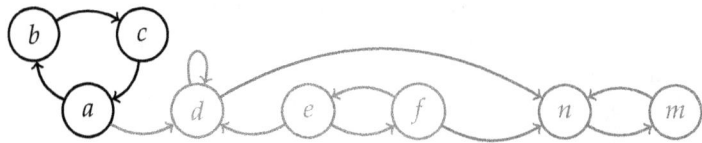

Figure 3.2: Weak Expansion Chains

Finally, we present a simplified method for checking whether an argument a is in some, respectively all extensions of an AF $\mathcal{F}$.

Proposition 3.6. *Given a semantics σ satisfying directionality and let $C = \langle \mathcal{F}_1, ..., \mathcal{F}_n \rangle$ be a weak expansion chain of $\mathcal{F}$. If i is the smallest integer, such that $a \in A(\mathcal{F}_i)$, then*

1. $a \in \bigcup_{E \in \mathcal{E}_\sigma(\mathcal{F})} E \Leftrightarrow a \in \bigcup_{E' \in \mathcal{E}_\sigma(\mathcal{F}_i)} E'$ *(status decidable in $\mathcal{F}_i$),*

2. $a \in \bigcap_{E \in \mathcal{E}_\sigma(\mathcal{F})} E \Leftrightarrow a \in \bigcap_{E' \in \mathcal{E}_\sigma(\mathcal{F}_i)} E'$ *(status decidable in $\mathcal{F}_i$).*

Proof. 1. ($\Rightarrow$) Given $a \in \bigcup_{E \in \mathcal{E}_\sigma(\mathcal{F})} E$. Consequently, there exists an $E \in \mathcal{E}_\sigma(\mathcal{F})$, $a \in E$. We have $a \in A(\mathcal{F}_i)$ and $\mathcal{F}_i \leq_W \mathcal{F}$. With the help of statement 3, Theorem 3.2 we conclude $E = E' \cup C$ where $E' \in \mathcal{E}_\sigma(\mathcal{F}_i)$ and $C \subseteq A(\mathcal{F}) \setminus A(\mathcal{F}_i)$. Since i is the smallest integer, such that $a \in A(\mathcal{F}_i)$ we deduce $a \in E'$ and finally, $a \in \bigcup_{E' \in \mathcal{E}_\sigma(\mathcal{F}_i)} E'$. ($\Leftarrow$) Given $a \in \bigcup_{E' \in \mathcal{E}_\sigma(\mathcal{F}_i)} E'$. In the light of statement 1, Proposition 3.3 we immediately get $a \in \bigcup_{E \in \mathcal{E}_\sigma(\mathcal{F})} E$.
2. ($\Rightarrow$) Let $a \in \bigcap_{E \in \mathcal{E}_\sigma(\mathcal{F})} E$. Thus, for any $E \in \mathcal{E}_\sigma(\mathcal{F})$, $a \in E$. Assume that there is an $E' \in \mathcal{E}_\sigma(\mathcal{F}_i)$, such that $a \notin E'$. Applying statement 2, Theorem 3.2 we deduce the existence of an $E \in \mathcal{E}_\sigma(\mathcal{F})$, such that $E = E' \cup C$ where $C \subseteq A(\mathcal{F}) \setminus A(\mathcal{F}_i)$. Since i is the smallest integer, such that $a \in A(\mathcal{F}_i)$ we deduce $a \notin C$ and thus, $a \notin E$ in contrast to the assumption. ($\Leftarrow$) This direction follows immediately by statement 2, Proposition 3.3. □

Chapter 3. Monotonic Aspects

The proposition shows that it is sufficient to check the acceptability of an argument a in the chain-member $\mathcal{F}_i$ which is the first AF in which a appears. The following example illustrates this property and concludes the section.

Example 3.7. Consider again the revised version $\mathcal{F}_{rev}$ of the initial framework $\mathcal{F}$ (compare Example 3.1). Since $\{f\} \in \mathcal{E}_{pr}(\mathcal{F})$ we deduce f is credulously accepted in $\mathcal{F}_{rev}$ and thus, e is not sceptically accepted in $\mathcal{F}_{rev}$.

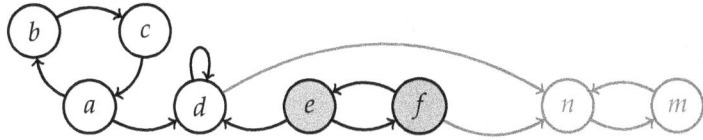

Figure 3.3: Computation of Justification States

3.3 Application: Reasoning about Action

Reasoning about action is a subfield of artificial intelligence that is concerned with representing and reasoning about dynamic domains. We propose to employ abstract argumentation for this purpose. Specifically, we present a translation of action domains from a specification language into Dung-style argumentation frameworks. As the key advantage of our approach, we use existing semantics for argumentation to make predictions about the domain in various manners and utilize monotonicity results to show that the approach can be efficiently implemented. This demonstrates the practical value not only of its theoretical results, but also of abstract argumentation itself.

3.3.1 Background: Action Theories

Action theories are used to represent agents' knowledge about dynamic domains. Envisioned by pioneer John McCarthy as early as 1959 [McCarthy, 1959], one of their tasks is to predict how the world will evolve over time. Such predictions are highly relevant in artificial intelligence, for example to enable an agent to plan ahead the course of actions suitable to meet its goals. A major representational and inferential problem any action theory must solve is the so-called *frame problem* of specifying the world properties that do *not* change when an action is performed [McCarthy and Hayes, 1969]. Having been discovered in 1969, it took until 1991 that the frame problem was solved in a generally accepted way [Reiter, 2001]. Today, the field of reasoning about actions and change has brought forth quite a number of logic-based formalisms, each with their own solution to the frame problem (see [Thielscher, 2011] for a more comprehensive treatment).

Empowered by these advances in classical action theories, researchers began to recognize that agents typically have only incomplete knowledge about their environment, and started to address this issue. Mueller [Mueller, 2006] described a general method for default reasoning about (linear) time based on the circumscription of abnormality predicates. Kakas et al. [Kakas et al., 2008] sketched an integration of temporal and default reasoning and in subsequent works developed an argumentation-based semantics for the approach using linear time [Michael and Kakas, 2009; Michael and Kakas, 2011]. Lakemeyer and Levesque [Lakemeyer and Levesque, 2009] gave a definition of progression [Reiter, 2001] in the presence of state defaults for a modal fragment of the Situation Calculus that uses branching time. Baumann et al. [Baumann et al., 2010] approached the state default problem of inferring what usually holds in an abstract formalism that is independent of a particular time structure [Thielscher, 2011].

In parallel to the development of the highly expressive, logic-based formalisms for reasoning about changing worlds, researchers have proposed so-called *action languages* [Gelfond and Lifschitz, 1998] for describing domains. They are simpler, much closer to natural language and usually have a semantics based on state transition systems. The approach presented in this paper is in effect also based on this paradigm since we define an action language with argumentation-based semantics.

3.3.2 Specifying Action Domains

This section describes our approach for reasoning about actions and change via abstract argumentation. Roughly, the approach works as follows. The user specifies an action domain in an action language that we define next. Our definitions construct an argumentation framework from a description of a domain in this action language. The obtained argumentation framework can be used to answer queries about the domain using various semantics.

The vocabulary for speaking about dynamic domains consists of three components. First fluents, properties that may change over time. Second actions (also called events), that happen and initiate those changes. Third a time structure, that specifies the time points at which we are interested in the state of the world and how these time points relate to each other. The first two can be viewed as constant symbols, the third element is given by a directed tree whose nodes are time points and whose edges induce a reachability relation among the time points.

Definition 3.8. A *domain vocabulary* is a tuple $(\mathbf{F}, \mathbf{A}, \mathbf{T}, <)$, where

- $\mathbf{F}$ is a set of fluents. A *fluent literal* is of the form f or $\neg f$ for some $f \in \mathbf{F}$. Define $\overline{f} \stackrel{\text{def}}{=} \neg f$ and $\overline{\neg f} \stackrel{\text{def}}{=} f$, and for a set L of fluent literals set $\overline{L} \stackrel{\text{def}}{=} \{\overline{l} \mid l \in L\}$. The set of all fluent literals is then $\mathbf{F}^{\pm} \stackrel{\text{def}}{=} \mathbf{F} \cup \overline{\mathbf{F}}$.

- $\mathbf{A}$ is a set of actions.

- The pair $(\mathbf{T}, <)$ is a *time structure*, a directed graph with the properties

    - $\mathbf{T}$ is a countable set of time points, each with a finite degree,
    - $\mathbf{0} \in \mathbf{T}$ is the root of the tree (called the *least time point*),
    - for any $t \in \mathbf{T}$ there is a unique finite, directed path from $\mathbf{0}$ to t.

Chapter 3. Monotonic Aspects

The intuition behind trees as time structures is that edges lead to direct successor time points and the direction of the edges express the flow of time. Hence our time structures may be branching into the future, but they are always linear with respect to the past. This is a very abstract view of time that can accommodate different notions of time that are used in the literature: For example, the pair $(\mathbb{N}, \{(n, n+1) \mid n \in \mathbb{N}\})$ of the natural numbers with the usual successor relation defines a discrete linear time structure [Mueller, 2006]. The second major time structure used in the reasoning about actions community, the branching time of *situations* [Reiter, 2001], can be modeled using terms.

Taking **0** to denote the initial situation and the binary function do to indicate action application at a time point, we inductively define $\mathbf{T}^0 \stackrel{\text{def}}{=} \{\mathbf{0}\}$, and for $i \geq 0$

$$\mathbf{T}^{i+1} \stackrel{\text{def}}{=} \mathbf{T}^i \cup \left\{ do(a, s) \mid a \in \mathbf{A}, s \in \mathbf{T}^i \right\} \quad \text{and} \quad \mathbf{T} \stackrel{\text{def}}{=} \bigcup_{i=0}^{\infty} \mathbf{T}^i.$$

The ordering on situations with least element **0** is then defined as usual [Reiter, 2001].[1]

The state of the world at a time point of any time structure is described by providing the truth values of the relevant aspects of the environment represented by fluents. Since some world aspects might be unknown, we allow the representation of incomplete knowledge about a time point. Formally, a so-called *state* is modeled as a consistent set of fluent literals. This prevents an agent from believing contradictory propositions, but allows for incompleteness of its knowledge.

Actions are formalized by stating their *action preconditions* – world properties that must hold in order for the action to be executable – and direct effects, that express how actions change the state of the world. Since world states are modeled using fluents, the changes initiated by actions are modeled by fluent literals that become true whenever certain fluent literals – the *effect preconditions* – hold. Finally, to formalize how the world normally behaves we use *defaults* – which say that a fluent literal normally holds whenever all literals in a set of *prerequisites* hold. Several action languages offer additional expressiveness, for example indirect action effects. We however want to keep it simple here since our main goal is to show how abstract argumentation can be used to reason about dynamic domains.

A specification of an action domain in our action language consists of two parts: the first part contains such general knowledge about the domain – action preconditions, action effects, state defaults –, the second part contains information about what holds and happens at various time points of a specific domain instance. We begin with how to express knowledge about the general workings of a domain.

Definition 3.9. Consider a fixed domain vocabulary $(\mathbf{F}, \mathbf{A}, \mathbf{T}, <)$, and let $a \in \mathbf{A}$ be an action, $l \in \mathbf{F}^{\pm}$ be a fluent literal and $C \subseteq \mathbf{F}^{\pm}$ be a finite set of fluent literals. A *statement* can be:

---

[1] For the interested reader, the predecessor relation for situations is then given by $s < do(a, s)$ for all $a \in \mathbf{A}, s \in \mathbf{T}$.

3.3. Application: Reasoning about Action

- a *precondition statement*: **possible** a **if** C
- a *direct effect statement*: **action** a **causes** l **if** C
- a *default statement*: **normally** l **if** C

In statements of the above form, we will refer to literal l as the *consequent*. If $C = \emptyset$ for a statement, we omit the **if** part in writing. For illustration, we use the following running example throughout this section.

Example 3.10. (Machine supervision) In this simplified domain, the agent's task is to supervise the operation of a machine. If the temperature of the machine becomes too high, it has to be shut down. Normally, however, the machine operates within temperature. To make statements about this domain, we use the set of fluents $\mathbf{F} = \{\mathsf{On}, \mathsf{Cool}\}$ to express that the machine is on and within acceptable temperature, and the set of actions $\mathbf{A} = \{\mathsf{Switch}\}$ for toggling the machine's power button.

How fluents and actions interrelate is now given through the following statements. When switching the machine on/off, the status of fluent On flips

action Switch **causes** On **if** $\{\neg\mathsf{On}\}$, **action** Switch **causes** $\neg\mathsf{On}$ **if** $\{\mathsf{On}\}$

and turning it off furthermore causes the machine to cool down, formalized as

action Switch **causes** Cool **if** $\{\mathsf{On}\}$.

The usual state of affairs in normal operation mode is expressed by the default statement

normally Cool **if** $\{\mathsf{On}\}$.

For a specific domain instance, we will also assume given a *narrative* consisting of observations of the status of fluents and occurrences of actions at various time points. Although we introduce a more general notation, in this paper, we restrict our attention to actions that end in the direct successor time point.

Definition 3.11. For a fixed domain vocabulary $(\mathbf{F}, \mathbf{A}, \mathbf{T}, <)$, let $a \in \mathbf{A}$ be an action, $l \in \mathbf{F}^{\pm}$ a fluent literal and $s, t \in \mathbf{T}$ time points with $s < t$. An *axiom* can be:

- an *observation axiom*: **observed** l **at** t
- an *occurrence axiom*: **happens** a **from** s **to** t

So when we use the linear time structure shown earlier, action occurrences are always of the form **happens** a **from** n **to** $n+1$ for some $n \in \mathbb{N}$, which means that a has a fixed duration of one time-step. However, we allow the possibility of concurrency, that is, multiple actions happening at the same time.

The combination of general information about the domain and information about a specific instance is now called an action domain specification.

Chapter 3. Monotonic Aspects

Definition 3.12. An *action domain specification*, or *domain* for short, is a set $\Sigma = Y \cup \Omega$ where Y is a finite set of statements and Ω is a set of axioms.

Example 3.13. (Example 3.10 continued) We extend the partial domain signature of the machine supervision domain by the time structure $(\mathbf{T}, <)$ where $\mathbf{T} = \{0, 1, 2, 3\}$ and $<$ is given by $\{(0,1), (1,2), (2,3)\}$. (Note that in this case $0 = 0$.) Now we can express a narrative where the machine is initially off, then switched on at time point 1 and nothing further happens:

$$\Omega = \{\mathbf{observed}\ \neg\text{On}\ \mathbf{at}\ 0, \mathbf{happens}\ \text{Switch}\ \mathbf{from}\ 1\ \mathbf{to}\ 2\}.$$

Whenever using situations as underlying time structure of a domain, we tacitly assume included the set $\{\mathbf{happens}\ a\ \mathbf{from}\ s\ \mathbf{to}\ do(a, s) \mid a \in \mathbf{A}, s \in \mathbf{T}\}$ of occurrence axioms (expressing the meaning of situations as hypothetical future time points) and restrict the user specification to observation axioms.

For any action domain specification – be it linear-time or branching-time – we now want to make predictions about how the domain will normally evolve over time. We do this by translating the specifications into argumentation frameworks and using argumentation semantics to reason about the domains.

3.3.3 Encoding

In the previous section, we introduced the syntax of a language for describing dynamic domains. Now we present the argumentation-based semantics for that language. To this end, we define a translation function from action domain specifications into abstract argumentation frameworks.

This translation function will be mostly *modular*, which means that most of the constituents of a domain description can be translated in isolation, that is, without considering other parts of the domain. It is *mostly* modular because there will be one exception for modularity: to correctly express the effects of actions, we need access to effect statements when translating occurrence axioms. The rest of the translation will however be modular.

The basic intuition underlying our translation is the following. Each piece of knowledge about the domain is modeled as an argument [Michael and Kakas, 2009; Michael and Kakas, 2011]. The most important arguments will express whether a fluent holds at a time point. There will be other arguments, that state various causes for fluents to have a certain truth value. For example, there will be arguments stating persistence as a reason for a fluent being true or false, arguments about direct action effects, default conclusions and lastly observations. To resolve conflicts between these causes – e.g. persistence says a fluent should be false while a direct effect says it should be true – we use a fixed natural priority ordering that prefers observations over action effects, which are in turn preferred over defaults, that on their part trump persistence. This priority ordering is expressed in the defined framework via attacks, that (along with the remaining attacks) encode the relation between different pieces of knowledge in the domain and eventually determine the semantics of the given specification.

3.3. Application: Reasoning about Action

For the rest of this subsection, we assume a given action domain specification Σ over a domain vocabulary $(\mathbf{F}, \mathbf{A}, \mathbf{T}, <)$. Although the translation can be defined in a strictly formal way, we have chosen a less rigorous presentation that we hope is much easier to read. In the paragraphs below, we define arguments and attacks that are created from elements of Σ. To express that argument a attacks argument b, we will write $a \longrightarrow b$. The resulting argumentation framework $\mathcal{F}_\Sigma$ is understood to contain all arguments and attacks that we define below. Along the way, we will illustrate most of the definitions with the relevant parts of the argumentation framework of our running example domain.

Fluents and time points. First of all, we create arguments that express whether a fluent is true or false at a time point, or alternatively whether a fluent literal holds at a time point. For a fluent literal $l \in \mathbf{F}^\pm$ and time point $t \in \mathbf{T}$ they are of the form $holds(l, t)$. Obviously, a fluent cannot be both true and false at any one time point, so for all $f \in \mathbf{F}$ and $t \in \mathbf{T}$, we create the attacks $holds(f, t) \longrightarrow holds(\neg f, t)$ and $holds(\neg f, t) \longrightarrow holds(f, t)$.

Persistence. To solve the frame problem, we define arguments $frame(l, t_1, t_2)$ for all $l \in \mathbf{F}^\pm$ and $t_1 < t_2$. These arguments say "the truth value of fluent literal l persists from time point t_1 to its direct successor t_2." First, we express that l holding at t_1 is evidence against its negation $\bar{l}$ persisting from t_1 to t_2: $holds(l, t_1) \longrightarrow frame(\bar{l}, t_1, t_2)$. Also, l persisting from t_1 to t_2 is evidence against it being false at t_2: $frame(l, t_1, t_2) \longrightarrow holds(\bar{l}, t_2)$.

Example 3.14. (Example 3.13 continued) The AF about fluent On and time points $0, 1$ looks thus:

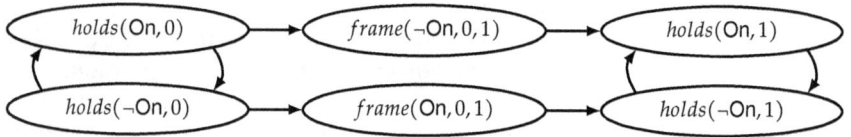

Defaults. Now we encode default conclusions: for **normally** l **if** $C \in \Sigma$ and $t \in \mathbf{T}$ we create arguments $def(l, t)$ and $def(\bar{l}, t)$ with the intended meaning that l normally holds (normally does not hold) at t. The argument for l being normally true at t attacks the arguments for l being false or normally false at t: We add $def(l, t) \longrightarrow holds(\bar{l}, t)$ and $def(l, t) \longrightarrow def(\bar{l}, t)$. A default is inapplicable if some prerequisite $c \in C$ is false at t, expressed by $holds(\bar{c}, t) \longrightarrow def(l, t)$ for each $c \in C$. Defaults generally override persistence, so a default **normally** l **if** C will attack persistence of literal $\bar{l}$ to time point t. If there is a time point $s < t$ we add the attack $def(l, t) \longrightarrow frame(\bar{l}, s, t)$.

We also create special arguments that express whether the world is abnormal with regard to a specific default, that is, whether the default was violated at the time point. A state default **normally** l **if** C is *violated* whenever all literals in C hold but l does not hold, which hints at an abnormality of the world. For a default **normally** l **if** $C \in \Sigma$ and time points $s, t \in \mathbf{T}$ with $s < t$ we create the argument $viol(l, s)$ and the attacks detailed below. First, we require that abnormal situations do not go away by default, therefore a

Chapter 3. Monotonic Aspects

violated default blocks its own application at the successor time point by the attack $viol(l,s) \rightarrow def(l,t)$. Conversely, a default is not violated at s iff one of: (1) it was applied, hence the attack $def(l,s) \rightarrow viol(l,s)$; (2) its consequent holds (the world is normal), thus we add $holds(l,s) \rightarrow viol(l,s)$; or (3) one of its prerequisites is false, hence we include the attack $holds(\bar{c},s) \rightarrow viol(l,s)$ for each $c \in C$.

Example 3.15. (Example 3.14 continued) Here is a part of the argumentation (sub-)framework expressing that the machine is usually cool when on. (The lower part about persistence of Cool is isomorphic to the above graph for persistence of On.) Below we can see that $holds(\neg On, 0)$ defends $def(Cool, 1)$, which in turn defends $holds(Cool, 1)$.

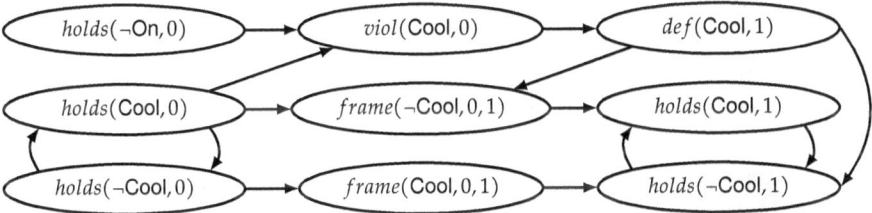

Action effects. Now for modeling the direct effects of actions. Let action occurrence **happens** a **from** s **to** $t \in \Sigma$ and effect statement **action** a **causes** l **if** $C \in \Sigma$. We devise an argument $dir(l,a,s,t)$ that encodes occurrence of effect l through a from s to t. First, the action effect l can never materialize if some precondition in C is false: we add $holds(\bar{c},s) \rightarrow dir(l,a,s,t)$ for all $c \in C$. If there is a precondition statement **possible** a **if** $C_a \in \Sigma$, we add the attacks[2] $holds(\bar{c},s) \rightarrow dir(l,a,s,t)$ for all $c \in C_a$. As usual, to derive effect l at t we attack its negation $\bar{l}$ at t, $dir(l,a,s,t) \rightarrow holds(\bar{l},t)$. Since effects override both defaults and persistence, a direct effect l also attacks persistence of and default conclusions about its negation $\bar{l}$, as well as possible conflicting effects. We add $dir(l,a,s,t) \rightarrow frame(\bar{l},s,t)$, $dir(l,a,s,t) \rightarrow def(\bar{l},t)$ and finally $dir(l,a,s,t) \rightarrow dir(\bar{l},a',s,t)$ for all $a' \in A$.

Example 3.16. (Example 3.15 continued) If the machine was turned off, it would cool down.

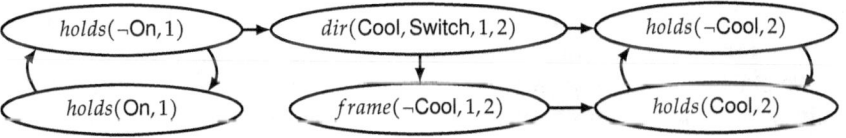

Observations. For observations **observed** l **at** $t \in \Sigma$ we create arguments $obs(l,t)$ and $obs(\bar{l},t)$ saying that l (resp. $\bar{l}$) has been observed. Observations are the strongest causes, i.e. they attack all other causes, including observations to the contrary: $obs(l,t) \rightarrow holds(\bar{l},t)$, $obs(l,t) \rightarrow frame(\bar{l},s,t)$ for $s < t$,

---

[2]If there is no precondition statement for an action, we assume that it is always possible.

3.3. Application: Reasoning about Action

$obs(l, t) \twoheadrightarrow def(\bar{l}, t)$, $obs(l, t) \twoheadrightarrow dir(\bar{l}, a, s', t)$ for $s' < t$ and all $a \in \mathbf{A}$, finally $obs(l, t) \twoheadrightarrow obs(\bar{l}, t)$.

Example 3.17. (Example 3.16 continued) Initially, the machine is observed to be off.

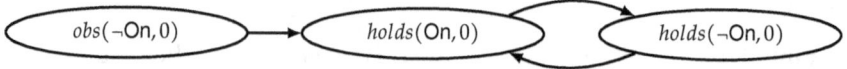

This concludes our definition of the argumentation framework $\mathcal{F}_\Sigma$ associated with an action domain specification Σ.

Example 3.18. (Example 3.17 continued) The full argumentation framework $\mathcal{F}_\Sigma$ constructed from the machine supervision domain is too large to show here, we can however have a look at its extensions and what they predict about the domain. The grounded extension of $\mathcal{F}_\Sigma$ is given by

{$obs(\neg On, 0)$, $holds(\neg On, 0)$,
$frame(\neg On, 0, 1)$, $holds(\neg On, 1)$,
$dir(On, Switch, 1, 2)$, $holds(On, 2)$, $def(Cool, 2)$, $holds(Cool, 2)$,
$frame(On, 2, 3)$, $holds(On, 3)$, $frame(Cool, 2, 3)$, $holds(Cool, 3)$}

In words, the observation that the machine is off lets us conclude it is indeed off; this persists to time point 1. Then the machine is switched on, fluent On becomes true and persists that way. Meanwhile, there is no information about the temperature of the machine; however, after it is switched on, the default can be applied and the machine is henceforth assumed to be cool.

There are two stable extensions, which contain everything that is grounded and additionally make a commitment toward the initial status of Cool, where the first one contains $holds(Cool, 0)$ and the second one $holds(\neg Cool, 0)$. There are three complete extensions, the grounded one and the two stable extensions.

3.3.4 Encoded Priorities

The example framework just seen illustrated the workings of our approach, yet allows no general conclusions about beneficiary properties of the constructed frameworks. We will now proceed to prove several nice properties that are highly relevant for knowledge representation and ultimately pave the way for an efficient implementation of our approach.

We start with some general observations. First, the AFs we construct in Section 3.3.3 are always finitary, that is, every argument has only finitely many attackers, although they may of course be infinite due to an infinite set of time points. Second, it follows from the definition that the argumentation frameworks are free of self-loops. According to results about expansion equivalence [Oikarinen and Woltran, 2011], this means that all attacks encoded therein are actually meaningful.

In this subsection we will show that the causes influencing the truth values of world properties respect a certain priority ordering. Intuitively, this

Chapter 3. Monotonic Aspects

means that the AFs we construct "do the right thing" with respect to domain behaviour.

A fluent f can hold, respectively not hold, at a time point t for different reasons: because an observation says so, it is a direct effect of an action, it is the conclusion of an applicable state default or simply because of persistence. Whenever two of these reasons are in conflict, for instance persistence says f should be true at t while an action effect says f should be false, we somehow have to decide for a truth value of f. We now want to show that the extensions of the translated AFs $\mathcal{F}_\Sigma$ satisfy a certain priority ordering among these reasons, namely (from most to least preferred)

observations < direct effects < default conclusions < persistence.

Hence when persistence says f should be true and a direct effect says f should be false, then the preferred cause "direct effect" takes precedence and f will indeed not hold. We will show that the above ordering is indeed established. We begin with showing the superiority of observations. In the sequel, we call an observation o, an action a or a default δ contrary to l at t in Σ iff they justify $\bar{l}$, that is **observed** $\bar{l}$ **at** $t \in \Sigma$, **happens** a **from** s **to** $t \in \Sigma$ and **action** a **causes** $\bar{l}$ **if** $C^a \in \Sigma$ or **normally** $\bar{l}$ **if** $C \in \Sigma$, respectively.

Proposition 3.19 (*Observations override everything*). *Let Σ be a domain and $\mathcal{F}_\Sigma$ the associated AF. If **observed** l **at** $t \in \Sigma$ and **observed** $\bar{l}$ **at** $t \notin \Sigma$, then for any semantics $\sigma \in \{stb, pr, co, gr\}$ and any $E \in \mathcal{E}_\sigma(\mathcal{F}_\Sigma)$, we find $holds(l,t) \in E$ and $holds(\bar{l},t) \notin E$.*

Proof. First, observe that $obs(l,t)$ is unattacked in $\mathcal{F}_\Sigma$, i.e. $obs(l,t) \in \Gamma_{\mathcal{F}_\Sigma}(\emptyset)$. Furthermore, it attacks the argument $holds(\bar{l},t)$ and the possibly existing argument(s) $def(\bar{l},t)$, $frame(\bar{l},s,t)$, $dir(\bar{l},a,s,t)$ for any $a \in \mathbf{A}$. These arguments are exactly the (possible) attackers of $holds(l,t)$. This means $holds(l,t)$ is defended by $obs(l,t)$. Hence, $holds(l,t) \in C_{\mathcal{F}_\Sigma}(C_{\mathcal{F}_\Sigma}(\emptyset))$ and thus, $holds(l,t) \in E$ where E is the unique grounded extension of $\mathcal{F}_\Sigma$. Note that the grounded extension E satisfies $E \subseteq E'$ for any $E' \in \mathcal{E}_\sigma(\mathcal{F}_\Sigma)$ where $\sigma \in \{stb, pr, co\}$. Consequently, $holds(l,t)$ is contained in any stable, preferred and complete extension of $\mathcal{F}_\Sigma$ too. Finally, $holds(\bar{l},t)$ cannot be included in an extension E since every considered semantics σ is based on conflict-freeness.[3] □

Proposition 3.20 (*Direct effects override default conclusions*). *Let Σ be a domain and $\mathcal{F}_\Sigma$ the associated AF. If*

1. **observed** $\bar{l}$ **at** $t \notin \Sigma$,   (*absence of contrary observation $\bar{l}$*)

2. **happens** a **from** s **to** $t \in \Sigma$,   (*attempting to execute action a*)

3. **action** a **causes** l **if** $C^a \in \Sigma$ and   (*l is a direct effect of action a*)

4. **possible** a **if** $C_a \in \Sigma$,   (*action preconditions of action a*)

---

[3] In the following proofs we will leave the last two comments out. This means we will only show that $holds(l,t)$ is contained in the unique grounded extension of $\mathcal{F}_\Sigma$.

3.3. Application: Reasoning about Action

then for any semantics $\sigma \in \{stb, pr, co, gr\}$ and any $E \in \mathcal{E}_\sigma(\mathcal{F}_\Sigma)$, such that

- $holds(C_a, s) \in E,$[4]   (action a is executable)

- $holds(C^a, s) \in E$ and   (effect statement is applicable)

- for any action $\bar{a}$ contrary to l at t in Σ there is a $c_{\bar{a}} \in C^{\bar{a}} \cup C_{\bar{a}}$ such that $holds(\overline{c_{\bar{a}}}, t) \in E$,   (contrary actions are blocked)

we find $holds(l, t) \in E$ and $holds(\bar{l}, t) \notin E$.

Proof. Let E be the unique grounded extension of $\mathcal{F}_\Sigma$. Since we assumed that $holds(C_a, s), holds(C^a, s) \in E$ and furthermore, $holds(\overline{c_{\bar{a}}}, t) \in E$ for any contrary actions $\bar{a}$ we state that $dir(l, a, s, t) \in E$ because the possible attackers of $dir(l, a, s, t)$, namely $holds(\overline{C_a}, s)$, $holds(\overline{C^a}, s)$ and $dir(\bar{l}, \bar{a}, s, t)$ are counterattacked by $holds(C_a, s)$, $holds(C^a, s)$ or $holds(\overline{c_{\bar{a}}}, t)$, respectively. Thus, $dir(l, a, s, t) \in E$. Observe that the argument $dir(l, a, s, t)$ attacks the arguments $holds(\bar{l}, t)$ and $frame(\bar{l}, s, t)$ as well as the possibly existing arguments $dir(\bar{l}, \bar{a}, s, t)$ and $def(\bar{l}, t)$. This means $holds(l, t)$ is defended by $dir(l, a, s, t)$ and thus, $holds(l, t) \in E$. □

Proposition 3.21 (*Default conclusions override persistence*). *Let Σ be a domain and $\mathcal{F}_\Sigma$ the associated AF. If*

1. **observed** $\bar{l}$ **at** $t \notin \Sigma$,   (absence of contrary observation $\bar{l}$)

2. **normally** l **if** $C \in \Sigma$,   (default δ concludes l)

then for any semantics $\sigma \in \{stb, pr, co, gr\}$ and any $E \in \mathcal{E}_\sigma(\mathcal{F}_\Sigma)$, such that

- $def(l, t) \in E$ and   (default δ is applied)

- for any action $\bar{a}$ contrary to l at t in Σ there is a $c_{\bar{a}} \in C^{\bar{a}} \cup C_{\bar{a}}$ such that $holds(\overline{c_{\bar{a}}}, t) \in E$,   (contrary actions are blocked)

we find $holds(l, t) \in E$ and $holds(\bar{l}, t) \notin E$.

Proof. Let $E \in \mathcal{E}_{gr}(\mathcal{F}_\Sigma)$, $def(l, t) \in E$ and furthermore, $holds(\overline{c_{\bar{a}}}, t) \in E$ for any contrary action $\bar{a}$. The attackers of $holds(l, t) \in E$ are: First, $holds(\bar{l}, t)$, $frame(\bar{l}, s, t)$ and possibly $def(\bar{l}, t)$ which are counterattacked by $def(l, t)$ and second, in case of existence, the arguments $dir(\bar{l}, \bar{a}, s, t)$ which are counterattacked by $holds(\overline{c_{\bar{a}}}, t)$. Thus, $holds(l, t)$ is defended by the set $A \subseteq E$ containing $def(l, t)$ and $holds(\overline{c_{\bar{a}}}, t) \in E$ for any contrary action $\bar{a}$. Consequently, $holds(l, t) \in E$ concluding the proof. □

Proposition 3.22 (*The frame problem is solved*). *Let Σ be a domain and $\mathcal{F}_\Sigma$ the associated AF. If* **observed** $\bar{l}$ **at** $t \notin \Sigma$ *(absence of contrary observation $\bar{l}$), then for any semantics $\sigma \in \{stb, pr, co, gr\}$ and any $E \in \mathcal{E}_\sigma(\mathcal{F}_\Sigma)$, such that*

---

[4] For a set $C = \{c_1, \ldots, c_n\}$ of fluent literals, the expression $holds(C, s) \in S$ abbreviates the more complex $\{holds(c_1, s), \ldots, holds(c_n, s)\} \subseteq S$.

Chapter 3. Monotonic Aspects

- $holds(l,s) \in E$,   *(fluent literal holds before)*

- *for any action $\bar{a}$ contrary to l at t in Σ there is a $c_{\bar{a}} \in C^{\bar{a}} \cup C_{\bar{a}}$ such that* $holds(\overline{c_{\bar{a}}}, t) \in E$,   *(contrary actions are blocked)*

- *for any default $\bar{\delta}$ contrary to l at t in Σ there is a $c_{\bar{\delta}} \in C_{\bar{\delta}}$ such that $holds(\overline{c_{\bar{\delta}}}, t) \in E$ or $viol(\bar{l}, s) \in E$,*   *(contrary defaults are blocked)*

we find $holds(l,t) \in E$ and $holds(\bar{l}, t) \notin E$.

Proof. Let $E \in \mathcal{E}_{gr}(\mathcal{F}_\Sigma)$, $holds(l,s) \in E$ and furthermore, $holds(\overline{c_{\bar{a}}}, t) \in E$ and $holds(\overline{c_{\bar{\delta}}}, t) \in E$ or $viol(\bar{l}, s) \in E$ for any contrary action $\bar{a}$ or contrary default $\bar{\delta}$, respectively. At first, it can be checked that $frame(l,s,t) \in E$ since all (possible) attackers, namely $dir(\bar{l}, \bar{a}, s, t)$, $def(\bar{l}, t)$ and $holds(\bar{l}, s) \in E$ are counterattacked by at least one argument mentioned above. Since $frame(l,s,t)$ attacks $holds(\bar{l}, s)$, $holds(l,s)$ attacks $frame(\bar{l}, s, t)$ and we assumed that contrary actions or defaults are blocked we deduce $holds(l,t) \in E$. □

Theorem 3.23. *Let Σ be a domain and $\mathcal{F}_\Sigma$ the associated AF, and consider reasoning in the stable, complete, preferred and grounded semantics. Concerning the causes that influence whether fluents hold at a time point, we have the following:*

1. *observations override action effects,*
2. *action effects override default conclusions,*
3. *default conclusions override persistence and*
4. *the frame problem is solved.*

Proof. Follows from Propositions 3.19–3.22. □

Hence when, for example, persistence says f should be true and a direct effect says f should be false, then the preferred cause "direct effect" takes precedence and f will indeed not hold. Above, default conclusions refers to the consequents of applicable state defaults, where we recall that violated state defaults are inapplicable at the next time point. So if, in our running example, the machine is On and ¬Cool at one time point, the default statement **normally** Cool **if** {On} is violated and ¬Cool will persist.

3.3.5 Stratification and Computing Extensions

The results of the previous section show that the argumentation semantics considered in this section respect a suitable priority ordering among causes. If this ordering is the same across the semantics, the reader may ask, then where lies the difference between them? Roughly, the different semantics are used to model different types of knowledge (complete v. incomplete) and modes of reasoning (well-founded v. hypothetical). The grounded semantics, for example, accepts only conclusions that are well-founded with respect to definite knowledge. However, this may lead to the agent having only incomplete knowledge about the domain. The stable semantics, on the other hand,

3.3. Application: Reasoning about Action

may make unproven assumptions about the world, but provides complete knowledge about the domain in each extension:

Proposition 3.24. *Let Σ be a domain with associated AF $\mathcal{F}_\Sigma = (A, R)$. For each $E \in \mathcal{E}_{stb}(\mathcal{F}_\Sigma)$, $t \in \mathbf{T}$ and $f \in \mathbf{F}$ we have either $holds(f, t) \in E$ or $holds(\neg f, t) \in E$.*

Proof. Let $E \in \mathcal{E}_{stb}(\mathcal{F}_\Sigma)$. Clearly $\{holds(f, t), holds(\neg f, t)\} \not\subseteq E$ since E is conflict-free. Now assume to the contrary that there are $t \in \mathbf{T}$ and $f \in \mathbf{F}$ with $holds(f, t) \notin E$ and $holds(\neg f, t) \notin E$. Since E is stable, we have $e_1, e_2 \in E$ with $(e_1, holds(f, t)) \in R$ and $(e_2, holds(\neg f, t)) \in R$. Roughly, e_1 and e_2 must be causes among observations, direct effects, defaults and persistence. If they have different priorities, say, e_1 overrides e_2, then $(e_1, e_2) \in R$ and E is not conflict-free, contradiction. If they have the same priority, we have $(e_1, e_2), (e_2, e_1) \in R$ and E is again not conflict-free, contradiction. □

While this result hints at the benefits of using the stable semantics for reasoning about actions with our frameworks, this conclusion has to be qualified: the existence of stable extensions cannot be guaranteed in general.

Example 3.25. Consider the domain Σ consisting only of the default statements

$$\textbf{normally } f \textbf{ if } \{\neg f\} \text{ and } \textbf{normally } \neg f \textbf{ if } \{f\}$$

over the vocabulary $(\{f\}, \emptyset, \{0\}, \emptyset)$. It leads to the argumentation framework $\mathcal{F}_\Sigma$ containing (among others) the odd cycle $def(f, 0) \rightarrow holds(\neg f, 0) \rightarrow holds(f, 0) \rightarrow def(f, 0)$. The single admissible set $\emptyset$ is a preferred extension, but there is no stable extension. The example is however contrived in that the specified defaults make no intuitive sense. How can f be normally false if it is true and vice versa?

We already observed that the constructed AF $\mathcal{F}_\Sigma$ may be infinite but is in any case finitary. Hence, using a result of Dung, we may obtain the unique grounded extension of $\mathcal{F}_\Sigma$ by iteratively applying the characteristic function on the empty set [Dung, 1995]. Unfortunately, there is no similar (constructive) method for the other semantics we are interested in. Nevertheless we will show that the evaluation of an argument from $\mathcal{F}_\Sigma$ can be implemented in principle since we only have to compute extensions in finite subframeworks. Here, the concept of *stratification* plays the lead. A stratification divides the arguments of an AF into layers that satisfy a simple syntactic dependency criterion: a layer of an argumentation framework is any subset of its arguments that is not attacked from the outside. If there is an increasing sequence of layers whose union is the set of all arguments, then the argumentation framework is called stratified. We will relate the new concept to the already introduced weak expansion chains (compare Definition 3.5) and thus may apply monotonicity results.

Definition 3.26. *Let $\mathcal{F} = (A, R)$ be an argumentation framework. A set $L \subseteq A$ is a layer of $\mathcal{F}$ iff for all $a \in L$ and $b \in A \setminus L$ we have $(b, a) \notin R$. A strictly increasing sequence of layers $L_0 \subset L_1 \subset \ldots$ is a stratification of $\mathcal{F}$ iff $L_0 \cup L_1 \cup \ldots = A$.*

Chapter 3. Monotonic Aspects

We call an stratification *infinite (finite)* iff it possesses infinitely (finitely) many layers. For a layer L_i, the argumentation framework *associated to layer* L_i is $\mathcal{F}_i \stackrel{\text{def}}{=} (L_i, R \cap (L_i \times L_i))$. An AF $\mathcal{F}_\Sigma$ automatically constructed from a domain Σ allows for a fairly straightforward definition of a stratification:

Definition 3.27. Let Σ be a domain over vocabulary $(\mathbf{F}, \mathbf{A}, \mathbf{T}, <)$ and $\mathcal{F}_\Sigma = (A, R)$ be the argumentation framework obtained from it by the encoding specified in Section 3.3.3. For time points $s, t \in \mathbf{T}$ define

$$L(t) \stackrel{\text{def}}{=} \{holds(l,t), obs(l,t), viol(l,t), def(l,t) \mid l \in \mathbf{F}^\pm\} \cap A$$
$$L(s,t) \stackrel{\text{def}}{=} \{frame(l,s,t), dir(l,a,s,t) \mid l \in \mathbf{F}^\pm, a \in \mathbf{A}\} \cap A$$

For a set $B \subseteq A$ of arguments denote by $\mathbf{T}(B)$ the time points occurring in B. Now define by induction on natural numbers the sets $L_0 \stackrel{\text{def}}{=} L(0)$ and for $n \in \mathbb{N}$,

$$L_{n+1} \stackrel{\text{def}}{=} \bigcup_{s \in \mathbf{T}(L_n), s < t} (L(s) \cup L(s,t) \cup L(t))$$

For the stratification, we construct the bottom layer by taking all arguments about the least time point 0. The following layers are then defined inductively according to the time structure: in each step, we add the arguments about direct successors of the time points in the arguments in the previous layer. It is straightforward to prove that the construction as a matter of fact yields a stratification:

Proposition 3.28. Let Σ be a domain over vocabulary $(\mathbf{F}, \mathbf{A}, \mathbf{T}, <)$; let $\mathcal{F}_\Sigma = (A, R)$ be the AF constructed from Σ by the encoding from Section 3.3.3. Then the sets $L_0, L_1, \ldots$ obtained according to Definition 3.27 are a stratification for $\mathcal{F}_\Sigma$.

Proof. We first show that for all $n \in \mathbb{N}$, we have $L_n \subset L_{n+1}$. Let $a \in L_n$. We do a case distinction on the structure of a.

1. $a \in L(t)$ for some $t \in \mathbf{T}$. Then $t \in \mathbf{T}(L_n)$ and $a \in L(t) \subseteq L_{n+1}$.
2. $a \in L(s,t)$ for $s, t \in \mathbf{T}$. Then $s \in \mathbf{T}(L_n)$ and $s < t$, thus $a \in L(s,t) \subseteq L_{n+1}$.

This shows $L_n \subseteq L_{n+1}$. Assuming that L_n is about s, for any $t \in \mathbf{T}$ with $s < t$ and $f \in \mathbf{F}$ it is clear that we have $holds(f,t) \in L_{n+1}$ but $holds(f,t) \notin L_n$, thus $L_n \subset L_{n+1}$. It remains to show that there are no attacks from $A \setminus L_n$ into L_n, that is, from $L_{n+1} \cup L_{n+2} \cup \ldots$ into L_n. From the encoding in Section 3.3.3, it follows that an argument about $t \in \mathbf{T}$ in $\mathcal{F}_\Sigma$ only attacks arguments about t or about $t' \in \mathbf{T}$ with $t < t'$. Hence for any $m \in \mathbb{N}$, layer L_m only attacks $L_m \cup L_{m+1}$. □

Of course the same holds for finite stratifications, that end in some layer $L_m = A$. The real power of stratifications lies however in being able to decide the status of an argument after discarding a possibly infinite part of the framework. In the following we anticipate Proposition 4.14 in Section 4.1.3 showing that grounded, complete and preferred semantics satisfy directionality even in case of infinite AFs.

Proposition 3.29. *Let Σ be a domain and $\mathcal{F}_\Sigma = (A, R)$ its associated argumentation framework with stratification $L_0, L_1, \ldots$ according to Definition 3.27. For any semantics $\sigma \in \{gr, co, pr\}$ we have:*

1. $a \in \bigcup_{E \in \mathcal{E}_\sigma(\mathcal{F}_\Sigma)} E \Leftrightarrow a \in \bigcup_{E' \in \mathcal{E}_\sigma(\mathcal{F}_n)} E'$ *(status decidable in $\mathcal{F}_n$),*

2. $a \in \bigcap_{E \in \mathcal{E}_\sigma(\mathcal{F}_\Sigma)} E \Leftrightarrow a \in \bigcap_{E' \in \mathcal{E}_\sigma(\mathcal{F}_n)} E'$ *(status decidable in $\mathcal{F}_n$).*

where $\mathcal{F}_n$ is the AF associated to the least layer L_n such that $a \in L_n$.

Proof. First, we use that grounded, preferred and complete semantics satisfy the directionality principle (compare Theorem 4.13). By Definition 3.27 and Proposition 3.28 we obtain a (possibly infinite) stratification of $\mathcal{F}_\Sigma$ and assume that L_n is the smallest (with respect to subset relation) layer with $a \in L_n$. Now $\mathcal{F}_\Sigma$ admits the (finite) weak expansion chain $C_n = \langle \mathcal{F}_1, \mathcal{F}_2, \ldots, \mathcal{F}_n, \mathcal{F}_\Sigma \rangle$. Applying Proposition 3.6 yields the assertion. □

It is crucial to note that this result makes implementing our approach *feasible in principle*, since it reduces the relevant decision problems about infinite AFs to equivalent problems about finite AFs. What is more, translating action domains to stratified argumentation frameworks also provides an important step towards an *efficient* implementation: An intelligent agent that constantly executes actions has to decide upon future actions in a timely manner. Consider an agent that has executed m actions since having been switched on. To plan n time points into the future, it needs to consider an argumentation framework of a size that grows in $m + n$. As time passes, n (the lookahead into the future) might be kept constant but m (the history of past actions) will surely grow up to the point where despite only involving a linear blowup the associated AFs are too large for the agent to handle. The solution to this problem is known as *progression* in the reasoning about actions community: every once in a while, the agent replaces its knowledge base about the whole past by a smaller but equivalent one about the present, thereby effectively resetting m [Reiter, 2001]. That way, the size of the knowledge base the agent handles can be significantly reduced while keeping the information it contains. For grounded semantics, the obvious technical approach to progression is then to replace the sub-framework $\mathcal{F}'_\Sigma$ speaking about the past until time point t by its unique grounded extension E', thereby obtaining a splitting. Splittings will be the object of study in Chapter 4. Empirical evidence about extensions of AFs admitting a splitting [Baumann et al., 2012] shows that this provides a promising basis for implementing the approach presented in this paper.

3.4 Conclusions and Related Work

We provided theoretical insights about the impact of further arguments and attack relations. In particular, we showed that the class of weak expansions

Chapter 3. Monotonic Aspects

behaves monotonically with respect to the cardinality of extensions and justification state of arguments if the considered semantics satisfies directionality. Note that the established results (restricted to the class of finite AFs) were already published in [Baumann and Brewka, 2010]. One key advantage of the shown relations is the use of the directionality principle which makes our results general enough to cover semantics which may be defined in the future.

There are several papers analyzing the dynamics in abstract argumentation. In [Cayrol et al., 2008] a typology of revisions is proposed for the case that one new argument along with one new interaction is added. Furthermore they proved sufficient conditions for being of a certain revision type. A further work [Cayrol et al., 2010] from the same authors continues this line including monotonicity and cardinality results. Furthermore, in [Boella et al., 2009] so-called *attack refinement principles* are introduced stating whether the extension stays the same if a single attack is added. They studied the fulfillment of these principles with respect to grounded semantics.

In the third section we presented an encoding of action domains into abstract argumentation frameworks published in [Baumann and Strass, 2012]. In being independent of a particular notion of time and able to do default reasoning in dynamic domains, the approach goes well beyond the capabilities of current action languages. We used theoretical results from argumentation to show how our approach can be put to use and argue how it can be implemented.

In continuation of their preliminary earlier work [Kakas et al., 2008], Michael and Kakas developed an approach that combines default reasoning with temporal reasoning and is based on *assumption-based* argumentation [Michael and Kakas, 2009; Michael and Kakas, 2011]. The approach uses a fixed linear time structure and a tailor-made definition of argumentation semantics. The work presented in this paper uses a more general notion of time, furthermore we employ *abstract* argumentation with the standard definitions of its semantics, and can immediately use existing results from this area, in particular existing solvers. Earlier, Kakas et al. [Kakas et al., 1999] translated domains in an action language based on linear time into the argumentation framework of logic programming without negation as failure, a contribution much closer to the original work on action languages [Gelfond and Lifschitz, 1998].

Chapter 4

Splitting Results

Splitting is a fundamental principle and has been investigated for several nonmonotonic formalisms. For instance, in an important and much cited paper Vladimir Lifschitz and Hudson Turner [Lifschitz and Turner, 1994] have shown how, under certain conditions, logic programs under answer set semantics can be split into two disjoint parts, a "bottom" part and a "top" part. The bottom part can be evaluated independently of the top part. Results of the evaluation, i.e., answer sets of the bottom part, are then used to simplify the top part. To obtain answer sets of the original program one simply has to combine an answer set of the simplified top part with the answer set which was used to simplify this part. Similar results were obtained for default logic [Turner, 1996] as well as auto-epistemic logic [Gelfond and Przymusinska, 1992]. The possibility of splitting has important implications, both from the theoretical and from the practical point of view. On the theoretical side, splitting allows for simplification of proofs showing properties of a particular formalism. On the practical side, splitting results may yield more efficient computations.

In this chapter we study the concept of splitting in Dung's abstract argumentation frameworks. We first consider so-called *classical splittings* which are intimately connected to the already introduced concept of weak expansions. In the second part we show how the conditions under which splitting is possible can be relaxed. The main idea is to encode the possibility of attack in form of new arguments. This way we can split in cases where classical splitting is not possible. Finally, we convey our idea of a non-classical splitting to logic programs generalizing results proven in [Lifschitz and Turner, 1994].

4.1 Classical Splitting Results

We start with the consideration of partitions into two parts such that the remaining attacks are restricted to a single direction. It turns out that for stable semantics the result is similar to logic programs. However, for admissible,

4.1. Classical Splitting Results

preferred, complete and grounded semantics, a more sophisticated modification is needed which takes into account that arguments may be neither accepted nor refuted in extensions. As a byproduct we will prove that admissible, preferred, complete and grounded semantics satisfy directionality even in the case of infinite AFs. Furthermore, the specific way of partitioning allows us to transfer the obtained results into the field of dynamical argumentation. In particular, we will strengthen a former monotonicity result (Theorem 3.2).
We proceed with the formal foundation, i.e. we develop the technical tools which are needed to prove the splitting results.

4.1.1 Classical Splitting, Reduct, Undefined Set and Modification

We start with the definition of classical splittings.

Definition 4.1. *Let $\mathcal{F}_1 = (A_1, R_1)$ and $\mathcal{F}_2 = (A_2, R_2)$ be AFs such that $A_1 \cap A_2 = \emptyset$. Let $R_3 \subseteq A_1 \times A_2$. We call the tuple $(\mathcal{F}_1, \mathcal{F}_2, R_3)$ a classical splitting of the argumentation framework $\mathcal{F} = (A_1 \cup A_2, R_1 \cup R_2 \cup R_3)$.*

For short, a classical splitting of a given AF $\mathcal{F}$ is a partition in two disjoint AFs $\mathcal{F}_1$ and $\mathcal{F}_2$ such that the remaining attacks between $\mathcal{F}_1$ and $\mathcal{F}_2$ are restricted to a single direction. The following proposition establishes the connection between splittings and weak expansions. It states that weak expansions and the introduced splitting definition are in a sense two sides of the same coin. Note that this property is pretty obvious. Being aware of this fact, we still present it in the form of a proposition.

Proposition 4.2. *If $(\mathcal{F}_1, \mathcal{F}_2, R_3)$ is a splitting of $\mathcal{F}$, then $\mathcal{F}$ is a weak expansion of $\mathcal{F}_1$. Vice versa, if $\mathcal{F} = (A, R)$ is a weak expansion of $\mathcal{F}_1 = (A_1, R_1)$, then $(\mathcal{F}_1, \mathcal{F}_2, R_3)$ with $\mathcal{F}_2 = (A \smallsetminus A_1, R \cap (A \smallsetminus A_1 \times A \smallsetminus A_1))$ and $R_3 = R \cap (A_1 \times (A \smallsetminus A_1))$ is a splitting of $\mathcal{F}$.*

Now we turn to the main question. Given a splitting $(\mathcal{F}_1, \mathcal{F}_2, R_3)$ of an AF $\mathcal{F}$. How are the extensions of $\mathcal{F}$ and the extensions of $\mathcal{F}_1$ and $\mathcal{F}_2$ related?

Example 4.3. Consider the following AF $\mathcal{F}$ and its splitting $(\mathcal{F}_1, \mathcal{F}_2, R_3)$.

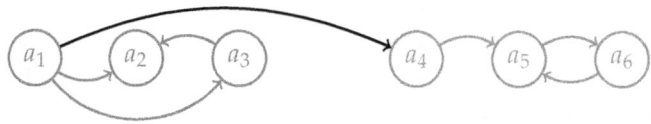

Figure 4.1: Classical Splitting

There are two stable extensions of $\mathcal{F}$, namely $E' = \{a_1, a_5\}$ and $E'' = \{a_1, a_6\}$. Furthermore we observe that $E_1 = \{a_1\}$ and $E_2 = \{a_4, a_6\}$ are the unique stable extensions of $\mathcal{F}_1$ and $\mathcal{F}_2$, respectively. Observe that we cannot *reconstruct* the extensions E' and E'' out of the extensions E_1 and E_2. This is not very surprising because we do not take into account the attack (a_1, a_4).

Chapter 4. Splitting Results

If we delete the argument a_4 in $\mathcal{F}_2$ which is attacked by E_1 and then compute the stable extensions in the reduced AF $(\{a_5, a_6\}, \{(a_5, a_6), (a_6, a_5)\})$ we get the "missing" singletons $\{a_5\}$ and $\{a_6\}$. In fact, E' and E'' can be reconstructed as unions, namely $E' = \{a_1\} \cup \{a_5\}$ and $E'' = \{a_1\} \cup \{a_6\}$.

Figure 4.2: The Reduct of $\mathcal{F}_2$ with respect to E_1 and R_3

We will see that this observation holds in general for the stable semantics. The following definition of a reduct captures the intuitive idea.

Definition 4.4. Let $\mathcal{F} = (A, R)$ be an AF, A' a set disjoint from A, $S \subseteq A'$ and $L \subseteq A' \times A$. The (S, L)-reduct of $\mathcal{F}$, denoted $\mathcal{F}^{S,L}$ is the AF

$$\mathcal{F}^{S,L} = (A^{S,L}, R^{S,L})$$

where

$$A^{S,L} = \{a \in A \mid (S, a) \notin L)\} \text{ and } R^{S,L} = \{(a, b) \in R \mid a, b \in A^{S,L}\}.$$

The intuitively described reduced version of the AF $\mathcal{F}_2$ depicted in Figure 4.2 can be formalized in the following way: $(\{a_5, a_6\}, \{(a_5, a_6), (a_6, a_5)\}) = \mathcal{F}_2^{E_1, R_3}$. Unfortunately it turns out that the reduct used above does not obtain the desired properties for other semantics we are interested in. Here is a counterexample.

Example 4.5. Given the following AF $\mathcal{F}$ and its splitting $(\mathcal{F}_1, \mathcal{F}_2, R_3)$.

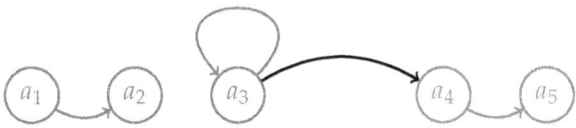

Figure 4.3: Counterexample Reduct ($\sigma \in \{pr, co, gr\}$)

We observe that $E = \{a_1\}$ is the unique preferred, complete and grounded extension of $\mathcal{F}$. The same holds for the AF $\mathcal{F}_1$, i.e. $E_1 = \{a_1\}$. Furthermore, the (E_1, R_3)-reduct of $\mathcal{F}_2$ equals $\mathcal{F}_2$ since a_1 does not attack arguments in $\mathcal{F}_2$. The reduct of $\mathcal{F}_2$ establishes the unique preferred, complete and grounded extension $E_2 = \{a_4\}$. Yet the union of E_1 and E_2 differs from E.

Given a splitting $(\mathcal{F}_1, \mathcal{F}_2, R_3)$ of $\mathcal{F}$. The problem described above stems from the fact that the distinction between those arguments which are not in the extension *because they are refuted* (attacked by an accepted argument) and those not in the extension *without being refuted* is not taken care of. The former

4.1. Classical Splitting Results

have no influence on $\mathcal{F}_2$. However, the latter - which we will call undefined in contrast to the refuted ones - indeed have an influence on $\mathcal{F}_2$, as illustrated in Example 4.5. To overcome this problem, we introduce a simple modification. Whenever there is an undefined argument a in the extension of the first AF $\mathcal{F}_1$ which attacks an argument b in the second (maybe reduced) AF $\mathcal{F}_2$, we modify the latter so that b is both origin and goal of the attack.

Example 4.6 (Example 4.5 continued). The fact that a_3 is undefined with respect to E_1 leads to the undefinedness of both a_4 and a_5, and this is not captured by the (E_1, R_3)-reduct of $\mathcal{F}_2$. We can enforce undefinedness of a_4 (and thus of a_5) in $\mathcal{F}_2$ by introducing a self-attack for a_4.

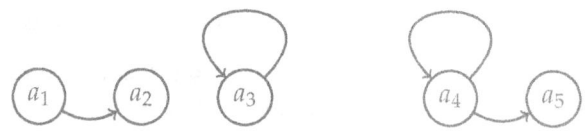

Figure 4.4: The Modification of $\mathcal{F}_2$ with respect to E_1 and R_3

Now, the modified AF $(\{a_4, a_5\}, \{(a_4, a_4), (a_4, a_5)\})$ establishes the empty set as the unique extension under preferred, complete and grounded semantics. Thus, E can be reconstructed as union, namely $E = \{a_1\} \cup \emptyset$.

Here are the formal definitions capturing the ideas outlined above.

Definition 4.7. Let $\mathcal{F} = (A, R)$ be an AF, E an extension of $\mathcal{F}$. The set of arguments *undefined with respect to E* is

$$U_E = \{a \in A \mid a \notin E, (E, a) \notin R\}.$$

Definition 4.8. Let $\mathcal{F} = (A, R)$ be an AF, A' a set disjoint from A, $S \subseteq A'$ and $L \subseteq A' \times A$. The (S, L)-modification of $\mathcal{F}$, denoted $mod_{S,L}(\mathcal{F})$, is the AF

$$mod_{S,L}(\mathcal{F}) = (A, R \cup \{(b, b) \mid (S, b) \in L\}).$$

Given a splitting $(\mathcal{F}_1, \mathcal{F}_2, R_3)$ of $\mathcal{F}$, an extension E of $\mathcal{F}_1$ which leaves the set of arguments U_E undefined, we will use $mod_{U_E, R_3}\left(\mathcal{F}_2^{E, R_3}\right)$ to compute what is missing from E. In case of Example 4.6 we computed the extensions of $(\{a_4, a_5\}, \{(a_4, a_4), (a_4, a_5)\})$ which can be equivalently formalized as $mod_{\{a_3\}, R_3}\left(\mathcal{F}_2^{E_1, R_3}\right)$. Note that, although links are added, under all standard measures of the size of a graph (e.g. number of links plus number of vertices) we have $|\mathcal{F}_1| + \left|mod_{U_E, R_3}\left(\mathcal{F}_2^{E, R_3}\right)\right| \leq |\mathcal{F}|$.

In the next section we will present our formal results using the technical tools defined here. Before doing so we will prove some simple properties which are frequently used throughout the proofs.

Proposition 4.9. Given an AF $\mathcal{F} = (A, R)$ which possesses a splitting $(\mathcal{F}_1, \mathcal{F}_2, R_3)$ such that $\mathcal{F}_1 = (A_1, R_1)$ and $\mathcal{F}_2 = (A_2, R_2)$, the following hold:

Chapter 4. Splitting Results

1. $E_1 \in \mathcal{E}_{stb}(\mathcal{F}_1) \Rightarrow mod_{U_{E_1},R_3}\left(\mathcal{F}_2^{E_1,R_3}\right) = \mathcal{F}_2^{E_1,R_3}$,

(neutrality of the modification with respect to the stable reduct)

2. $E \in cf(\mathcal{F}) \Rightarrow E \cap A_1 \in cf(\mathcal{F}_1) \wedge E \cap A_2 \in cf\left(\mathcal{F}_2^{E \cap A_1, R_3}\right)$,

(preserving conflict-freeness [intersection])

3. $E_1 \in cf(\mathcal{F}_1) \wedge E_2 \in cf\left(mod_{U_{E_1},R_3}\left(\mathcal{F}_2^{E_1,R_3}\right)\right) \Rightarrow E_1 \cup E_2 \in cf(\mathcal{F})$.

(preserving conflict-freeness [union])

Proof. 1. Given $E_1 \in \mathcal{E}_{stb}(\mathcal{F}_1)$. Consequently, all arguments in $A_1 \setminus E_1$ are attacked by E_1 and thus, $U_{E_1} = \emptyset$. This means, $mod_{U_{E_1},R_3}\left(\mathcal{F}_2^{E_1,R_3}\right) = \mathcal{F}_2^{E_1,R_3}$.
2. Subsets of conflict-free sets are conflict-free. Furthermore, $R_1 \subseteq R$ and $R_2^{E \cap A_1, R_3} \subseteq R_2 \subseteq R$. Hence, it suffices to show that $E \cap A_1 \subseteq A_1$ (obvious) and $E \cap A_2 \subseteq A_2^{E \cap A_1, R_3}$. The latter can be seen as follows: the assumption $E \in cf(\mathcal{F})$ guarantees $(E \cup A_1, E \cup A_2) \notin R_3$. Thus, $E \cup A_2 \subseteq A_2^{E \cap A_1, R_3}$ concluding the proof.
3. First, $E_1 \in cf(\mathcal{F})$ since we have assumed that $E_1 \in cf(\mathcal{F}_1)$ and $(\mathcal{F}_1, \mathcal{F}_2, R_3)$ is a splitting of $\mathcal{F}$. Furthermore, $E_2 \in cf\left(mod_{U_{E_1},R_3}\left(\mathcal{F}_2^{E_1,R_3}\right)\right)$ implies $E_2 \in cf\left(\mathcal{F}_2^{E_1,R_3}\right)$ (less attacks). Thus, $E_2 \in cf(\mathcal{F})$ and $(E_1, E_2) \notin R_3$ has to hold. Finally, $(E_2, E_1) \notin R_3$ because $(\mathcal{F}_1, \mathcal{F}_2, R_3)$ is a splitting of $\mathcal{F}$. Altogether, $E_1 \cup E_2 \in cf(\mathcal{F})$ is implied. □

4.1.2 Splitting Theorem

Given a splitting of an AF $\mathcal{F}$, the general idea is to compute an extension E_1 of $\mathcal{F}_1$, reduce and modify $\mathcal{F}_2$ depending on what extension we got, and then compute an extension E_2 of the modification of the reduct of $\mathcal{F}_2$. The resulting union of E_1 and E_2 is an extension of $\mathcal{F}$. The second part of the theorem proves the completeness of this method, i.e. all extensions are constructed this way.

Theorem 4.10. Let $\mathcal{F} = (A, R)$ be an AF which possesses a splitting $(\mathcal{F}_1, \mathcal{F}_2, R_3)$ with $\mathcal{F}_1 = (A_1, R_1)$ and $\mathcal{F}_2 = (A_2, R_2)$. Furthermore, let $\sigma \in \{stb, ad, pr, co, gr\}$.

1. If E_1 is an extension of $\mathcal{F}_1$ and E_2 is an extension of the (U_{E_1}, R_3)-modification of $\mathcal{F}_2^{E_1, R_3}$, then $E = E_1 \cup E_2$ is an extension of $\mathcal{F}$.

$$E_1 \in \mathcal{E}_\sigma(\mathcal{F}_1) \wedge E_2 \in \mathcal{E}_\sigma\left(mod_{U_{E_1},R_3}\left(\mathcal{F}_2^{E_1,R_3}\right)\right) \Rightarrow E_1 \cup E_2 \in \mathcal{E}_\sigma(\mathcal{F})$$

2. If E is an extension of $\mathcal{F}$, then $E_1 = E \cap A_1$ is an extension of $\mathcal{F}_1$ and $E_2 = E \cap A_2$ is an extension of the (U_{E_1}, R_3)-modification of $\mathcal{F}_2^{E_1, R_3}$.

$$E \in \mathcal{E}_\sigma(\mathcal{F}) \Rightarrow E \cap A_1 \in \mathcal{E}_\sigma(\mathcal{F}_1) \wedge E \cap A_2 \in \mathcal{E}_\sigma\left(mod_{U_{E \cap A_1},R_3}\left(\mathcal{F}_2^{E \cap A_1, R_3}\right)\right)$$

4.1. Classical Splitting Results

Proof. **(stable semantics)** 1. Statement 3 of Proposition 4.9 guarantees conflict-freeness of $E_1 \cup E_2$ in $\mathcal{F}$. We will show now that $E_1 \cup E_2$ attacks all arguments in $(A_1 \cup A_2) \setminus (E_1 \cup E_2)$. First, for any $a \in A_1 \setminus E_1$ we have $(E_1, a) \in R_1 \subseteq R$ because $E_1 \in \mathcal{E}_{stb}(\mathcal{F}_1)$ is assumed. On the other hand, if $a \in A_2 \setminus E_2$ we have either $(E_2, a) \in R_2$ or $(E_1, a) \in R_3$. This can be seen as follows: by statement 1 of Proposition 4.9 we have $mod_{U_{E_1}, R_3}\left(\mathcal{F}_2^{E_1, R_3}\right) = \mathcal{F}_2^{E_1, R_3}$. Thus, $E_2 \in \mathcal{E}_{stb}\left(\mathcal{F}_2^{E_1, R_3}\right)$ and consequently, if $a \in A_2^{E_1, R_3}$, then $(E_2, a) \in R_2^{E_1, R_3} \subseteq R_2 \subseteq R$. If not, then $a \in \{a \in A_2 \mid (E_1, a) \in R_3\}$ (compare Definition 4.4). Consequently, $(E_1, a) \in R_3 \subseteq R$.

2. By statement 2 of Proposition 4.9 we have $E \cap A_1 = E_1 \in cf(\mathcal{F}_1)$. Furthermore, any $a \in A_1 \setminus E_1$ is attacked by E_1 in $\mathcal{F}_1$ because we assumed that $E \in \mathcal{E}_{stb}(\mathcal{F})$ and $(\mathcal{F}_1, \mathcal{F}_2, R_3)$ is a splitting of $\mathcal{F}$. This means, $E_1 \in \mathcal{E}_{stb}(\mathcal{F}_1)$. Given statements 1 and 2 of Proposition 4.9 we deduce $E \cap A_2 = E_2 \in cf\left(mod_{U_{E_1}, R_3}\left(\mathcal{F}_2^{E_1, R_3}\right)\right)$. Since $A_2^{E_1, R_3} = \{a \in A_2 \mid (E_1, a) \notin R_3\}$ and $E \in \mathcal{E}_{stb}(\mathcal{F})$ we deduce $(E_2, a) \in R_2$ for any $a \in A_2^{E_1, R_3} \setminus E_2$. Finally, using statement 1 of Proposition 4.9 justifies $E_2 \in \mathcal{E}_{stb}\left(mod_{U_{E \cap A_1}, R_3}\left(\mathcal{F}_2^{E \cap A_1, R_3}\right)\right)$.

(admissible semantics) 1. Since admissible sets are conflict-free we deduce $E_1 \cup E_2 \in cf(\mathcal{F})$ (statement 3 of Proposition 4.9). We have to show that $E_1 \cup E_2$ is even admissible in $\mathcal{F}$. We show this by case distinction. Assume $(a, E_1) \in R$. Thus, $a \in A_1$ and $(a, E_1) \in R_1$ since $(\mathcal{F}_1, \mathcal{F}_2, R_3)$ is assumed to be a splitting of $\mathcal{F}$. Furthermore, $E_1 \in \mathcal{E}_{ad}(\mathcal{F}_1)$ implies $(E_1, a) \in R_1 \subseteq R$. Assume now $(a, E_2) \in R$. If $a \in A_1$ or more precisely $a \in A_1 \setminus E_1$, then $(E_1, a) \in R_1 \subseteq R$ has to hold because assuming the contrary, i.e. $(E_1, a) \notin R_1$ implies $(E_2, E_2) \in R\left(mod_{U_{E_1}, R_3}\left(\mathcal{F}_2^{E_1, R_3}\right)\right)$ (added self-loop) in contrast to the assumed admissibility and hence conflict-freeness of E_2 in $mod_{U_{E_1}, R_3}\left(\mathcal{F}_2^{E_1, R_3}\right)$. Let $a \in A_2$. If $a \in \{a \in A_2 \mid (E_1, a) \in R_3 \subseteq R\}$, then nothing is to show. If not, we have $a \in A_2^{E_1, R_3} = \{a \in A_2 \mid (E_1, a) \notin R_3\}$. Since $E_2 \in \mathcal{E}_{ad}\left(mod_{U_{E_1}, R_3}\left(\mathcal{F}_2^{E_1, R_3}\right)\right)$ is assumed it follows $E_2 \in \mathcal{E}_{ad}\left(\mathcal{F}_2^{E_1, R_3}\right)$ and hence, $(E_2, a) \in R_2^{E_1, R_3} \subseteq R_2 \subseteq R$. Altogether, $E_1 \cup E_2 \in \mathcal{E}_{ad}(\mathcal{F})$ is shown.

2. First, $E \cap A_1 = E_1 \in cf(\mathcal{F}_1)$ and $E \cap A_2 = E_2 \in cf\left(\mathcal{F}_2^{E \cap A_1, R_3}\right)$ is given by statement 2 of Proposition 4.9. Furthermore, $E_2 \in cf\left(mod_{U_{E_1}, R_3}\left(\mathcal{F}_2^{E_1, R_3}\right)\right)$ because an additional self-loop in E_2 resulting from an attack of an undefined argument with respect to E_1 in $\mathcal{F}_1$ would contradict the admissibility of E in $\mathcal{F}$ because we assumed that $(\mathcal{F}_1, \mathcal{F}_2, R_3)$ is an splitting of $\mathcal{F}$. For the same reasons we conclude that $E_1 \in \mathcal{E}_{ad}(\mathcal{F}_1)$. Consider now E_2. Let $a \in A\left(mod_{U_{E_1}, R_3}\left(\mathcal{F}_2^{E_1, R_3}\right)\right) \setminus E_2$ and $(a, E_2) \in R\left(mod_{U_{E_1}, R_3}\left(\mathcal{F}_2^{E_1, R_3}\right)\right)$. Remember that $A\left(mod_{U_{E_1}, R_3}\left(\mathcal{F}_2^{E_1, R_3}\right)\right) = A_2^{E_1, R_3} = \{a \in A_2 \mid (E_1, a) \notin R_3\}$. Furthermore, since $a \notin E_2$ we have $(a, E_2) \in R_2^{E_1, R_3} \subseteq R_2$. Since $E \in \mathcal{E}_{ad}(\mathcal{F})$ is assumed we conclude $(E_2, a) \in R_2$ and even $(E_2, a) \in R_2^{E_1, R_3} \subseteq R\left(mod_{U_{E_1}, R_3}\left(\mathcal{F}_2^{E_1, R_3}\right)\right)$ because $a \in A_2^{E_1, R_3}$. Thus, $E_2 \in \mathcal{E}_{ad}\left(mod_{U_{E \cap A_1}, R_3}\left(\mathcal{F}_2^{E \cap A_1, R_3}\right)\right)$ is shown.

Chapter 4. Splitting Results

(**preferred semantics**) 1. Due to the previous proof and since any preferred extension is admissible we have $E_1 \cup E_2 \in \mathcal{E}_{ad}(\mathcal{F})$. Assume $E_1 \cup E_2$ is not maximal with respect to set inclusion, i.e. there is an $E' \in \mathcal{E}_{ad}(\mathcal{F}) : E_1 \cup E_2 \subset E'$. Consequently, at least one of the following two cases hold: $E_1 \subset E' \cap A_1$ or $E_2 \subset E' \cap A_2$. The first case contradicts the maximality of E_1 in $\mathcal{F}_1$ because $E' \cap A_1$ is an admissible extension of $\mathcal{F}_1$ (statement 2 of Theorem 4.10 for $\sigma = ad$). Consider now the second case, namely $E_2 \subset E' \cap A_2$. Without loss of generality we may assume $E' \cap A_1 = E_1$. In combination with statement 2 of Theorem 4.10 we deduce $E' \cap A_2 \in \mathcal{E}_{ad}\left(mod_{U_{E_1},R_3}\left(\mathcal{F}_2^{E_1,R_3}\right)\right)$ in contradiction to the maximality of E_2 in $mod_{U_{E_1},R_3}\left(\mathcal{F}_2^{E_1,R_3}\right)$. Thus, $E_1 \cup E_2 \in \mathcal{E}_{pr}(\mathcal{F})$ has to hold.

2. Let $E \in \mathcal{E}_{pr}(\mathcal{F})$. By statement 2 of Theorem 4.10 $E_1 = E \cap A_1 \in \mathcal{E}_{ad}(\mathcal{F}_1)$ and $E_2 = E \cap A_2 \in \mathcal{E}_{ad}\left(mod_{U_{E_1},R_3}\left(\mathcal{F}_2^{E_1,R_3}\right)\right)$ are guaranteed. Assume E_1 is not preferred in $\mathcal{F}_1$, i.e. there is an $E'_1 \in \mathcal{E}_{ad}(\mathcal{F}_1)$, such that $E_1 \subset E'_1$. We will show now that E_2 is admissible in $mod_{U_{E'_1},R_3}\left(\mathcal{F}_2^{E'_1,R_3}\right)$ and thus, $E'_1 \cup E_2 \in \mathcal{E}_{ad}(\mathcal{F})$ (statement 1 of Theorem 4.10) in contradiction to the maximality of E in $\mathcal{F}$. First, $E_2 \subseteq A_2^{E'_1,R_3} = A\left(mod_{U_{E'_1},R_3}\left(\mathcal{F}_2^{E'_1,R_3}\right)\right)$ because assuming the contrary would imply $(E'_1, E_2) \in R_3$. Since E is assumed to be admissible in $\mathcal{F}$ and furthermore, $(\mathcal{F}_1, \mathcal{F}_2, R_3)$ is an splitting of $\mathcal{F}$ we conclude $(E_1, E'_1) \in R_1$ in contradiction to the conflict-freeness of $E'_1 \in \mathcal{F}_1$. Second, observe that $U_{E'_1} \subseteq U_{E_1}$ and $R_2^{E'_1,R_3} \subseteq R_2^{E_1,R_3}$. Consequently, $E_2 \in cf\left(mod_{U_{E_1},R_3}\left(\mathcal{F}_2^{E_1,R_3}\right)\right)$ justifies $E_2 \in cf\left(mod_{U_{E'_1},R_3}\left(\mathcal{F}_2^{E'_1,R_3}\right)\right)$. Finally, using the admissibility of E_2 in $mod_{U_{E_1},R_3}\left(\mathcal{F}_2^{E_1,R_3}\right)$ and the observations $U_{E'_1} \subseteq U_{E_1}$ and $R_2^{E'_1,R_3} \subseteq R_2^{E_1,R_3}$ justify $E_2 \in \mathcal{E}_{ad}\left(mod_{U_{E'_1},R_3}\left(\mathcal{F}_2^{E'_1,R_3}\right)\right)$. Consequently, $E'_1 \cup E_2 \in \mathcal{E}_{ad}(\mathcal{F})$ (statement 1 of Theorem 4.10) contradicting the assumption $E \in \mathcal{E}_{pr}(\mathcal{F})$. Finally, assuming that $E \cap A_2 \notin \mathcal{E}_{pr}\left(mod_{U_{E \cap A_1},R_3}\left(\mathcal{F}_2^{E \cap A_1,R_3}\right)\right)$ contradicts the assumption $E \in \mathcal{E}_{pr}(\mathcal{F})$. This can be seen as follows: First, $E \cap A_2 \in \mathcal{E}_{ad}\left(mod_{U_{E \cap A_1},R_3}\left(\mathcal{F}_2^{E \cap A_1,R_3}\right)\right)$ is given by statement 2 of Theorem 4.10 ($\sigma = ad$). Thus, there exists a set E'_2 being a proper superset of E_2, such that $E'_2 \in \mathcal{E}_{ad}\left(mod_{U_{E \cap A_1},R_3}\left(\mathcal{F}_2^{E \cap A_1,R_3}\right)\right)$. Hence, by statement 1 of Theorem 4.10 $E_1 \cup E'_2 \in \mathcal{E}_{ad}(\mathcal{F})$ in contradiction to $E \in \mathcal{E}_{pr}(\mathcal{F})$.

(**complete semantics**) 1. We have to show $E_1 \cup E_2 \in \mathcal{E}_{co}(\mathcal{F})$. Note that admissibility is already given by statement 1 of Theorem 4.10 ($\sigma = ad$). Thus, it suffices to show that for each $a \in A_1 \cup A_2$ defended by $E_1 \cup E_2$ in $\mathcal{F}$, $a \in E_1 \cup E_2$. Assume not, hence there is an $a \in (A_1 \cup A_2) \setminus (E_1 \cup E_2)$, such that a is defended by $E_1 \cup E_2$ in $\mathcal{F}$. Supposing $a \in A_1 \setminus E_1$ contradicts $E_1 \in \mathcal{E}_{co}(\mathcal{F}_1)$ because $(\mathcal{F}_1, \mathcal{F}_2, R_3)$ is assumed to be a splitting of $\mathcal{F}$. Consider now $a \in A_2 \setminus E_2$. Obviously, $a \in A_2^{E_1,R_3}$ because assuming the contrary would contradict the

53

conflict-freeness of E_1 in $\mathcal{F}_1$. We have to consider three attack-scenarios: First, a is attacked by arguments in $A_2 \smallsetminus A_2^{E_1,R_3}$ (and obviously defended by E_1 in $\mathcal{F}$). The reduct-relation $R_2^{E_1,R_3}$ do not contain such attacks and hence, every "attack" is counter-attacked by E_2 in $mod_{U_{E_1},R_3}\left(\mathcal{F}_2^{E_1,R_3}\right)$. Second, a is attacked by arguments in $A_2^{E_1,R_3} \smallsetminus E_2$. Hence, it must be defended by elements of E_2 in $\mathcal{F}_2^{E_1,R_3}$. Furthermore, it is even defended by E_2 in $mod_{U_{E_1},R_3}\left(\mathcal{F}_2^{E_1,R_3}\right)$ because the modified attack-relation still retains such counter-attacks. Finally, a may be attacked by arguments $A_1 \smallsetminus E_1$. Thus, it has to be defended by E_1 in $\mathcal{F}$ since $(\mathcal{F}_1,\mathcal{F}_2,R_3)$ is a splitting of $\mathcal{F}$. Obviously, the reduct-relation $R_2^{E_1,R_3}$ do not contain such attacks. Consequently, every "attack" is counter-attacked by E_2 in $mod_{U_{E_1},R_3}\left(\mathcal{F}_2^{E_1,R_3}\right)$. This means, in any case we conclude $E_2 \notin \mathcal{E}_{co}\left(mod_{U_{E_1},R_3}\left(\mathcal{F}_2^{E_1,R_3}\right)\right)$ in contradiction to the assumption.

2. Given $E \in \mathcal{E}_{co}(\mathcal{F})$. By statement 2 of Theorem 4.10 $E_1 = E \cap A_1 \in \mathcal{E}_{ad}(\mathcal{F}_1)$ and $E_2 = E \cap A_2 \in \mathcal{E}_{ad}\left(mod_{U_{E_1},R_3}\left(\mathcal{F}_2^{E_1,R_3}\right)\right)$ are guaranteed. Assume $E_1 \notin \mathcal{E}_{co}(\mathcal{F}_1)$. Thus, there exists an $a \in A_1 \smallsetminus E_1$ defended by E_1 in $\mathcal{F}_1$. Since $(\mathcal{F}_1,\mathcal{F}_2,R_3)$ is assumed to be a splitting of $\mathcal{F}$ we conclude a is defended by E in $\mathcal{F}$ in contrast to $E \in \mathcal{E}_{co}(\mathcal{F})$. Thus, $E_1 \in \mathcal{E}_{co}(\mathcal{F}_1)$. Consider now E_2. Suppose there is an argument $a \in A_2^{E_1,R_3} \smallsetminus E_2$, such that a is defended by E_2 in $mod_{U_{E_1},R_3}\left(\mathcal{F}_2^{E_1,R_3}\right)$. Clearly, the potential attackers of a in $A_2^{E_1,R_3} \smallsetminus E_2$ are still counter-attacked by E_2 in $\mathcal{F}$. Furthermore, since $A_2^{E_1,R_3} = \{b \in A_2 \mid (E_1,b) \notin R_3\}$ we have: if a is attacked by arguments in $A_2 \smallsetminus A_2^{E_1,R_3}$, then a is defended by E_1 in $\mathcal{F}$. Finally, note that a is not self-defeating in $mod_{U_{E_1},R_3}\left(\mathcal{F}_2^{E_1,R_3}\right)$ because E_2 has to be conflict-free. Thus, a is not attacked by arguments in U_{E_1} in $\mathcal{F}$. This means, if a is attacked by arguments in $A_1 \smallsetminus E_1$ in $\mathcal{F}$, then a is defended by E in $\mathcal{F}$. Altogether, $E \notin \mathcal{E}_{co}(\mathcal{F})$ in contradiction to the assumption.

(grounded semantics) 1. Since grounded extensions are complete we have $E_1 \cup E_2 \in \mathcal{E}_{co}(\mathcal{F})$ (statement 1 of Theorem 4.10 for $\sigma = co$). Thus, it suffices to show minimality with respect to subset relation. Assume not, i.e. there is a set $E' \in \mathcal{E}_{co}(\mathcal{F})$, such that $E' \subset E_1 \cup E_2$. Consequently, at least one of the following two cases has to hold: $E'_1 = E' \cap A_1 \subset E_1$ or $E'_2 = E' \cap A_2 \subset E_2$. The first case contradicts the minimality of E_1 in $\mathcal{F}_1$ because E'_1 is complete in $\mathcal{F}_1$ (statement 2 of Theorem 4.10 for $\sigma = co$). Consider now the second case and assume $E'_1 = E_1$. Using statement 2 of Theorem 4.10 we deduce $E'_2 \in \mathcal{E}_{co}\left(mod_{U_{E_1},R_3}\left(\mathcal{F}_2^{E_1,R_3}\right)\right)$ in contradiction to the minimality of E_2 in $mod_{U_{E_1},R_3}\left(\mathcal{F}_2^{E_1,R_3}\right)$. Thus, $E_1 \cup E_2 \in \mathcal{E}_{co}(\mathcal{F})$ has to hold.

2. Let $E \in \mathcal{E}_{gr}(\mathcal{F})$. By statement 2 of Theorem 4.10 $E_1 = E \cap A_1 \in \mathcal{E}_{co}(\mathcal{F}_1)$ and $E_2 = E \cap A_2 \in \mathcal{E}_{co}\left(mod_{U_{E_1},R_3}\left(\mathcal{F}_2^{E_1,R_3}\right)\right)$ are implied. This means, $\Gamma_{\mathcal{F}}(E) = E$ and $\Gamma_{\mathcal{F}_1}(E_1) = E_1$. Assume E_1 is not grounded in $\mathcal{F}_1$. This means, there is an $E'_1 \in \mathcal{E}_{co}(\mathcal{F}_1)$, such that $E'_1 \subset E_1$ and $\Gamma_{\mathcal{F}_1}(E'_1) = E'_1$. Since E_1 and E'_1 are fixpoints of $\Gamma_{\mathcal{F}_1}$ and $(\mathcal{F}_1,\mathcal{F}_2,R_3)$ is assumed to be a splitting of $\mathcal{F}$ we have:

Chapter 4. Splitting Results

$\Gamma^i_\mathcal{F}(E_1) \setminus E_1 \subseteq A_2$ and $\Gamma^i_\mathcal{F}(E'_1) \setminus E'_1 \subseteq A_2$ for any ordinal i. Furthermore, by assumption we have $E'_1 \subset E_1 \subseteq E$. Consequently, $\Gamma^i_\mathcal{F}(E'_1) \subset \Gamma^i_\mathcal{F}(E_1) \subseteq \Gamma^i_\mathcal{F}(E) = E$ for any ordinal i since $\Gamma_\mathcal{F}$ is monotonic. Infinite iteration (exceeding the number of arguments) stabilizes the chain, i.e. $\Gamma^j_\mathcal{F}(E'_1) = \Gamma(\Gamma^j_\mathcal{F}(E'_1))$ for some ordinal j (see [Grossi and Gabbay, 2013, Remark 1] for further information). This contradicts $E \in \mathcal{E}_{gr}(\mathcal{F})$ since $\Gamma^j_\mathcal{F}(E'_1) \subset E$ has to hold by construction. This means, $E_1 \in \mathcal{E}_{gr}(\mathcal{F}_1)$. Finally, assuming that $E_2 \notin \mathcal{E}_{gr}\left(mod_{U_{E_1},R_3}\left(\mathcal{F}_2^{E_1,R_3}\right)\right)$ contradicts the assumption $E \in \mathcal{E}_{gr}(\mathcal{F})$. This can be seen as follows: Assuming the existence of an $E'_2 \in \mathcal{E}_{co}\left(mod_{U_{E_1},R_3}\left(\mathcal{F}_2^{E_1,R_3}\right)\right)$, such that $E'_2 \subset E_2$ implies $E_1 \cup E'_2 \in \mathcal{E}_{co}(\mathcal{F})$ (statement 2 of Theorem 4.10). Consequently, $E \in \mathcal{E}_{gr}(\mathcal{F})$ cannot hold in contrast to the assumption. □

We want to emphasize that the proofs do not make use of finiteness. Hence, splitting results can be used for the whole class of AFs if considered under stable, admissible, preferred, complete or grounded semantics.

What about the other semantics we consider in this book? The following examples show that the splitting theorem 4.10 does not hold in general in case of semi-stable, stage, eager, ideal and naive semantics. The development of the *right* technical tools required to prove similar splitting results is part of future work.

Example 4.11. Given the following AF $\mathcal{F}$ and its splitting $(\mathcal{F}_1, \mathcal{F}_2, R_3)$. We have $\mathcal{E}_{ss}(\mathcal{F}) = \mathcal{E}_{stg}(\mathcal{F}) = \{\{a_2\}, \{a_3\}\}$. Thus, $\mathcal{E}_{eg}(\mathcal{F}) = \{\emptyset\}$. Furthermore, $\mathcal{E}_{ss}(\mathcal{F}_1) = \mathcal{E}_{stg}(\mathcal{F}_1) = \{\{a_2\}\}$. Hence, $\mathcal{E}_{eg}(\mathcal{F}_1) = \{\{a_2\}\}$. This means, $\{a_3\} \cap \{a_1, a_2, a_3\} = \{a_3\} \notin \mathcal{E}_\sigma(\mathcal{F}_1)$ for $\sigma \in \{ss, stg, eg\}$ in contrast to the splitting theorem 4.10, statement 2.

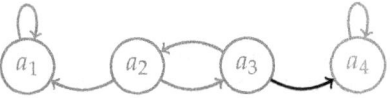

Figure 4.5: *Counterexample for $\sigma \in \{ss, stg, eg\}$*

Now we turn to ideal semantics. We want to mention that one may prove a similar result as statement 2, Theorem 4.10, namely

$$E \in \mathcal{E}_{id}(\mathcal{F}) \Rightarrow E \cap A_1 \in \mathcal{E}_{id}(\mathcal{F}_1) \wedge E \cap A_2 \in \mathcal{E}_{ad}\left(mod_{U_{E \cap A_1},R_3}\left(\mathcal{F}_2^{E \cap A_1,R_3}\right)\right).$$

Additionally, one may even prove that

$$E \in \mathcal{E}_{id}(\mathcal{F}) \Rightarrow E \cap A_2 \subseteq \bigcap_{P \in \mathcal{E}_{pr}\left(mod_{U_{E \cap A_1},R_3}\left(\mathcal{F}_2^{E \cap A_1,R_3}\right)\right)} P$$

holds but in general, it cannot be shown that $E \cap A_2$ is maximal with respect to this property as required for ideal extensions. Consider therefore the following counterexample.

4.1. Classical Splitting Results

Example 4.12. Given the following AF $\mathcal{F}$ and its splitting $(\mathcal{F}_1, \mathcal{F}_2, R_3)$. We have $\mathcal{E}_{pr}(\mathcal{F}) = \{\{a_1, a_3, a_5\}, \{a_2, a_4\}, \{a_2, a_5\}\}$. Thus, $\mathcal{E}_{id}(\mathcal{F}) = \{\emptyset\}$. Obviously, $\emptyset \cap \{a_1, a_2\} = \emptyset \in \mathcal{E}_{id}(\mathcal{F}_1)$ since $\mathcal{E}_{pr}(\mathcal{F}_1) = \{\{a_1\}, \{a_2\}\}$.

Figure 4.6: Counterexample for $\sigma = id$

Clearly, the empty set does not attack anything. Thus, the reduct of $\mathcal{F}_2$ with respect to $\emptyset$ and R_3 equals $\mathcal{F}_2$. Consider now the modification of $\mathcal{F}_2$ with respect to $U_\emptyset = \{a_1, a_2\}$ and R_3 depicted in the following figure:

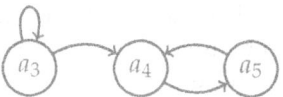

Figure 4.7: Modification of $\mathcal{F}_2$ with respect to $\{a_1, a_2\}$ and R_3

On the one hand, we have $\mathcal{E}_{pr}\left(mod_{U_\emptyset, R_3}\left(\mathcal{F}_2^{\emptyset, R_3}\right)\right) = \{\{a_5\}\}$ and hence, $\mathcal{E}_{id}\left(mod_{U_\emptyset, R_3}\left(\mathcal{F}_2^{\emptyset, R_3}\right)\right) = \{\{a_5\}\}$. On the other hand, $\emptyset \cap \{a_3, a_4, a_5\} = \emptyset$ in contrast to statement 2 of the splitting theorem 4.10.

Finally, for naive semantics one may use the AF $\mathcal{F} = (\{a_1, a_2, a_3\}, \{(a_2, a_3)\})$ to show that classical splitting results do not hold.

4.1.3 Splittings, Directionality and Monotonicity

In this section we will draw benefit from the classical splitting results proven in the section before. First, we will establish a general relation between universally definedness, the splitting property and the directionality principle. In particular, we prove that admissible, preferred, complete and grounded semantics satisfy directionality even in case of infinite AFs. As a second application we will strengthen the former monotonicity results (Theorem 3.2).

Theorem 4.13. *For any (**possible**) semantics σ we have: If σ is universally defined and satisfies classical splitting results (Theorem 4.10), then σ satisfies directionality principle $\mathcal{DI}$ (Definition 2.11).*

Proof. Given a universally defined semantics σ satisfying classical splitting results. We have to show that for any $U \in \mathcal{US}(\mathcal{F}) = \{U \mid U \in A(\mathcal{F}), (A(\mathcal{F}) \setminus U, U) \notin R(\mathcal{F})\}$ we have: $\mathcal{E}_\sigma(\mathcal{F}|_U) = \{E \cap U \mid E \in \mathcal{E}_\sigma(\mathcal{F})\}$. Assume not, hence there is an $U \in \mathcal{US}(\mathcal{F})$, such that $\mathcal{E}_\sigma(\mathcal{F}|_U) \neq \{E \cap U \mid E \in \mathcal{E}_\sigma(\mathcal{F})\}$. Observe that $\mathcal{F}|_U \leq_W \mathcal{F}$ and thus, $(\mathcal{F}|_U, \mathcal{F}_2, R_3)$ is a splitting of $\mathcal{F}$ where $\mathcal{F}_2$ and R_3 are given as specified in Proposition 4.2. First, consider $\mathcal{E}_\sigma(\mathcal{F}|_U) \not\subseteq \{E \cap U \mid E \in \mathcal{E}_\sigma(\mathcal{F})\}$. This means, there exists an $E_1 \in \mathcal{E}_\sigma(\mathcal{F}|_U)$, such that $E_1 \notin \{E \cap U \mid E \in \mathcal{E}_\sigma(\mathcal{F})\}$. Since we assumed that σ warrants the existence of extensions for any AF we deduce $E_2 \in \mathcal{E}_\sigma\left(mod_{U_{E_1}, R_3}\left(\mathcal{F}_2^{E_1, R_3}\right)\right)$. Consequently, $E_1 \cup E_2 \in$

$\mathcal{E}_\sigma(\mathcal{F})$ by statement 1 of the splitting theorem 4.10. Obviously, $(E_1 \cup E_2) \cap U = E_1$ in contradiction to $E_1 \notin \{E \cap U \mid E \in \mathcal{E}_\sigma(\mathcal{F})\}$. Consider now $\{E \cap U \mid E \in \mathcal{E}_\sigma(\mathcal{F})\} \nsubseteq \mathcal{E}_\sigma(\mathcal{F}|_U)$. Hence, there exists an $E \in \mathcal{E}_\sigma(\mathcal{F})$, such that $E \cap U \notin \mathcal{E}_\sigma(\mathcal{F}|_U)$. This is impossible because we have that $(\mathcal{F}|_U, \mathcal{F}_2, R_3)$ is a splitting of $\mathcal{F}$ and thus, statement 2 of the splitting theorem 4.10 guarantees $E \cap U \in \mathcal{E}_\sigma(\mathcal{F}|_U)$ concluding the proof. □

Just as an aside, note that universally definedness and satisfying the directionality principle is not sufficient for the fulfillment of classical splitting results (consider ideal semantics as a counter-example). The following proposition generalizes a former result in [Baroni and Giacomin, 2007] depicted in Figure 2.4 stating that admissible, preferred, complete and grounded semantics satisfy directionality provided that the considered frameworks are finite.

Proposition 4.14. *Admissible, preferred, complete and grounded semantics satisfy directionality for the whole class of AFs.*

Proof. In consideration of Theorems 4.13 and 4.10 it suffices to show that admissible, preferred, complete and grounded semantics are universally defined. Clearly, the empty set is always admissible whether or not the considered AF is finite. Furthermore, Dung itself has already shown that every AF possess at least one preferred extension [Dung, 1995, Corollary 12]. Consequently, complete semantics warrants the existence of at least one extension since $pr \subseteq co$ holds for any AF (compare Proposition 2.7). Finally, due to the monotonicity of the characteristic function the existence of the unique grounded extension is guaranteed [Tarski, 1955] (see [Grossi and Gabbay, 2013, Remark 1] for further information). □

Let us compare the splitting theorem 4.10 with the former monotonicity result (Theorem 3.2). The splitting theorem obviously strengthens the outcome of the monotonicity result for the admissible, preferred, grounded and complete semantics which all satisfy the directionality principle. We do not only know that an old belief set is contained in a new one and furthermore every new belief set is the union of an old one and a (possibly empty) set of new arguments but rather that every new belief set is the union of an old one and an extension of the corresponding modified reduct and vice versa. Furthermore, the cardinality inequality of the monotonicity result can be strengthened in the sense that we may provide a precise value of the number of extensions of the expanded AF through counting extensions of certain modified and reduced parts of the entire framework only.

Corollary 4.15. *Let $(\mathcal{F}, \mathcal{F}_2, R_3)$ be a splitting of the argumentation framework $\mathcal{G} = (A_1 \cup A_2, R_1 \cup R_2 \cup R_3)$. This means, $\mathcal{F} \leq_W \mathcal{G}$. For any semantics $\sigma \in \{ad, pr, co, gr\}$ we have:*[1]

1. $|\mathcal{E}_\sigma(\mathcal{F})| \leq \sum_{E_i \in \mathcal{E}_\sigma(\mathcal{F})} \left| \mathcal{E}_\sigma \left(mod_{U_{E_i}, R_3} \left(\mathcal{F}_2^{E_i, R_3} \right) \right) \right| = |\mathcal{E}_\sigma(\mathcal{G})|,$ (cardinality)

---

[1] The red-highlighted parts illustrate the strengthening compared to the monotonicity result 3.2 obtained by the splitting theorem 4.10.

2. $\forall E \in \mathcal{E}_\sigma(\mathcal{F})\ \exists E' \in \mathcal{E}_\sigma(\mathcal{G})\ (\exists C \subseteq B \smallsetminus A :) \exists C \in \mathcal{E}_\sigma \left(mod_{U_E,R_3} \left(\mathcal{F}_2^{E,R_3} \right) \right) : E' = E \cup C$,

(subset)

3. $\forall E' \in \mathcal{E}_\sigma(\mathcal{G})\ \exists E \in \mathcal{E}_\sigma(\mathcal{F})\ (\exists C \subseteq B \smallsetminus A :) \exists C \in \mathcal{E}_\sigma \left(mod_{U_E,R_3} \left(\mathcal{F}_2^{E,R_3} \right) \right) : E' = E \cup C$

(representation)

4.1.4 Dynamic Scenario: Reusing Extensions

Argumentation is a dynamic process. Obviously the set of extensions of an AF may change if new arguments and their corresponding interactions are added. Computing the justification state of an argument or even extensions from scratch each time new information is added is very inefficient. In subsection 3.2 we showed that in case of weak expansions and semantics satisfying directionality (compare Proposition 3.3) credulously as well as sceptically justified arguments persist. In this section we go a step further. Now we not only use already computed extensions to draw conclusions about the justification state of an argument but also to precisely determine the new set of extensions. Consider therefore the following example.

Example 4.16. Given an AF $\mathcal{F} = (\{a_1, ..., a_n\}, R)$ and its set of extensions $\mathcal{E}_\sigma(\mathcal{F}) = \{E_1, ..., E_m\}$ ($\sigma \in \{pr, co, gr\}$). Consider now a weak expansion with two additional arguments b_1 and b_2, such that b_1 is attacked by the *old* arguments a_1 and a_2 and furthermore, b_2 is defeated by b_1. Let $\mathcal{G}$ be the resulting AF depicted below.[2]

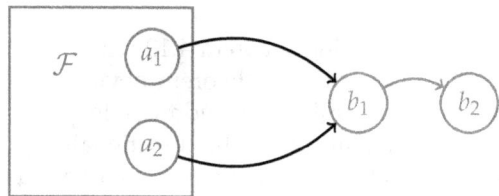

Figure 4.8: Reusing Extensions - The AF $\mathcal{G}$

What are the σ-extensions of the expanded AF $\mathcal{G}$. Since $\mathcal{G}$ is a weak expansion of $\mathcal{F}$ we immediately obtain a splitting (compare Proposition 4.2) and thus, may apply the splitting theorem 4.10. Given an extension E_i of $\mathcal{F}$ we have to distinguish three cases:

1. a_1 or a_2 is an element of E_i,

2. a_1 and a_2 are not in E_i and not in U_{E_i},

3. a_1 and a_2 are not in E_i and at least one of them is in U_{E_i}.

---

[2]We omit the arguments $a_3, ...a_n$ as well as the attack relation R since they are unimportant for the purpose of this example.

The AFs below are the resulting modifications in these three cases. In the first case the argument b_1 disappears because b_1 is attacked by at least one element of the extension E_i. In the second and third case the arguments b_1 and b_2 survive because they are not attacked by E_i. Furthermore in the last case we have to add a self-loop for b_1 since a_1 or a_2 are undefined with regard to E_i.

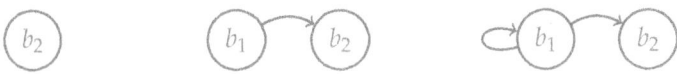

Figure 4.9: *Reusing Extensions - Different Cases*

The resulting preferred, complete and grounded extensions of the modified frameworks depicted above are $\{b_2\}$ in the first case, $\{b_1\}$ in the second and $\emptyset$ in the last case. Now we may construct (Theorem 4.10, statement 1) σ-extensions of the expanded AF $\mathcal{G}$ by using the already computed σ-extensions of $\mathcal{F}$, namely

1. $E_i \cup \{b_2\} \in \mathcal{E}_\sigma(\mathcal{G})$,
2. $E_i \cup \{b_1\} \in \mathcal{E}_\sigma(\mathcal{G})$ and
3. $E_i \in \mathcal{E}_\sigma(\mathcal{G})$.

Due to the completeness (Theorem 4.10, statement 2) of the splitting method we constructed all extensions of $\mathcal{G}$ concerning E_i.

Note that in general, new arguments occur as a response, i.e., an attack, to a former argument. In this situation the former extensions are usually not reusable because we do not operate within the class of weak expansions. Being aware of this fact, we emphasize that there are formalisms, like Value Based AFs [Bench-Capon, 2003] where weak expansions naturally occur. Former arguments may be arguments which advance higher values than the further arguments. Consequently, the new arguments cannot attack the former (compare the idea of "attack-succeed" in [Bench-Capon, 2003]).

4.1.5 Static Scenario: Computing Extensions

In [Baumann et al., 2011] we perform a systematic empirical evaluation of the effects of splitting on the computation of extensions. Our study shows that the performance of algorithms may dramatically improve when splitting is applied. In this section we sketch the main idea and present some performance results for stable, preferred and grounded semantics. For readers interested in more details we refer to the original paper [Baumann et al., 2011].

The main idea is as follows:

1. given an AF $\mathcal{F}$
2. generate a splitting $(\mathcal{F}_1, \mathcal{F}_2, R_3)$ of $\mathcal{F}$

3. compute extensions $E_1, ..., E_n$ of $\mathcal{F}_1$ via existing algorithm Alg

4. reduce and modify $\mathcal{F}_2$ with respect to E_i, i.e. compute $mod_{U_{E_i}, R_3}\left(\mathcal{F}_2^{E_i, R_3}\right)$

5. compute extensions $E_i^1, ..., E_i^m$ of $mod_{U_{E_i}, R_3}\left(\mathcal{F}_2^{E_i, R_3}\right)$ via existing algorithm Alg

6. combine E_i and E_i^j to an extension of $\mathcal{F}$

4.1.5.1 How to generate a splitting?

To generate a splitting we use the related graph-theoretic concept of *strongly connected components (SCC)*. A directed graph is strongly connected if there is a path from each vertex to every other vertex. The SCCs of a graph $\mathcal{F}$ are its maximal strongly connected subgraphs. Contracting every SCC to a single vertex leads to an acyclic graph. It is well-known that an acyclic graph induces a partial order on the set of vertices. Based on this order every SCC-decomposition can be easily transformed into a splitting. The following figure exemplifies the idea. We sketch three different splittings, namely S_1, S_2 and S_3. Note that S_1 corresponds to the most obvious possibility, namely taking the union of the initial nodes of the decomposition (= $\mathcal{F}_1$) and the union of the remaining subgraph (= $\mathcal{F}_2$).

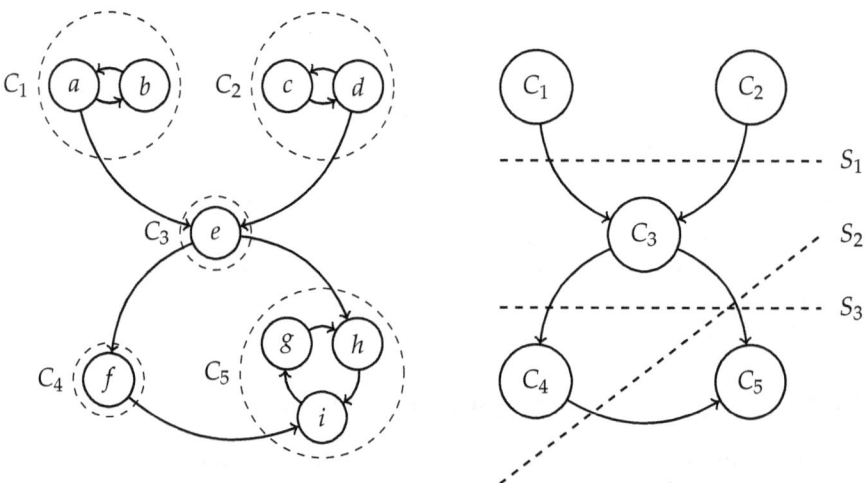

Figure 4.10: SCCs and Splittings

For computing strongly connected components we used standard Tarjan algorithm from [Tarjan, 1972]. The running time of it is linear in the number of arguments and attacks of $\mathcal{F}$. As depicted in Figure 4.10 there are usually more than one possibilities to generate a splitting. In the implementation we tried to achieve a splitting $(\mathcal{F}_1, \mathcal{F}_2, R)$ of $\mathcal{F}$, such that the number of arguments of $\mathcal{F}_1$ is approximately the same as or slightly more than the number of arguments of $\mathcal{F}_2$. By choosing this range we attempted to balance the number of arguments and at the same time keeping in mind that any further computations may slow down the splitting process. According to those requirements an optimal splitting for the example depicted above would run along the line S_3.

4.1.5.2 Labelling Approach and Splitting Results

The evaluation was based on an implementation of Caminada's labelling algorithm [Modgil and Caminada, 2009], arguably the standard genuine algorithm for computing extensions. We will not show how this algorithm works but we will briefly introduce the labelling approach [Caminada, 2006] and show how to reformulate splitting results in terms of labellings.

Given an AF $\mathcal{F} = (A, R)$, a *labelling* is a total function $L : A \to \{in, out, undec\}$. We use $x(L)$ for $L^{-1}(\{x\})$, i.e. $x(L) = \{a \in A \mid L(a) = x\}$. Analogously to $\mathcal{E}_\sigma(\mathcal{F})$ we write $\mathcal{L}_\sigma(\mathcal{F})$ for the set of all labellings prescribed by semantics σ for an AF $\mathcal{F}$ also referred to as σ-labellings of $\mathcal{F}$ or simply (if clear from context), labellings of $\mathcal{F}$. The following concept of a *complete* labelling play a central role within the context of labelling-based semantics.

Definition 4.17. Given an AF $\mathcal{F} = (A, R)$ and a labelling L of it. L is called a *complete* labelling ($L \in \mathcal{L}_{co}(\mathcal{A})$) iff for any $a \in A$ the following holds:

1. If $a \in in(L)$, then for each $b \in A$ such that $(b, a) \in R$, $b \in out(L)$,

2. If $a \in out(L)$, then there is an $b \in A$ such that $(b, a) \in R$ and $b \in in(L)$,

3. If $a \in undec(L)$, then there is an $b \in A$ such that $(b, a) \in R$ and $b \in undec(L)$ and there is no $b \in A$ such that $(b, a) \in R$ and $b \in in(L)$.

Now we are ready to define the remaining counterparts of the extension-based semantics in terms of complete labellings.

Definition 4.18. Given an AF $\mathcal{F} = (A, R)$ and a labelling $L \in \mathcal{L}_{co}(\mathcal{F})$. L is a

1. *stable* labelling ($L \in \mathcal{L}_{st}(\mathcal{F})$) iff $undec(L) = \emptyset$,

2. *preferred* labelling ($L \in \mathcal{L}_{pr}(\mathcal{F})$) iff for each $L' \in \mathcal{L}_{co}(\mathcal{F})$, $in(L) \not\subset in(L')$,

3. *grounded* labelling ($L \in \mathcal{L}_{gr}(\mathcal{F})$) iff for each $L' \in \mathcal{L}_{co}(\mathcal{F})$, $in(L') \not\subset in(L)$.

The following theorem opens the door to make splitting results applicable in case of labelling-based semantics.

Theorem 4.19. *[Modgil and Caminada, 2009] Given an AF $\mathcal{F}$. For each $\sigma \in \{co, stb, pr, gr\}$,*

1. $E \in \mathcal{E}_\sigma(\mathcal{F})$ iff $\exists L \in \mathcal{L}_\sigma(\mathcal{F}) : in(L) = E$ and

2. $|\mathcal{E}_\sigma(\mathcal{F})| = |\mathcal{L}_\sigma(\mathcal{A})|$ holds.

Now we may reformulate the splitting theorem 4.10 in terms of labellings.

Theorem 4.20. *Let $\mathcal{F} = (A, R)$ be an AF which possesses a splitting $(\mathcal{F}_1, \mathcal{F}_2, R_3)$ with $\mathcal{F}_1 = (A_1, R_1)$ and $\mathcal{F}_2 = (A_2, R_2)$. Furthermore, let $\sigma \in \{stb, pr, co, gr\}$.*

1. If L_1 is a labelling of $\mathcal{F}_1$ and L_2 is a labelling of the $(undec(L_1), R_3)$-modification of $\mathcal{F}_2^{in(L_1),R_3}$, then there is exactly one labelling L of $\mathcal{F}$, such that $in(L) = in(L_1) \cup in(L_2)$.

$$L_1 \in \mathcal{L}_\sigma(\mathcal{F}_1) \wedge L_2 \in \mathcal{L}_\sigma\left(mod_{undec(L_1),R_3}\left(\mathcal{F}_2^{in(L_1),R_3}\right)\right) \Rightarrow$$
$$\exists! \, L \in \mathcal{L}_\sigma(\mathcal{F}) : in(L) = in(L_1) \cup in(L_2)$$

2. If L is a labelling of $\mathcal{F}$, then there is exactly one labelling L_1 of $\mathcal{F}_1$, such that $in(L_1) = in(L) \cap A_1$ and furthermore, exactly one labelling L_2 of the $(undec(L) \cap A_1, R_3)$-modification of $\mathcal{F}_2^{in(L) \cap A_1, R_3}$.

$$L \in \mathcal{L}_\sigma(\mathcal{F}) \Rightarrow \exists! \, L_1 \in \mathcal{L}_\sigma(\mathcal{F}_1) : in(L_1) = in(L) \cap A_1 \wedge$$
$$\exists! \, L_2 \in \mathcal{L}_\sigma\left(mod_{undec(L) \cap A_1, R_3}\left(\mathcal{F}_2^{in(L) \cap A_1, R_3}\right)\right) : in(L_2) = in(L) \cap A_2$$

4.1.5.3 Empirical Results

The evaluation was mainly focused on preferred and stable semantics. We also included results for grounded semantics, but as this semantics is known to be polynomial an improvement of performance here was never expected, and our results confirm this.

Our evaluation of runtime for the grounded, preferred and stable semantics was based on the sampling of 100 generated frameworks where 20 examples randomly extracted from each of the following *arguments/attacks* combinations: 10/30, 50/100, 100/175, 200/375 and 500/750.

The following items summarize the most important observations:

1. execution with splitting was always faster than the one without

2. a gain of around 60% on average for both preferred and stable semantics

3. splitting may significantly improve runtime for stable semantics in frameworks where no stable labellings exist

In a nutshell, the results confirm that the additional overhead introduced by splitting is negligible. On the contrary it is worthwhile to use splittings.

4.2 Generalized Splitting Results

In the previous section we studied the concept of a classical splitting which allows to obtain the extensions of the original framework by computing and combining extensions of two disjoint and partially modified subframeworks. The restriction we made was that the attacks between the splitted parts are restricted to a single direction. The aim of this section is to make this technique also applicable in cases where classical splits are not possible. This means, we present a simple modification approach allowing for *arbitrary* splits. We present generalized splitting results for stable semantics representing one of

Chapter 4. Splitting Results

the most important semantics for Dung frameworks. All results are already published in [Baumann et al., 2012].

4.2.1 k-Splitting, Conditional Extension and Match

We start with the definition of *k-splittings*.

Definition 4.21. Let $F = (A, R)$ be an AF. Given a set $S \subseteq A$ we define $\bar{S} = A \smallsetminus S$, $R^S_\rightarrow = R \cap (S \times \bar{S})$ and $R^S_\leftarrow = R \cap (\bar{S} \times S)$. The set S is called a *k-splitting* of $\mathcal{F}$ if $k = |R^S_\leftarrow|$.

A *k*-splitting S of $\mathcal{F}$ induces two sub-frameworks of $\mathcal{F}$, namely $\mathcal{F}^S_1 = \mathcal{F}|_S$ and $\mathcal{F}^S_2 = \mathcal{F}|_{\bar{S}}$, together with the sets of links $R^S_\rightarrow$ and $R^S_\leftarrow$ connecting the sub-frameworks in the two possible directions. If $|R^S_\leftarrow| = 0$, then $(\mathcal{F}^S_1, \mathcal{F}^S_2, R^S_\rightarrow)$ is a classical splitting of $\mathcal{F}$ (compare Definition 4.1). This means, there is a one-to-one correspondence between 0-splittings and classical splittings.

Example 4.22. Consider the following AF $\mathcal{F}$ and its 4-splitting $S = \{a_1, a_2, a_3\}$. Observe that $\mathcal{E}_{stb}(\mathcal{F}) = \{\{a_3, b_2, b_5\}, \{a_1, b_2, b_3, b_4\}\}$.

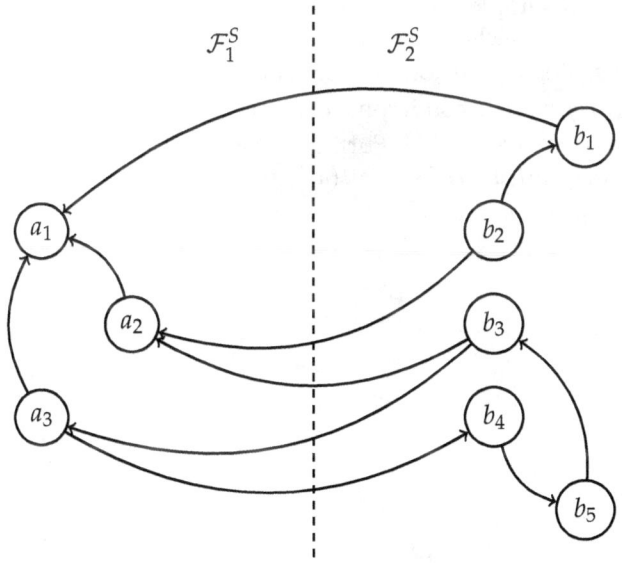

Figure 4.11: A 4-splitting of $\mathcal{F}$

How to obtain or reconstruct the stable extensions of $\mathcal{F}$? The basic idea is as follows. We first find a stable extension of $\mathcal{F}^S_1$. However, we have to take into account that one of the elements of S may be attacked by an argument of $\mathcal{F}^S_2$. For this reason we must provide, for each argument a in $\mathcal{F}^S_1$ attacked by an argument in $\mathcal{F}^S_2$, the possibility *to assume it is attacked*. To this end we add for a a new argument $att(a)$ such that a and $att(a)$ attack each other. Now we may choose to include the new argument in an extension which

4.2. Generalized Splitting Results

corresponds to the assumption that a is attacked. Later, when extensions of $\mathcal{F}_2^S$ are computed, we need to check whether the assumptions we made actually are satisfied. Only if they are, we can safely combine the extensions of the sub-frameworks we have found.

Definition 4.23. Let $\mathcal{F} = (A, R)$ be an AF and S a k-splitting of $\mathcal{F}$. A *conditional extension* of F_1^S is a stable extension of the modified AF $[\mathcal{F}_1^S] = (A_S, R_S)$ where

- $A_S = S \cup \left\{ att(a) \mid a \in A_{R_\leftarrow^S}^{\oplus} \right\}$ and

- $R_S = (R \cap (S \times S)) \cup \left\{ (att(a), a), (a, att(a)) \mid a \in A_{R_\leftarrow^S}^{\oplus} \right\}$.

We use $A_{R_\leftarrow^S}^{\oplus} = \{ a \mid (A, a) \in R_\leftarrow^S \}$.

In other words, $[F_1^S]$ is obtained from F_1^S by adding a copy $att(a)$ for each argument a attacked from F_2^S, and providing for each such a a mutual attack between a and $att(a)$. For a k-splitting we thus add k nodes and $2k$ links to F_1^S. In case of a 0-splitting S, note that $[F_1^S] = F_1^S$ as intended.

Example 4.24 (Example 4.22 continued). Consider again the 4-splitting $S = \{a_1, a_2, a_3\}$ of $\mathcal{F}$ depicted in Figure 4.11. According to Definition 4.23 we obtain $[\mathcal{F}_1^S]$ as depicted below. The set of the grey highlighted arguments $E = \{att(a_1), att(a_2), a_3\}$ is a conditional extension of $\mathcal{F}_1^S$, i.e. a stable extension of the modified framework $[\mathcal{F}_1^S]$. Further stable extensions of $[\mathcal{F}_1^S]$ are given by $E_1 = \{a_1, att(a_2), att(a_3)\}$, $E_2 = \{att(a_1), a_2, att(a_3)\}$, $E_3 = \{att(a_1), a_2, a_3\}$ and $E_4 = \{att(a_1), att(a_2), att(a_3)\}$.

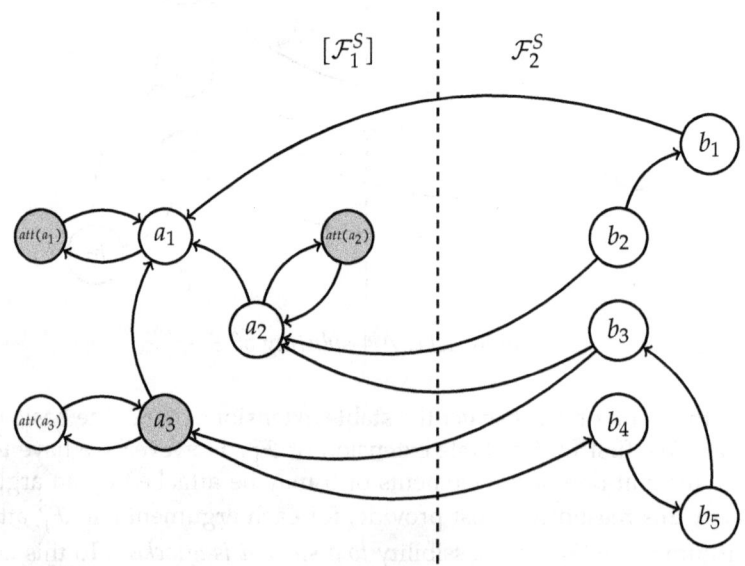

Figure 4.12: Conditional Extensions of $\mathcal{F}_1^S$

Chapter 4. Splitting Results

Each conditional extension E of $\mathcal{F}_1^S$ may contain meta-information in the form of $att(x)$ elements. This information will later be disregarded, but is important to verify the assumptions extensions of $\mathcal{F}_2^S$ need to fulfill so that E can be augmented to an extension of the entire argumentation framework $\mathcal{F}$. As in case of classical splitting, E will be used to modify $\mathcal{F}_2^S$ accordingly, thus propagating effects of elements in E on $\mathcal{F}_2^S$. In the generalized case we also need to take the meta-information in E into account to make sure the assumptions we made are valid. In particular,

- if $att(a)$ is in E yet a is not attacked by another element in $E \cap S$, then we know that a must be externally attacked from an element of F_2^S. In other words, we are only interested in extensions of F_2^S which contain at least one attacker of a.

- if b is in E yet it is attacked by some argument in F_2^S, then we are only interested in extensions of F_2^S not containing any of the attackers of b.

Before we turn these ideas into a definition we present a useful lemma which helps to understand our definition. In the following, given an AF $\mathcal{F} = (A, R)$ and a set $E \subseteq A$ we use $E_R^+ = E \cup \{b \mid (E, b) \in R\}$. Observe that any conflict-free set such that $E_R^+ = A$ is a stable extension of $\mathcal{F}$ and vice versa.

Lemma 4.25. *Let $\mathcal{F} = (A, R)$ be an AF. Let further B and $C_1, \ldots, C_n$ be sets such that $B, C_1, \ldots, C_n \subseteq A$, and $D = \{d_1, \ldots, d_n\}$ such that $D \cap A = \emptyset$. The stable extensions of the AF*

$$\mathcal{F}' = (A \cup D, R \cup \{(b, b) \mid b \in B \text{ or } b \in D\} \cup \{(c, d_j) \mid c \in C_j, 1 \leq j \leq n\})$$

are exactly the stable extensions E of $\mathcal{F}$ containing no element of B and at least one element of every C_i, i.e. $C_i \cap E \neq \emptyset$ for every $i \in \{1, \ldots, n\}$.

Proof. Let $E \in \mathcal{E}_{stb}(\mathcal{F})$, such that E does not contain an element of B (+) and $C_i \cap E \neq \emptyset$ for every $i \in \{1, \ldots, n\}$ (*). We observe $E \in cf(\mathcal{F}')$ because of $E \subseteq A$ and (+), and furthermore, $E_{R(\mathcal{F}')}^+ = E_{R(\mathcal{F})}^+ \cup D = A \cup D$ because of (*). Thus, $E \in \mathcal{E}_{stb}(\mathcal{F}')$.

Assume now $E \in \mathcal{E}_{stb}(\mathcal{F}')$. We observe that $E \cap B = \emptyset$ since conflict-freeness has to be fulfilled. Furthermore, $C_i \cap E \neq \emptyset$ for every $i \in \{1, \ldots, n\}$ has to hold because $E_{R(\mathcal{F}')}^+ = A \cup \{d_1, \ldots, d_n\}$ and only arguments in C_i attack the argument d_i by construction. Obviously, $E \subseteq A$ since the elements of D are self-attacking. Furthermore, $E \in cf(F)$ because $R(\mathcal{F}) \subseteq R(\mathcal{F}')$. Consider now the attack-relation $R(\mathcal{F}')$. We obtain $E_{R(\mathcal{F})}^+ = E_{R(\mathcal{F}')}^+ \setminus D = (A \cup D) \setminus D = A$ which proves $E \in \mathcal{E}_{stb}(\mathcal{F})$. □

Based on the lemma we can now define the modification of $\mathcal{F}_2^S$ that is needed to compute those extensions which comply with a conditional extension E, capturing also the assumptions made in E. First, we can eliminate all arguments attacked by an element of E. This step corresponds to the usual propagation needed for classical splittings as well. In addition, we make sure that only those extensions of the resulting framework are generated which

4.2. Generalized Splitting Results

(1) contain an attacker for all externally attacked nodes of S, and (2) do not contain an attacker for any element in E. For this purpose the techniques of the lemma are applied.

Definition 4.26. Let $\mathcal{F} = (A, R)$ be an AF, S a k-splitting of $\mathcal{F}$ and let E be a conditional extension of $\mathcal{F}_1^S$. Furthermore, let

$$EA(S, E) = \{a \in S \setminus E \mid a \notin (S \cap E)_R^\oplus\}$$

denote the set of arguments from F_1^S not contained in E because they are externally attacked. An (E, S)-*match* of $\mathcal{F}$ is a stable extension of the AF $[\mathcal{F}_2^S]_E = (A', R')$ where

- $A' = \left(\bar{S} \setminus E^+_{R^S_\rightarrow}\right) \cup \{in(a) \mid a \in EA(S, E)\}$ and

- $R' = (R \cap (A' \times A')) \cup \{(in(a), in(a)), (b, in(a)) \mid a \in EA(S, E), (b, a) \in R^S_\leftarrow\}$
  $\cup \{(c, c) \mid (c, E) \in R^S_\leftarrow\}$.

In other words, we take the framework $\mathcal{F}_2^S$ and modify it with respect to a given conditional extension E of $\mathcal{F}_1^S$. To this end, we remove those arguments from $\mathcal{F}_2^S$ which are attacked by E via $R^S_\rightarrow$, but we make a copy of each argument a in $\mathcal{F}_1^S$ externally attacked by $\mathcal{F}_2^S$ via $R^S_\leftarrow$. These additional self-attacking arguments $in(a)$ are used to represent the forbidden situation where an externally attacked argument a actually remains unattacked. Finally, we exclude those arguments in $\mathcal{F}_2^S$ from potential extensions which attack an argument in E located in $\mathcal{F}_1^S$; these are the self-loops (c, c) for arguments s with $(c, E) \in R^S_\leftarrow$. Again the size of the modification is small whenever k is small: we add at most k nodes and $2k$ links to $\mathcal{F}_2^S$.

Example 4.27 (Example 4.24 continued). We continue. On the right-hand side we have the modification of $\mathcal{F}_2^S$ with respect to the conditional extension $E = \{att(a_1), att(a_2), a_3\}$, i.e. the AF $[\mathcal{F}_2^S]_E$. Observe that $EA(S, E) = \{a_2\}$ because a_2 is not an element of E and furthermore, it is not attacked by an argument in $E \cap S = \{a_3\}$. Hence, we have to add a self-attacking node $in(a_2)$ to $\mathcal{F}_2^S$ which is attacked by the attackers of a_2, namely the arguments b_2 and b_3. The argument a_3 (which belongs to the extension E) is attacked by the argument b_3 and attacks the argument b_4. Hence, we have to add a self-loop for b_3 and further, we have to delete b_4 and its corresponding attacks.

Chapter 4. Splitting Results

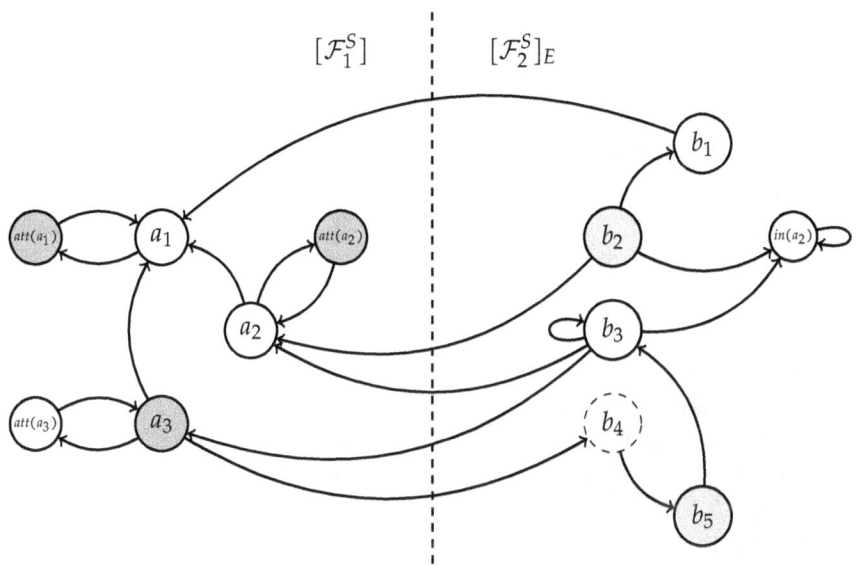

Figure 4.13: (E,S)-match of $\mathcal{F}$

The set of the light-grey highlighted arguments $E' = \{b_2, b_5\}$ is an (E,S)-match of $\mathcal{F}$, i.e. a stable extension of $[\mathcal{F}_2^S]_E$; in fact, it is the only (E,S)-match of $\mathcal{F}$. Recall that $(E \cap S) \cup E' = \{a_3, b_2, b_5\}$ is a stable extension of the initial AF $\mathcal{F}$. One can further check that $\{b_2, b_3, b_4\}$ is an (E_1, S)-match of $\mathcal{F}$ with E_1 as given in the previous example. On the other hand, for the remaining conditional extensions of F_1^S, no corresponding matches exist. Take, for instance, $E_2 = \{att(a_1), a_2, att(a_3)\}$; here we have to put self-loops for b_2 and b_3, but b_2 remains unattacked in $[\mathcal{F}_2^S]_{E_2}$. Thus no stable extension for $[\mathcal{F}_2^S]_{E_2}$ exists.

4.2.2 Generalized Splitting Theorem

Now we are prepared to prove a splitting theorem for k-splittings in case of stable semantics.

Theorem 4.28. *Let $\mathcal{F} = (A, R)$ be an AF and let S be a k-splitting of $\mathcal{F}$.*

1. *If E is a conditional extension of $\mathcal{F}_1^S$ and E' an (E,S)-match of $\mathcal{F}$, then $(E \cap S) \cup E'$ is a stable extension of $\mathcal{F}$.*

2. *If H is an extension of $\mathcal{F}$, then there is a set $X \subseteq \left\{att(a) \mid a \in A_{R^S_\leftarrow}^+\right\}$ such that $E = (H \cap S) \cup X$ is a conditional extension of $\mathcal{F}_1^S$ and $H \cap \bar{S}$ is an (E,S)-match of $\mathcal{F}$.*

Proof. ad 1. First we will show that $(E \cap S) \cup E' \in cf(\mathcal{F})$. Given that E is a conditional extension of $\mathcal{F}_1^S$ we deduce $E \in cf([\mathcal{F}_1^S])$. Consequently, $E \cap S \in cf([\mathcal{F}_1^S])$ (less arguments) and thus, $E \cap S \in cf(\mathcal{F}_1^S)$ (less attacks). Finally,

$E \cap S \in cf(\mathcal{F})$ since $(\mathcal{F}_1^S)|_S = \mathcal{F}|_S$ holds. Let E' be an (E,S)-match of $\mathcal{F}$, i.e. $E' \in \mathcal{E}_{stb}([\mathcal{F}_2^S]_E)$. According to Lemma 4.25, $E' \in stb(\mathcal{F}')$ where $\mathcal{F}' = \left(\bar{S} \setminus E^\oplus_{R^S_\rightarrow}, R(\mathcal{F}) \setminus \left(E^\oplus_{R^S_\rightarrow}, E^\oplus_{R^S_\rightarrow}\right)\right)$. Thus, $E' \in cf(\mathcal{F})$. Obviously, $(E \cap S, E') \notin R(\mathcal{F})$ since $E \cap S \subseteq S$ and $E' \subseteq \bar{S} \setminus E^\oplus_{R^S_\rightarrow}$. Assume now $(E', E \cap S) \in R(\mathcal{F})$. This means, there are arguments $e' \in E'$ and $e \in E \cap S$, such that $(e', e) \in R$. This contradicts the conflict-freeness of E' in $[\mathcal{F}_2^S]_E$ because $R([\mathcal{F}_2^S]_E)$ contains the set $\{(c, c) \mid (c, E) \in R^S_\leftarrow\}$. Thus, $(E', E \cap S) \notin R(\mathcal{F})$ and $(E \cap S) \cup E' \in cf(\mathcal{F})$ is shown.

We now show that $((E \cap S) \cup E')^+_{R(\mathcal{F})} = S \cup \bar{S} = A$. Let us consider an argument s, such that $s \notin ((E \cap S) \cup E')^+_{R(\mathcal{F})}$. Assume $s \in S$. Consequently, $s \in EA(S, E)$. Since E' is an (E, S)-match of $\mathcal{F}$, by Lemma 4.25, $E \in \mathcal{E}_{stb}(\mathcal{F}')$ such that $(E, in(s)) \in R(\mathcal{F}')$. By definition of $\mathcal{F}'$ we have $R(\mathcal{F}') \subseteq R(\mathcal{F})$ and thus $(E, in(s)) \in R(\mathcal{F})$ contradicting the assumption. Assume now $s \in \bar{S}$. Obviously, $s \in \bar{S} \setminus E^\oplus_{R^S_\rightarrow}$. Since E' is an (E, S)-match of $\mathcal{F}$ we deduce by Lemma 4.25 that E' is a stable extension of $\mathcal{F}'$ as defined above. Thus, $s \in (E')^+_{R(\mathcal{F})}$ contradicting the assumption. Altogether, we have shown that $(E \cap S) \cup E'$ is a stable extension of $\mathcal{F}$.

ad 2. Let $(H \cap S)^+_{R(\mathcal{F})} \dot{\cup} B = S$. Since $H \in \mathcal{E}_{stb}(\mathcal{F})$ is assumed it follows $B \subseteq (H \cap \bar{S})^+_{R(\mathcal{F})}$. This means, $B \subseteq A^+_{R^S_\leftarrow}$. Consider now $X = \{att(b) \mid b \in B\}$. It can be easily seen that $E = (H \cap S) \cup X$ is a conditional extension of $[\mathcal{F}_1^S]$. Note that $B = EA(S, E)$. Since $H \in \mathcal{E}_{stb}(\mathcal{F})$ it follows $H \cap \bar{S} \in \mathcal{E}_{stb}(\mathcal{F}')$ where $\mathcal{F}'$ is as above. Furthermore, there is no argument $c \in H \cap \bar{S}$, such that $(c, d) \in R^S_\leftarrow$ with $d \in E$. Remember that $B = EA(S, E)$. Hence, for every $b \in B$, $(H \cap \bar{S}) \cap \{b\}^\oplus_{R^S_\leftarrow} \neq \emptyset$, since $H \in \mathcal{E}_{stb}(\mathcal{F})$. Thus, Lemma 4.25 is applicable which implies that $H \cap \bar{S}$ is an (E, S)-match of $\mathcal{F}$. □

Note that k-splittings, on the one hand, divide the search space into smaller fractions in many cases, but on the other hand this might result in the computation of more stable extensions which turn out to be "useless" when propagated from the first to the second part of the split, i.e. they do not contribute to the evaluation of the entire framework. So from the theoretical side it is not clear how k-splitting effects the computation times. Recently, in [Wong, 2013] an empirical evaluation of the effects of k-splitting on the average computation time was performed. The results therein provide empirical support for our claim that using k-splittings can improve the performance of existing algorithms similar to classical splittings as discussed in Section 4.1.5.

4.3 Splittings for Logic Programs

In an important and much cited paper [Lifschitz and Turner, 1994] Vladimir Lifschitz and Hudson Turner have shown how, under certain conditions, logic programs under answer set semantics can be split into two disjoint parts, a "bottom" part and a "top" part. The bottom part can be evaluated independently of the top part. Results of the evaluation, i.e., answer sets of the bottom part, are then used to simplify the top part. To obtain answer sets of the original program one simply has to combine an answer set of the simplified top part with the answer set which was used to simplify this part. In this section we present a simple generalization of these results very similar to the generalized splitting results in case of argumentation frameworks (Theorem 4.28).

We point out that generalizations for logic programs have been investigated in depth in [Janhunen et al., 2009]. In fact, Janhunen et al. describe an entire rather impressive module theory for logic programs based on modules consisting of rules together with input, output and hidden variables. To a certain extent some of our results can be viewed as being "implicit" already in that paper. Nevertheless, we believe that the constructions we describe here complement the abstract theory in a useful manner.

4.3.1 Background: Logic Programs

We briefly introduce the theory of logic programs. Thereby we restrict the discussion to normal logic programs. Such programs are sets of rules of the form

$$a \leftarrow b_1, \ldots, b_m, \text{not } c_1, \ldots, \text{not } c_n \tag{4.1}$$

where a and all b_i's and c_j's are atoms. Intuitively, the rule is a justification to "establish" or "derive" that a (the so called *head*) is true, if all default literals to the right of $\leftarrow$ (the so called *body*) are true in the following sense: a non-negated atom b_i is true if it has a derivation, a negated one, not c_j, is true if c_j does not have one. Variables can appear in rules, however they are just convenient abbreviations for the collection of their ground instantiations.

The semantics of (ground) normal logic programs [Gelfond and Lifschitz, 1988; Gelfond and Lifschitz, 1991] is defined in terms of answer sets, also called stable models for this class of programs. Programs without negation in the bodies have a unique answer set, namely the smallest set of atoms closed under the rules. Equivalently, this set can be characterized as the least model of the rules, reading $\leftarrow$ as classical implication and the comma in the body as conjunction.

For programs with negation in the body, one needs to guess a candidate set of atoms S and then verifies the choice. This is achieved by evaluating negation with respect to S and checking whether the "reduced" negation-free program corresponding to this evaluation has S as answer set. If this is the case, it is guaranteed that all applicable rules were applied, and that each atom in S has a valid derivation based on appropriate rules. Here is the formal definition:

Definition 4.29. Let P be a normal logic program, S a set of atoms. The *Gelfond/Lifschitz-reduct* (*GL-reduct*) of P and S is the negation-free program P^S obtained from P by

1. deleting all rules $r \in P$ with not c_j in the body for some $c_j \in S$,

4.3. Splittings for Logic Programs

2. deleting all negated atoms from the remaining rules.

S is an answer set of P iff S is the answer set of P^S. We denote the collection of answer sets of a program P by $AS(P)$.

For convenience we will use rules without head (constraints) of the form

$$\leftarrow b_1, \ldots, b_m, \text{ not } c_1, \ldots, \text{ not } c_n \quad (4.2)$$

as abbreviation for

$$f \leftarrow \text{not } f, b_1, \ldots, b_m, \text{ not } c_1, \ldots, \text{ not } c_n \quad (4.3)$$

where f is an atom not appearing anywhere else in the program. Adding rule 4.2 to a program has the effect of eliminating those answer sets of the original program which contain all of the b_is and none of the c_js.

For other types of programs and an introduction to answer set programming, a problem solving paradigm based on the notion of answer sets, the reader is referred to [Brewka et al., 2011].

4.3.2 k-Splitting, Conditional Answer Set and Match

A classical splitting S can be defined as a set of atoms dividing a program P into two disjoint subprograms P_1^S and P_2^S such that no head of a rule in P_2^S appears anywhere in P_1^S.

One way to visualize this is via the dependency graph of P. The nodes of the dependency graph are the atoms in P. In addition, there are two kinds of links, positive and negative ones: whenever b appears positively (negatively) in a rule with head c, then there is a positive (negative) link from b to c in the dependency graph. A (proper) splitting S now is a set of atoms such that the dependency graph has no links - positive or negative - to an atom in S from any atom outside S.

Now let us consider the situation where a (small) number k of atoms in the heads of rules in P_2^S appear negatively in bodies of P_1^S. This means that the dependency graph of the program has a small number of negative links *pointing in the wrong direction*. As we will see methods similar to those we used for argumentation can be applied here as well. In the following $head(r)$ denotes the head of rule r, $At(r)$ the atoms appearing in r and $pos(r)$ (respectively $neg(r)$) the positive (respectively negative) atoms in the body of r. We also write $head(P)$ for the set $\{head(r) \mid r \in P\}$ and $At(P)$ for $\{At(r) \mid r \in P\}$.

Definition 4.30. Let P be a normal logic program. Let $S \subseteq At(P)$, such that for each rule $r \in P$, $head(r) \in S$ implies $pos(r) \subseteq S$. Furthermore, let $\bar{S} = At(P) \setminus S$ and $V_S = \{c \in \bar{S} \mid r \in P, head(r) \in S, c \in neg(r)\}$. The set S is called a k-splitting if $|V_S| = k$.

Analogously to AFs where a splitting induces two disjoint sub-frameworks, here the set S induces two disjoint sub-programs, P_1^S having heads in S and

Chapter 4. Splitting Results

P_2^S having heads in $\bar{S}$. Whenever $|V_S| \neq 0$, there are some heads in P_2^S which may appear in bodies of P_1^S, but only in negated form. This is not allowed in classical splittings (as defined in [Lifschitz and Turner, 1994]) which thus correspond to 0-splittings.

Note an important distinction between splittings for AFs and for logic programs: an arbitrary subset of arguments is an AF splitting, whereas a splitting S for a program P needs to fulfill the additional requirement that for each rule $r \in P$ with its head contained in S, also the entire positive body stems from S.

Example 4.31. Consider the following simple program

(1) $a \leftarrow$ not b
(2) $b \leftarrow$ not a
(3) $c \leftarrow a$

The program does not have a (nontrivial) classical splitting. However, it possesses the 1-splitting $S = \{a, c\}$ (together with the complementary 1-splitting $\bar{S} = \{b\}$). P_1^S consists of rules (1) and (3), P_2^S of rule (2). It is easily verified that $V_S = \{b\}$.

For the computation of answer sets we proceed in the same spirit as before by adding rules to P_1^S which allow answer sets to contain meta-information about the assumptions which need to hold for an answer set. We introduce atoms $ndr(b)$ to express the assumption that b will be underivable from P_2^S.

Definition 4.32. Let P be a normal logic program, let S be a k-splitting of P and let P_1^S, respectively P_2^S, be the sets of rules in P with heads in S, respectively in $\bar{S}$. Moreover, let V_S be the set of atoms in $\bar{S}$ appearing negatively in P_1^S.

$[P_1^S]$ is the program obtained from P_1^S by adding, for each $b \in V_S$, the following two rules:

(1_b) $b \leftarrow$ not $ndr(b)$
(2_b) $ndr(b) \leftarrow$ not b

E is called *conditional answer set* of P_1^S iff E is an extension of $[P_1^S]$.

Intuitively, $ndr(b)$ represents the assumption that b is not derivable from P_2^S (since $b \in V_S$ it cannot be derivable from P_1^S anyway). The additional rules allow us to assume b - a condition which we later need to verify. The construction is similar to the one we used for AFs where it was possible to assume that an argument is attacked (compare Definition 4.23).

Now, given an answer set E of $[P_1^S]$, we use E to modify P_2^S in the following way. We first use E to simplify P_2^S in exactly the same way this was done by Lifschitz and Turner: we replace atoms in S appearing positively (negatively) in rule bodies of P_2^S with *true* whenever they are (are not) contained in E. Moreover, we delete each rule with a positive (negative) occurrence of an S-atom in the body which is not (which is) in E. We call the resulting

4.3. Splittings for Logic Programs

program E-evaluation of P_2^S. Next we add adequate rules (constraints) which guarantee that answer sets generated by the modification of P_2^S match the conditions expressed in the meta-atoms of E. This is captured in the following definition:

Definition 4.33. Let P, S, and P_2^S be as in Def. 4.32, and let E be a conditional answer set of P_1^S. Let $[P_2^S]_E$, the E-modification of P_2^S, be obtained from the E-evaluation of P_2^S by adding the following k rules

$$\{\leftarrow not\ b \mid b \in E \cap V_S\} \cup \{\leftarrow b \mid ndr(b) \in E\}.$$

E' is called (E, S)-match iff E' is an answer set of $[P_2^S]_E$.

4.3.3 Generalized Splitting Theorem

Now we can verify that answer sets of P can be obtained by computing an answer set E_1 of $[P_1^S]$ and an answer set E_2 of the E_1-modification of P_2^S (we just have to eliminate the meta-atoms in E_1).

Theorem 4.34. *Let P be a normal logic program and let S be a quasi-splitting of P.*

1. *If E is a conditional answer set of P_1^S and F an (E, S)-match, then $(E \cap S) \cup F$ is an answer set of P.*

2. *If H is an answer set of P, then there is a set $X \subseteq \{ndr(a) \mid a \in V_S\}$ such that $E = (H \cap S) \cup X$ is a conditional answer set of P_1^S and $H \cap \bar{S}$ is an (E, S)-match of P.*

Proof. Thanks to the richer syntax of logic programs compared to AFs, the proof of this result is conceptually simpler than the one for Theorem 4.28. We sketch the proof which is done in two main steps.

First, given a normal program P and S a quasi-splitting of P, define the program $P_S = [P_1^S] \cup \bar{P}_2^S \cup \{\leftarrow not\ b', b;\ \leftarrow b', ndr(b) \mid b \in V_S\}$ where $\bar{P}_2^S$ results from P_2^S by replacing each atom $s \in \bar{S}$ with a fresh atom s'. One can show that P and P_S are equivalent in the following sense: (1) if E is an answer set of P then $X = (E \cap S) \cup \{s' \mid s \in E \cap \bar{S}\} \cup \{ndr(s) \mid s \in V_S \setminus E\}$ is an answer set of P_S; in particular, by the definition of X and $V_S \subseteq \bar{S}$, we have, for each $s \in V_S$, $s' \in X$ iff $s \in X$ iff $ndr(s) \notin X$; (2) if H is an answer set of P_S, then $(H \cap S) \cup \{s \mid s' \in H\}$ is an answer set of P.

Next, we observe that P_S has a proper split T with $T = S \cup \{ndr(s) \mid s \in V_S\}$. With this result at hand, it can be shown that after evaluating the bottom part of this split, i.e. $(P_S)_1^T = [P_1^S]$, the rules $\{\leftarrow not\ b', b;\ \leftarrow b', ndr(b) \mid b \in V_S\}$ in P_S play exactly the role of the replacements defined in $[P_2^S]_E$. In other words, we have for each $E \in AS((P_S)_1^T) = AS([P_1^S])$, the E-evaluation of $(P_S)_2^T$ is equal to $([P_2^S]_E)'$, where $([P_2^S]_E)'$ denotes $[P_2^S]_E$ replacing all atoms s by s'. Together with the above relation between answer sets of P and P_S the assertion is now shown in a quite straightforward way. □

Example 4.35 (Example 4.31 continued). To determine the conditional answer sets of P_1^S we use the program

(1) $a \leftarrow$ not b
(3) $c \leftarrow a$
(1_b) $b \leftarrow$ not $ndr(b)$
(2_b) $ndr(b) \leftarrow$ not b

This program has two answer sets, namely $E_1 = \{a, c, ndr(b)\}$ and $E_2 = \{b\}$.

The E_1-modification of P_2^S consists of the single rule $\leftarrow b$ and its single answer set $\emptyset$ is an (S, E_1)-match. We thus obtain the first answer set of P, namely $(E_1 \cap S) \cup \emptyset = \{a, c\}$.

The E_2-modification of P_2^S is the program:

b
$\leftarrow$ not b

Its single answer set $\{b\}$ is an (S, E_2)-match. We thus obtain the second answer set of P, namely $(E_2 \cap S) \cup \{b\} = \{b\}$.

Of course, the example is meant to illustrate the basic ideas, not to show the potential computational benefits of our approach. However, if we find a k-splitting with small k for large programs with hundreds of rules such benefits may be tremendous.

As the thoughtful reader may have recognized our results are stated for normal logic programs while Lifschitz and Turner define splittings for the more general class of disjunctive programs. However, our results can be extended to disjunctive programs in a straightforward way.

4.4 Conclusions and Related Work

Splitting results in nonmonotonic formalisms have a long tradition. On the one hand, these results can be used to improve existing computational procedures, and on the other hand they yield deeper theoretical insights into how a nonmonotonic approach works. In the 90's splitting results for autoepistemic logic [Gelfond and Przymusinska, 1992], logic programs [Lifschitz and Turner, 1994] and default theory [Turner, 1996] were established.

In this chapter we presented similar splitting results for argumentation frameworks and convey some ideas to normal logic programs generalizing results in [Lifschitz and Turner, 1994]. Most of the presented results are already published in [Baumann, 2011; Baumann et al., 2011; Baumann et al., 2012]. An unpublished and very exciting result is the general relation between universally definedness, the splitting property and the directionality principle as stated in Theorem 4.13. The study of such general relations, i.e. abstracting away from the concrete semantics, contributes to a better understanding of the universe of argumentation semantics and continues the line of abstract analysis initiated in [Baroni and Giacomin, 2007].

4.4. Conclusions and Related Work

We mention two works in the line of decomposition-based computation of extensions, namely [Baroni et al., 2005] and [Liao et al., 2011]. In the first one a general recursive schema for argumentation semantics were introduced. Furthermore they have shown that all admissibility-based semantics was covered by this definition. The great benefit of this approach is that the extensions of an AF $\mathcal{F}$ can be incrementally constructed by the extensions along its strongly connected components. In this sense our results are certainly related to the SCC-approach but there are some important differences on the technical level. Furthermore, we generalized our results, such that the parts into which an AF is split may be, but do not necessarily have to be SCCs. The second work by Liao et al. is very similar to our classical splitting approach both published in 2011. The paper addresses the problem of reconstructing the set of extensions of an AF $\mathcal{F}$ by computing and combining the extensions of certain subframeworks. They distinguish between "Dung-style sub-frameworks" which are unattacked by other subframeworks and "quasi sub-frameworks" which are attacked by arguments of other sub-frameworks. Finally, they group sub-frameworks in layers such that the sub-frameworks in a given layer are only restricted by sub-frameworks in lower layers and describe how to evaluate the different layers incrementally such that the extensions of the unpartitioned AF are reconstructable.

Chapter 5

Notions of Equivalence and Replacement

In general, equivalence tells us whether two syntactically different objects represent the same information - which is relevant, for instance, for simplification issues. In monotonic formalisms, there is usually one standard notion of equivalence which serves this purpose as well as the purpose of "equivalence for replacement", where one asks whether a replacement of equivalent objects within a larger one, does not change the meaning of the latter. For nonmonotonic formalisms these two purposes require, in general, different notions of equivalence, Take as an example logic programs under stable model semantics [Gelfond and Lifschitz, 1988]. The two programs $P_1 = \{a\}$ and $P_2 = \{a \leftarrow not\ b\}$ have the same stable model, namely $\{a\}$. However, if P_1 and P_2 are later extended with the fact b, then the stable models will no longer coincide: we obtain $\{a,b\}$ for the former, $\{b\}$ for the latter. This observation led to the investigation of stronger equivalence notions for logic programs [Lifschitz et al., 2001], and more recently also for argumentation.

Oikarinen and Woltran introduced the notion of *strong equivalence* for abstract AFs [Oikarinen and Woltran, 2011]. Two AFs $\mathcal{F}$ and $\mathcal{G}$ are strongly equivalent if for any AF $\mathcal{H}$, $\mathcal{F}$ conjoined with $\mathcal{H}$ and $\mathcal{G}$ conjoined with $\mathcal{H}$ possess the same extensions. From now on, for the sake of clarity we will use the term *expansion equivalence* instead of *strong equivalence* since any kind of expansions are allowed. The powerful notion of expansion equivalence was the starting point for our research. However, for several typical argumentation scenarios expansion equivalence seems too strong a notion. Just like in the case of other nonmonotonic formalisms where further equivalence notions in-between strong and standard equivalence were motivated, defined and studied (see [Woltran, 2010] for an excellent overview) we looked for corresponding notions for AFs which take the very nature of argumentation into account. For instance, let us consider a reasoning process about defeasible information stored in a knowledge base as illustrated in Figure 1.1. What happens on the abstract level if a new piece of information is added? It turns out that older arguments and their corresponding attacks survive and only new arguments which may interact with the previous ones arise. This means, in contrast to Oikarinen and Woltran which studied equivalence with respect to arbitrary expansions we are interested in equivalence relations with respect to specific expansions where the attack relationship between former arguments remains unchanged. Such kinds of dynamic scenarios correspond to the already defined concepts of *normal* or *strong expansions* firstly introduced in [Baumann and Brewka, 2010].

As a valuable tool for deciding expansion equivalence, Oikarinen and Woltran introduced the notion of a *kernel* of an AF. Informally speaking, kernels are frameworks without redundant attacks. The significant insight of [Oikarinen and Woltran, 2011] was that expansion equivalence for most semantics is exactly captured by syntactical equivalence of suitably chosen notions of kernels. All kernels found so far differed only in case where self-loops are permitted; in other words, if self-loops are omitted it turns out that each framework is its own kernel (a related observation for graph problems in general was given independently at the same time in [Lonc and Truszczyński, 2011]).

In this chapter we present characterization theorems with respect to normal and strong expansion equivalence and the ten prominent semantics considered in this book, namely stable, semi-stable, stage, preferred, admissible, complete, grounded, ideal, eager and naive semantics. Except for naive semantics, we will present kernel-based, i.e. purely syntactical characterizations of the mentioned equivalence notions. Depending on the semantics we either use already existing kernels or we have to introduce novel and more involved kernel definitions. Furthermore, we consider local equivalence with respect to stage semantics as well as weak expansion equivalence for stable and preferred semantics.

One main result is that expansion equivalence coincides with normal expansion equivalence for all considered semantics. Even more surprisingly we will see that stage, stable, semi-stable, eager and naive semantics do not "distinguish" between normal expansion and strong expansion equivalence. Both results are quite unexpected since the class of strong (normal) expansions is obviously a proper subset of the class of normal (arbitrary) expansions.

5.1 Recapitulation: Expansion Equivalence

As already mentioned in Section 2.1.5 as well as illustrated in Example 2.15 standard equivalence of two AFs is not sufficient for their mutual replaceability in dynamic argumentation scenarios. That means, possessing the same extensions does not guarantee to share the same acceptable sets of arguments with respect to all expansions. Expansion equivalence guarantees inter-substitutability in any dynamical scenario by definition. We briefly summarize concepts and results firstly presented in [Oikarinen and Woltran, 2011] since they are frequently used throughout the whole chapter.

Chapter 5. Notions of Equivalence and Replacement

5.1.1 σ-Kernel

One main result in [Oikarinen and Woltran, 2011] is that expansion equivalence can be decided via so-called *kernels*. A kernel of an AF $\mathcal{F}$ is itself an AF obtained from $\mathcal{F}$ by deleting certain attacks based on the actual structure of the given AF but without computing extensions or any related semantical objects. This makes kernels a purely syntactical — and thus efficiently computable — concept.

Here are the relevant kernel definitions.

Definition 5.1. Given a semantics $\sigma \in \{stb, ad, gr, co\}$ and an AF $\mathcal{F} = (A, R)$. We define the *σ-kernel* of $\mathcal{F}$ as $\mathcal{F}^{k(\sigma)} = (A, R^{k(\sigma)})$ where

1. $R^{k(stb)} = R \smallsetminus \{(a,b) \mid a \neq b, (a,a) \in R\}$,

2. $R^{k(ad)} = R \smallsetminus \{(a,b) \mid a \neq b, (a,a) \in R, \{(b,a),(b,b)\} \cap R \neq \emptyset\}$,

3. $R^{k(gr)} = R \smallsetminus \{(a,b) \mid a \neq b, (b,b) \in R, \{(a,a),(b,a)\} \cap R \neq \emptyset\}$,

4. $R^{k(co)} = R \smallsetminus \{(a,b) \mid a \neq b, (a,a), (b,b) \in R\}$.

For the purpose of illustration we present the following minimalistic example.

Example 5.2. Consider the AF $\mathcal{F}$ and its corresponding stable-kernel $\mathcal{F}^{k(stb)}$ and grounded-kernel $\mathcal{F}^{k(gr)}$.

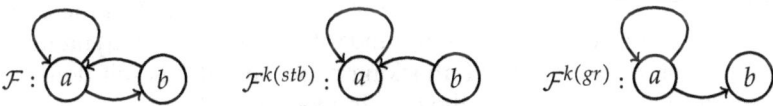

Figure 5.1: Stable and Grounded Kernel of $\mathcal{F}$

We want to mention three simple properties applying to any σ-kernel defined so far (as well as all novel kernel definitions which will be introduced in the following): First, $\mathcal{F}$ and $\mathcal{F}^{k(\sigma)}$ share exactly the same arguments. Second, the attack-relation of $\mathcal{F}^{k(\sigma)}$ is contained in the attack-relation of $\mathcal{F}$ and third, the kernel operation is idempotent, i.e. $\mathcal{F}^{k(\sigma)} = \left(\mathcal{F}^{k(\sigma)}\right)^{k(\sigma)}$.

5.1.2 Characterization Theorems

A kernel characterizes an equivalence relation as follows: two AFs are equivalent if and only if their corresponding kernels are identical. In other words, the kernel detects the redundant attacks with respect to a certain kind of expansion and semantics.

We list now some non-trivial results showing relations between the syntactical concept of σ-kernels and semantically defined equivalence relations.[1] The main aim of the rest of the chapter is to develop similar characterization theorems for normal and strong expansion equivalence with respect to all semantics considered in this book.

Lemma 5.3. *For any AF $\mathcal{F}$ and $\sigma \in \{stb, ad, gr, co\}$, $\mathcal{F} \equiv^\sigma \mathcal{F}^{k(\sigma)}$.*

Lemma 5.4. *For any AFs $\mathcal{F}, \mathcal{G}$ and $\sigma \in \{stb, ad, gr, co\}$ the following holds: If $\mathcal{F}^{k(\sigma)} = \mathcal{G}^{k(\sigma)}$, then $(\mathcal{F} \cup \mathcal{H})^{k(\sigma)} = (\mathcal{G} \cup \mathcal{H})^{k(\sigma)}$ for all AFs $\mathcal{H}$.*

Theorem 5.5. *For any AFs $\mathcal{F}, \mathcal{G}$ and $\sigma \in \{stb, ad, gr, co\}$:*

$$\mathcal{F}^{k(\sigma)} = \mathcal{G}^{k(\sigma)} \Leftrightarrow \mathcal{F} \equiv_E^\sigma \mathcal{G}.$$

Theorem 5.6. *For any AFs $\mathcal{F}$ and $\mathcal{G}$:*

$$\mathcal{F}^{k(co)} = \mathcal{G}^{k(co)} \Leftrightarrow \mathcal{F}^{k(ad)} = \mathcal{G}^{k(ad)} \text{ and } \mathcal{F}^{k(gr)} = \mathcal{G}^{k(gr)}.$$

Theorem 5.7. *For any AFs $\mathcal{F}$ and $\mathcal{G}$ the following holds:*

$$\mathcal{F} \equiv_E^{ad} \mathcal{G} \Leftrightarrow \mathcal{F} \equiv_E^{pr} \mathcal{G} \Leftrightarrow \mathcal{F} \equiv_E^{id} \mathcal{G} \Leftrightarrow \mathcal{F} \equiv_E^{ss} \mathcal{G} \Leftrightarrow \mathcal{F} \equiv_E^{eg} \mathcal{G}.$$

Theorem 5.8. *For any AFs $\mathcal{F}$ and $\mathcal{G}$:*

$$\mathcal{F} \equiv_E^{stb} \mathcal{G} \Leftrightarrow \mathcal{F} \equiv_E^{stg} \mathcal{G}.$$

Consider again Example 5.2. In case of stable semantics we may delete attacks from self-attacking arguments and still retain the same extensions ($\mathcal{F} \equiv^{stb} \mathcal{F}^{k(stb)}$, Lemma 5.3). Due to the idempotency of stb-kernel even the same extensional behavior with respect to arbitrary expansions is guaranteed ($\mathcal{F} \equiv_\leq^{stb} \mathcal{F}^{k(stb)}$, Theorem 5.5). In case of grounded semantics the conditions under which an attack (a, b) is redundant in the light of dynamics differ, namely: First, a and b are self-attacking or second, if b is self-defeating and counterattacks a. The latter is the case in Example 5.2.

5.2 Characterizing Strong Expansion Equivalence

In this section we will characterize strong expansion equivalence with respect to the ten semantics considered in this book. Analogously to the characterization of expansion equivalence (compare Section 5.1.2) we provide syntactical criteria to decide this notion of equivalence. In some cases we have to introduce more involved kernel definitions, so-called σ-*-kernels which allow more deletions than their counterparts for expansion equivalence. The main results can be summarized as follows:

---

[1] The following lemmata and theorems are mainly taken from [Oikarinen and Woltran, 2011]. Lemma 5.3 summarizes Lemmata 1, 4, 6 and 10 in [Oikarinen and Woltran, 2011]. Likewise, Lemma 5.4 is a summary of Lemmata 2, 5, 7 and 11. Furthermore, Theorem 1 combines Theorems 1, 2, 3 and 4. Theorem 5.6 and Theorem 5.7 correspond directly with Theorem 5 or Theorem 2 in [Oikarinen and Woltran, 2011], respectively. The last Theorem 5.8 corresponds to Theorem 2 in [Gaggl and Woltran, 2011].

- Strong expansion equivalence with respect to stable and stage semantics coincide and can be decided by the already defined *stb*-kernel [Oikarinen and Woltran, 2011]. This means, in case of these semantics, expansion and strong expansion equivalence coincide.

- In case of semi-stable and eager semantics we observe a similar behaviour. Both notions coincide and are characterizable via the *ad*-kernel [Oikarinen and Woltran, 2011]. Thus, there is no difference between expansion and strong expansion equivalence.

- The concepts of strong expansion equivalence with respect to admissible, preferred and ideal semantics coincide and can be adequately described by the newly introduced *ad-*-*kernel. Expansion equivalence with respect to these semantics implies strong expansion equivalence but not vice versa.

- The characterization of strong expansion equivalence with respect to grounded and complete semantics was the most difficult part. These notions can be decided by the newly introduced *gr-*-* or *co-*-*kernel, respectively. Both concepts are weaker than their corresponding expansion equivalence notions.

- Finally, possessing the same naive extensions and sharing the same arguments is sufficient and necessary for strong expansion equivalence with respect to naive semantics.

5.2.1 Splitting Results: A Tool for Simplifying Proofs

In Chapter 4 we presented splitting results for argumentation frameworks as well as logic programs. In general, splitting results are concerned with the question whether it is possible to divide a formal theory T in disjoint subtheories $S_1,...,S_n$ such that the formal semantics of the entire theory T can be obtained by constructing the semantics of $S_1,...,S_n$. Such results, especially in nonmonotonic formalisms [Gelfond and Przymusinska, 1992; Lifschitz and Turner, 1994; Turner, 1996; Baumann, 2011], are of great importance since first, they allow for simplification of proofs showing properties of a particular formalism and second, they may yield more efficient computations. In the following we will use splitting results for AFs as a tool for simplification.

An ongoing task in this section is the question whether it is possible to find an AF $\mathcal{H}$ such that, given two AFs $\mathcal{F}$ and $\mathcal{G}$, the semantics of $\mathcal{F} \cup \mathcal{H}$ and $\mathcal{G} \cup \mathcal{H}$ do not coincide. The difficulty is that we usually have very limited information about the AFs $\mathcal{F}$ and $\mathcal{G}$.

Example 5.9. Consider the following AFs $\mathcal{F}$ and $\mathcal{G}$. We have $A(\mathcal{F}) = A(\mathcal{G}) = \{a,b,c\} \cup B$ where B is a (possibly empty) set of further arguments. Furthermore, $R(\mathcal{F}) = \{(a,b),(b,b),(b,c)\} \cup R$ and $R(\mathcal{G}) = \{(a,b),(b,b)\} \cup S$ where R and S represent possible but unknown attacks (indicated by dashed arrows).

5.2. Characterizing Strong Expansion Equivalence

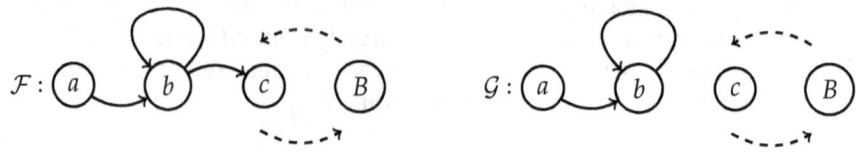

Figure 5.2: Partially Known AFs

Since we have only partial information about the AFs we cannot compute/compare their extensions. For instance, in case of $B = \emptyset$ we deduce $\mathcal{E}_{pr}(\mathcal{F}) = \mathcal{E}_{pr}(\mathcal{G}) = \{\{a,c\}\}$, i.e. they possess the same preferred extension. Consider now the AFs $\mathcal{F} \cup \mathcal{H}$ and $\mathcal{G} \cup \mathcal{H}$ where $\mathcal{H} = (A(\mathcal{F}) \cup \{d\}, \{(d,a)\} \cup \{(d,b) \mid b \in B\})$. Observe that $\mathcal{F} \cup \mathcal{H}$ and $\mathcal{G} \cup \mathcal{H}$ are strong expansions of $\mathcal{F}$ or $\mathcal{G}$, respectively.

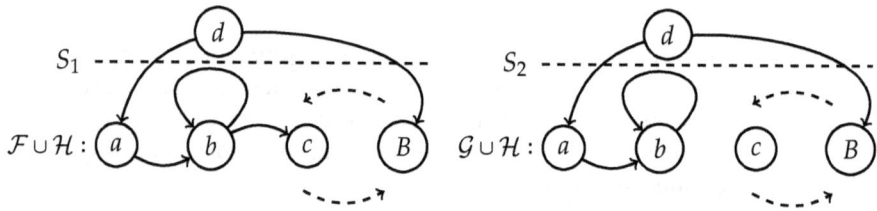

Figure 5.3: Strong Expansion and Classical Splitting

Using splitting results for AFs (Theorem 4.10) we are able to compute the extensions of the partially known AFs $\mathcal{F} \cup \mathcal{H}$ and $\mathcal{G} \cup \mathcal{H}$ in an iterative way. The procedure is as follows (consider Section 4.1 for detailed information): We split the initial AFs into two subframeworks, namely $\mathcal{F}_1 = \mathcal{G}_1 = (\{d\}, \emptyset)$ and $\mathcal{F}_2 = \mathcal{F}$ or $\mathcal{G}_2 = \mathcal{G}$, respectively (indicated by the dashed lines S_1 and S_2). We then take the unique preferred extension $\{d\}$ of $\mathcal{F}_1$ and $\mathcal{G}_1$ to reduce the AFs $\mathcal{F}_2$ or $\mathcal{G}_2$, respectively. In this case, "reducing" quite simply means deleting all arguments attacked by d. The following AFs $\mathcal{F}_2^{\{d\}}$ and $\mathcal{G}_2^{\{d\}}$ illustrate these reducts. Note that these reducts are uniquely determined!

Figure 5.4: Uniquely Determined Reducts

We now compute the preferred extensions of the reduced AFs, namely $\mathcal{E}_{pr}(\mathcal{F}_2^{\{d\}}) = \{\emptyset\}$ and $\mathcal{E}_{pr}(\mathcal{G}_2^{\{d\}}) = \{\{c\}\}$. Finally, the preferred extensions of $\mathcal{F} \cup \mathcal{H}$ and $\mathcal{G} \cup \mathcal{H}$ can be obtained by combining $\{d\}$ and $\emptyset$ or $\{c\}$, respectively. That means, $\{\{d\}\} = \mathcal{E}_{pr}(\mathcal{F} \cup \mathcal{H}) \neq \mathcal{E}_{pr}(\mathcal{G} \cup \mathcal{H}) = \{\{c,d\}\}$. In summary, we have shown that the partially known AFs $\mathcal{F}$ and $\mathcal{G}$ have strong expansions which possess different preferred and hence, semi-stable and ideal extensions

Chapter 5. Notions of Equivalence and Replacement

(Proposition 2.10) as well as admissible and complete extensions (Proposition 2.17).

5.2.2 Strong Expansion Equivalence for Stable and Stage Semantics

In case of stable and stage semantics it is already shown [Oikarinen and Woltran, 2011; Gaggl and Woltran, 2011] that attacks (a, b) where a is a self-attacking argument do not contribute in the evaluation of an AF $\mathcal{F}$, no matter how $\mathcal{F}$ is extended. Furthermore the syntactical equivalence of *stb*-kernels, which "delete" such attacks of a given AF, is necessary and sufficient for expansion equivalence between two AFs (Theorems 5.5, 5.8). Since the classes of strong and arbitrary expansions are in a proper subset relation we suspected that there are strong expansion equivalent AFs which are not expansion equivalent. The construction of such an example failed and we tried to prove that even strong expansion equivalence between two AFs is fulfilled if and only if they possess the same *stb*-kernels. The following theorem proves this conjecture.

Theorem 5.10. *For any AFs $\mathcal{F}$ and $\mathcal{G}$,*

$$\mathcal{F}^{k(stb)} = \mathcal{G}^{k(stb)} \Leftrightarrow \mathcal{F} \equiv_S^{stb} \mathcal{G} \Leftrightarrow \mathcal{F} \equiv_S^{stg} \mathcal{G}.$$

Proof. Let $\sigma \in \{stb, stg\}$. Throughout the proof we will use that possessing at least one stable extension is sufficient for the agreement of stable and stage semantics (Proposition 2.9). Moreover, we only have to show that $\mathcal{F} \equiv_S^{\sigma} \mathcal{G} \Rightarrow \mathcal{F}^{k(st)} = \mathcal{G}^{k(stb)}$ holds since $\mathcal{F}^{k(stb)} = \mathcal{G}^{k(stb)} \Rightarrow \mathcal{F} \equiv_E^{\sigma} \mathcal{G} \Rightarrow \mathcal{F} \equiv_N^{\sigma} \mathcal{G} \Rightarrow \mathcal{F} \equiv_S^{\sigma} \mathcal{G}$ is given by Theorems 5.5, 5.8 and statement 1 of Proposition 2.18. We will prove this implication by contraposition.

Suppose $\mathcal{F}^{k(stb)} \neq \mathcal{G}^{k(stb)}$. **1st case:** Consider $A(\mathcal{F}^{k(stb)}) \neq A(\mathcal{G}^{k(stb)})$. Consequently, without loss of generality there exists an argument $a \in A(\mathcal{F}) \setminus A(\mathcal{G})$. Let c be a fresh argument and $B = A(\mathcal{F} \cup \mathcal{G}) \setminus \{a\}$. We define

$$\mathcal{H} = (B \cup \{c\}, \{(c, c') \mid c' \in B\}).$$

If a is contained in some $E \in \mathcal{E}_\sigma(\mathcal{F} \cup \mathcal{H})$, then $E \notin \mathcal{E}_\sigma(\mathcal{G} \cup \mathcal{H})$ since $a \notin A(\mathcal{G} \cup \mathcal{H})$ was supposed. If not, consider $\mathcal{H}' = \mathcal{H} \cup (\{a\}, \emptyset)$. Then, $E = \{a, c\}$ is a stable extension in $\mathcal{G} \cup \mathcal{H}'$ and therefore, $E \in \mathcal{E}_\sigma(\mathcal{G} \cup \mathcal{H}')$. Furthermore, $E \notin \mathcal{E}_\sigma(\mathcal{F} \cup \mathcal{H}')$ since $\mathcal{F} \cup \mathcal{H}' = \mathcal{F} \cup \mathcal{H}$ holds. Consequently, $\mathcal{F} \not\equiv_S^{\sigma} \mathcal{G}$ is shown.

2nd case: Consider $A(\mathcal{F}^{k(stb)}) = A(\mathcal{G}^{k(stb)})$ and $R(\mathcal{F}^{k(stb)}) \neq R(\mathcal{G}^{k(stb)})$. Hence, there are $a, b \in A(\mathcal{F})$, such that $(a, b) \in R(\mathcal{F}^{k(stb)}) \setminus R(\mathcal{G}^{k(stb)})$. Let c be a fresh argument. We define

$$\mathcal{I} = (A(\mathcal{F}) \cup \{c\}, \{(c, c') \mid c' \in A(\mathcal{F}) \setminus \{a, b\}\}).$$

Case 2.1: Let $a = b$, therefore $(a, a) \in R(\mathcal{F}^{k(stb)}) \setminus R(\mathcal{G}^{k(stb)})$ and consequently, $(a, a) \in R(\mathcal{F}) \setminus R(\mathcal{G})$ because self-loops remain the same after applying the kernel-operator. Obviously, $\mathcal{E}_{stb}(\mathcal{G} \cup \mathcal{I}) = \{\{a, c\}\}$ and thus, $\{a, c\} \in$

5.2. Characterizing Strong Expansion Equivalence

$\mathcal{E}_{stg}(\mathcal{G} \cup \mathcal{I})$. On the other hand, $\{a,c\} \notin cf(\mathcal{F} \cup \mathcal{I})$ and therefore, $\{a,c\} \notin \mathcal{E}_\sigma(\mathcal{F} \cup \mathcal{I})$. This means, $\mathcal{F} \not\equiv_S^\sigma \mathcal{G}$ is shown. Thus from now on we assume that $R(\mathcal{F}^{k(stb)})$ and $R(\mathcal{G}^{k(stb)})$ contain the same self-loops.

Case 2.2: Let $a \neq b$. Since $(a,b) \in R(\mathcal{F}^{k(stb)}) \setminus R(\mathcal{G}^{k(stb)})$, it follows that $(a,b) \in R(\mathcal{F})$, $(a,a) \notin R(\mathcal{F})$, consequently $(a,a) \notin R(\mathcal{G})$ and $(a,b) \notin R(\mathcal{G})$. First, we assume $(b,b) \notin R(\mathcal{F})$. Thus, $(b,b) \notin R(\mathcal{G})$. In any case $\{a,c\} \in \mathcal{E}_{stb}(\mathcal{F} \cup \mathcal{I})$ and consequently, $\{a,c\} \in \mathcal{E}_{stg}(\mathcal{F} \cup \mathcal{I})$. On the other hand, we state $\{a,c\} \notin \mathcal{E}_{stg}(\mathcal{G} \cup \mathcal{I})$ because $\{\{a,b,c\}\} = \mathcal{E}_{stb}(\mathcal{G} \cup \mathcal{I})$ if $(b,a) \notin R(\mathcal{G})$ and $\{\{b,c\}\} = \mathcal{E}_{stb}(\mathcal{G} \cup \mathcal{I})$ if $(b,a) \in R(\mathcal{G})$. Second, let $(b,b) \in R(\mathcal{F})$. Consequently, $(b,b) \in R(\mathcal{G})$. In contrast to the other cases the AF $\mathcal{I}$ does not enforce different stage extensions (but it works for stable semantics since $\mathcal{E}_{stb}(\mathcal{F} \cup \mathcal{I}) = \{\{a,c\}\} \neq \emptyset = \mathcal{E}_{stb}(\mathcal{G} \cup \mathcal{I}))$. In particular, $\mathcal{E}_{stg}(\mathcal{F} \cup \mathcal{I}) = \mathcal{E}_{stg}(\mathcal{G} \cup \mathcal{I}) = \{\{a,c\}\}$ independently of the occurrence of the attack (b,a) in $\mathcal{F}$ or $\mathcal{G}$. We define

$$\mathcal{S} = (A(\mathcal{F}) \cup \{c,d\}, \{(e,f) \mid e \in \{c,d\} \wedge f \in A(\mathcal{F}) \setminus \{a,b\}\} \cup \{(c,d),(d,a),(d,c)\}).$$

For the purpose of illustration we omit further arguments different from a, b, c and d. Bear in mind that these arguments are attacked by c and d. The dashed arrows reflect the situation that (b,a) may or not be in $R(\mathcal{F})$ or $R(\mathcal{G})$.

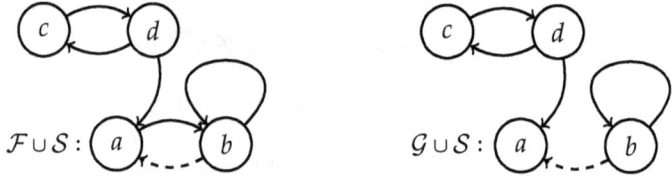

Note that $b \notin R^+_{\mathcal{G} \cup \mathcal{S}}(E)$ for any set $E \in cf(\mathcal{G} \cup \mathcal{S})$. Since $R^+_{\mathcal{G} \cup \mathcal{S}}(\{d\}) = A(\mathcal{G} \cup \mathcal{S}) \setminus \{b\}$ we deduce $\{d\} \in \mathcal{E}_{stg}(\mathcal{G} \cup \mathcal{S})$. On the other hand, $R^+_{\mathcal{F} \cup \mathcal{S}}(\{d\}) = A(\mathcal{F} \cup \mathcal{S}) \setminus \{b\} \subset A(\mathcal{F} \cup \mathcal{S}) = R^+_{\mathcal{F} \cup \mathcal{S}}(\{a,c\})$. Consequently, $\{d\} \notin \mathcal{E}_{stg}(\mathcal{F} \cup \mathcal{S})$ is shown. Thus, $\mathcal{F} \not\equiv_S^{stg} \mathcal{G}$ as well as $\mathcal{F} \not\equiv_S^{stb} \mathcal{G}$ concluding the proof. □

5.2.3 Strong Expansion Equivalence for Semi-Stable and Eager Semantics

Semi-stable semantics is, as the name suggests, very close to stable semantics. It can be shown that any stable extension is semi-stable and furthermore, if there is at least one stable extension then the set of stable and semi-stable extensions coincide (compare Propositions 2.7, 2.9). In spite of these similarities the characterizing kernels of the correspondent expansion equivalence notions differ. Oikarinen and Woltran showed that instead of the *stb*-kernel the equality of the more restrictive *ad*-kernel of two AFs adequately determine expansion equivalence with respect to semi-stable semantics (compare Theorems 5.5, 5.7). Apart from this, the following theorem states a similar result to Theorem 5.10, namely that the equality of the *ad*-kernels of two AFs is even necessary for their strong expansion equivalence with respect to semi-stable semantics. Interestingly, in contrast to expansion equivalence

Chapter 5. Notions of Equivalence and Replacement

the *ad*-kernel does not serve as a uniform characterization for strong expansion equivalence with respect to admissible, preferred, ideal, semi-stable and eager semantics. Only semi-stable and eager semantics are captured by this kernel as the following theorem shows.

Theorem 5.11. *For any AFs $\mathcal{F}$ and $\mathcal{G}$,*

$$\mathcal{F}^{k(ad)} = \mathcal{G}^{k(ad)} \Leftrightarrow \mathcal{F} \equiv_S^{ss} \mathcal{G} \Leftrightarrow \mathcal{F} \equiv_S^{eg} \mathcal{G}.$$

Proof. Let $\tau \in \{ss, eg\}$. Observe that Theorems 5.5, 5.7 and statement 1 of Proposition 2.18 guarantee $\mathcal{F}^{k(ad)} = \mathcal{G}^{k(ad)} \Rightarrow \mathcal{F} \equiv_S^{\tau} \mathcal{G}$. Hence, it suffices to show that $\mathcal{F}^{k(ad)} = \mathcal{G}^{k(ad)}$ is implied by $\mathcal{F} \equiv_S^{\tau} \mathcal{G}$. We show the contrapositive.

Assume $\mathcal{F}^{k(ad)} \neq \mathcal{G}^{k(ad)}$. We have to show $\mathcal{F} \not\equiv_S^{\tau} \mathcal{G}$ which we do by case analysis. In almost all cases (except for the case 2.2.4) we even prove that, given the assumption, $\mathcal{F} \not\equiv_S^{\sigma} \mathcal{G}$ for every $\sigma \in \{ad, pr, ss, id, eg\}$. This can be shown without extra effort.[2] **1st case:** Assume $A(\mathcal{F}^{k(ad)}) \neq A(\mathcal{G}^{k(ad)})$. Hence, $A(\mathcal{F}) \neq A(\mathcal{G})$ is implied and without loss of generality there exists an argument $a \in A(\mathcal{F}) \setminus A(\mathcal{G})$. Let c be a fresh argument, i.e. $c \notin A(\mathcal{F} \cup \mathcal{G})$ and $B = A(\mathcal{F} \cup \mathcal{G}) \setminus \{a\}$. We define

$$\mathcal{H} = (B \cup \{c\}, \{(c, c') \mid c' \in B\}).$$

Given $\sigma \in \{ad, pr, ss, id, eg\}$. If a is contained in some $E \in \mathcal{E}_\sigma(\mathcal{F} \cup \mathcal{H})$, then $E \notin \mathcal{E}_\sigma(\mathcal{G} \cup \mathcal{H})$ follows since $a \notin A(\mathcal{G} \cup \mathcal{H})$ was supposed. If not, consider $\mathcal{H}' = \mathcal{H} \cup (\{a\}, \emptyset)$. Applying splitting results (compare Section 5.2.1) it follows that $E = \{a, c\}$ is the unique preferred extension of $\mathcal{G} \cup \mathcal{H}'$. Consequently, E is admissible in $\mathcal{G} \cup \mathcal{H}'$ and the unique semi-stable, ideal and eager extension of $\mathcal{G} \cup \mathcal{H}'$ (Proposition 2.10). On the other hand, $E \notin \mathcal{E}_\sigma(\mathcal{F} \cup \mathcal{H}')$ since $\mathcal{F} \cup \mathcal{H}' = \mathcal{F} \cup \mathcal{H}$ holds. This means, $\mathcal{F} \not\equiv_S^{\sigma} \mathcal{G}$ for $\sigma \in \{ad, pr, ss, id, eg\}$ is shown since $\mathcal{F}$ and $\mathcal{G}$ combined with $\mathcal{H}$ or $\mathcal{H}'$ are strong expansions of $\mathcal{F}$ and $\mathcal{G}$.

2nd case: Consider now $R(\mathcal{F}^{k(ad)}) \neq R(\mathcal{G}^{k(ad)})$ and $A(\mathcal{F}^{k(ad)}) = A(\mathcal{G}^{k(ad)})$. Note that $A(\mathcal{F}) = A(\mathcal{G})$ is implied and furthermore, without loss of generality we may assume the existence of arguments $a, b \in A(\mathcal{F})$, such that $(a, b) \in R(\mathcal{F}^{k(ad)}) \setminus R(\mathcal{G}^{k(ad)})$. Let c be a fresh argument, i.e. $c \notin A(\mathcal{F})$. Furthermore we define

$$\mathcal{I} = (A(\mathcal{F}) \cup \{c\}, \{(c, c') \mid c' \in A(\mathcal{F}) \setminus \{a, b\}\}).$$

Case 2.1: Assume $a = b$. Therefore $(a, a) \in R(\mathcal{F}^{k(ad)}) \setminus R(\mathcal{G}^{k(ad)})$ and consequently $(a, a) \in R(\mathcal{F}) \setminus R(\mathcal{G})$ by definition of the *ad*-kernel. It can be checked (splitting results, Section 5.2.1) that $\mathcal{E}_{pr}(\mathcal{G} \cup \mathcal{I}) = \{\{a, c\}\}$ and $\mathcal{E}_{pr}(\mathcal{F} \cup \mathcal{I}) = \{\{c\}\}$. Hence, $\mathcal{F} \cup \mathcal{I} \not\equiv^{\sigma} \mathcal{G} \cup \mathcal{I}$ for $\sigma \in \{ad, pr, ss, id, eg\}$ can be obtained (Propositions 2.10, 2.17) and therefore, $\mathcal{F} \not\equiv_S^{\sigma} \mathcal{G}$ is shown. Thus from now on we assume that $R(\mathcal{F}^{k(ad)})$, $R(\mathcal{G}^{k(ad)})$, $R(\mathcal{F})$ and $R(\mathcal{G})$ contain the same self-loops.

---

[2] We will use these results in Section 5.3.2 to prove that the *ad*-kernel adequately describes normal expansion equivalence with respect to admissible, preferred and ideal semantics.

5.2. Characterizing Strong Expansion Equivalence

Case 2.2: Let $a \neq b$. Since $(a,b) \in R(\mathcal{F}^{k(ad)}) \setminus R(\mathcal{G}^{k(ad)})$, it follows $(a,b) \in R(\mathcal{F})$. Now we have to distinguish four cases with respect to the presence or absence of the self-loops (a,a) and (b,b). **Case 2.2.1:** Assume $(a,a), (b,b) \in R(\mathcal{F})$. This case is impossible because the definition of the ad-kernel (Definition 5.1) enforce the deletion of (a,b) in $R(\mathcal{F}^{k(ad)})$. **Case 2.2.2:** Consider $(a,a) \in R(\mathcal{F})$ and $(b,b) \notin R(\mathcal{F})$. We observe that $(b,a) \notin R(\mathcal{F})$ holds by kernel definition. Hence, $\mathcal{E}_{pr}(\mathcal{F} \cup \mathcal{I}) = \{\{c\}\}$ (splitting results, Section 5.2.1). For $\mathcal{G}$ three cases arise. First, $(a,b) \in R(\mathcal{G})$ and consequently $(b,a) \in R(\mathcal{G})$ because of the assumption $(a,b) \notin R(\mathcal{G}^{k(ad)})$. Second and third, $(a,b) \notin R(\mathcal{G})$ and (b,a) may or may not be in $R(\mathcal{G})$. Using splitting results (compare Section 5.2.1) it can be checked that $\mathcal{E}_{pr}(\mathcal{G} \cup \mathcal{I}) = \{\{b,c\}\}$ holds. Thus, using Propositions 2.10, 2.17 we obtain $\mathcal{F} \not\equiv_S^\sigma \mathcal{G}$ for $\sigma \in \{ad, pr, ss, id, eg\}$. **Case 2.2.3:** Let $(a,a), (b,b) \notin R(\mathcal{F})$. We deduce $(a,b) \notin \mathcal{G}$ since $(a,a) \notin R(\mathcal{G})$ and $(a,b) \notin R(\mathcal{G}^{k(ad)})$ was assumed. We have to distinguish four sub-cases with respect to the presence or absence of (b,a). Suppose $(b,a) \notin R(\mathcal{F})$. Hence, $\mathcal{E}_{pr}(\mathcal{F} \cup \mathcal{I}) = \{\{a,c\}\}$. If $(b,a) \notin R(\mathcal{G})$, $\mathcal{E}_{pr}(\mathcal{G} \cup \mathcal{I}) = \{\{a,b,c\}\}$. If not, $\mathcal{E}_{pr}(\mathcal{G} \cup \mathcal{I}) = \{\{b,c\}\}$. In both cases $\mathcal{E}_{pr}(\mathcal{F} \cup \mathcal{I}) \neq \mathcal{E}_{pr}(\mathcal{G} \cup \mathcal{I})$ holds. Consider now $(b,a) \in R(\mathcal{F})$. It can be checked that $\mathcal{E}_{pr}(\mathcal{F} \cup \mathcal{I}) = \{\{a,c\}, \{b,c\}\}$. Note that these sets are stable and therefore semi-stable extensions too. Furthermore, $\{c\}$ is admissible in $\mathcal{F} \cup \mathcal{I}$ and equals $\{a,c\} \cap \{b,c\}$. This means, $\mathcal{E}_{eg}(\mathcal{F} \cup \mathcal{I}) = \mathcal{E}_{id}(\mathcal{F} \cup \mathcal{I}) = \{\{c\}\}$. Again, if $(b,a) \notin R(\mathcal{G})$, $\mathcal{E}_{pr}(\mathcal{G} \cup \mathcal{I}) = \{\{a,b,c\}\}$. If not, $\mathcal{E}_{pr}(\mathcal{G} \cup \mathcal{I}) = \{\{b,c\}\}$. Hence, in all cases $\mathcal{F} \cup \mathcal{I} \not\equiv^\sigma \mathcal{G} \cup \mathcal{I}$ for $\sigma \in \{ad, pr, ss, id, eg\}$. Thus, $\mathcal{F} \not\equiv_S^\sigma \mathcal{G}$ for $\sigma \in \{ad, pr, ss, id, eg\}$ is shown. **Case 2.2.4:** Consider $(a,a) \notin R(\mathcal{F})$ and $(b,b) \in R(\mathcal{F})$. As described at the very beginning of the proof this sub-case is the decisive point where only semi-stable and eager semantics behave "well" with respect to the ad-kernel. This means, we will only show that $\mathcal{F}$ and $\mathcal{G}$ are not strong expansion equivalent with respect to semi-stable and eager semantics. In contrast to the other cases the AF $\mathcal{I}$ does not do the trick, i.e. $\mathcal{F} \cup \mathcal{I}$ and $\mathcal{G} \cup \mathcal{I}$ do not necessarily possess different semi-stable and eager extensions. We therefore consider the AF $\mathcal{S}$ already used in Theorem 5.10.

$$\mathcal{S} = (A(\mathcal{F}) \cup \{c,d\}, \{(e,f) \mid e \in \{c,d\} \land f \in A(\mathcal{F}) \setminus \{a,b\}\} \cup \{(c,d),(d,a),(d,c)\}).$$

The following figure illustrates $\mathcal{F} \cup \mathcal{S}$ and $\mathcal{G} \cup \mathcal{S}$. Note that $(a,b) \notin R(\mathcal{G})$ is implied since $(a,b) \notin R(\mathcal{G}^{k(ad)})$ and $(a,a) \notin R(\mathcal{G})$ is assumed. Remember that we already observed that in this case (b,a) may or not be in $R(\mathcal{F})$ or $R(\mathcal{G})$. The dashed arrows reflect this situation. The capital letter B is an abbreviation for the arguments in $A(\mathcal{F}) \setminus \{a,b\}$. Furthermore we left out possible attacks between B and $\{a,b\}$ since they are not important as we will see.

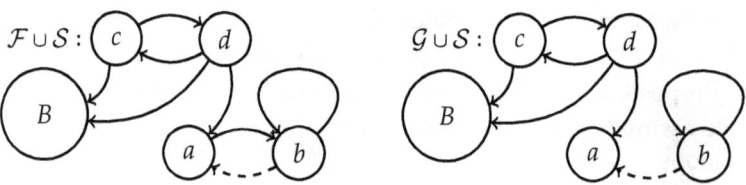

First notice that $\mathcal{E}_{ad}(\mathcal{F} \cup \mathcal{S}) = \mathcal{E}_{ad}(\mathcal{G} \cup \mathcal{S}) = \{\emptyset, \{a,c\}, \{c\}, \{d\}\}$. Remember that

84

semi-stable extensions are admissible too. It turns out that $\{d\} \in \mathcal{E}_{ss}(\mathcal{G} \cup \mathcal{S})$ and $\{d\} \notin \mathcal{E}_{ss}(\mathcal{F} \cup \mathcal{S})$ holds. This can be seen as follows: In both AFs the ranges of $\{d\}$ are identical, i.e. $R^+_{\mathcal{F} \cup \mathcal{S}}(\{d\}) = R^+_{\mathcal{G} \cup \mathcal{S}}(\{d\}) = A(\mathcal{F} \cup \mathcal{S}) \smallsetminus \{b\}$. Since $R^+_{\mathcal{F} \cup \mathcal{S}}(\{a,c\}) = A(\mathcal{F} \cup \mathcal{S})$ we deduce $\{d\} \notin \mathcal{E}_{ss}(\mathcal{F} \cup \mathcal{S})$ by definition of the semi-stable semantics. On the other hand, for any set $E \in \mathcal{E}_{ad}(\mathcal{G} \cup \mathcal{S})$, $b \notin R^+_{\mathcal{G} \cup \mathcal{S}}(E)$ because $b \notin E$ and $(E, b) \notin R(\mathcal{G} \cup \mathcal{S})$. Hence, $R^+_{\mathcal{G} \cup \mathcal{S}}(\{d\}) \not\subset R^+_{\mathcal{G} \cup \mathcal{S}}(E)$. Consequently, $\{d\} \in \mathcal{E}_{ss}(\mathcal{G} \cup \mathcal{S})$ is shown and thus $\mathcal{F} \not\equiv^{ss}_S \mathcal{G}$. More generally, one may easily check that $\{\{a,c\}\} = \mathcal{E}_{ss}(\mathcal{F} \cup \mathcal{S})$ and $\{\{c\},\{d\}\} = \mathcal{E}_{ss}(\mathcal{G} \cup \mathcal{S})$. Thus, $\mathcal{E}_{eg}(\mathcal{F} \cup \mathcal{S}) = \{\{a,c\}\} \neq \{\emptyset\} = \mathcal{E}_{eg}(\mathcal{G} \cup \mathcal{S})$. Altogether, $\mathcal{F} \not\equiv^{eg}_S \mathcal{G}$ concluding the proof. □

5.2.4 Strong Expansion Equivalence for Admissible, Preferred and Ideal Semantics

A special feature of strong expansions is that a former attack between old arguments will never become a counterattack to an added attack. In this sense, former attacks do not play a role with respect to being a potential defender of an added argument. Hence, in contrast to arbitrary expansions where such attacks might be relevant we may delete them without changing the behavior with respect to further evaluations. In the last two subsections we proved that in case of stable, stage, semi-stable and eager semantics there are no further redundant attacks if we consider strong expansion equivalence. In case of admissible, preferred and ideal semantics the situation becomes different. Consider the following example.

Example 5.12. The AFs $\mathcal{F}$ and $\mathcal{G}$ are not expansion equivalent with respect to admissible, preferred and ideal semantics since their corresponding ad-kernels $\mathcal{F}^{k(ad)} (= \mathcal{F})$ and $\mathcal{G}^{k(ad)} (= \mathcal{G})$ are different.

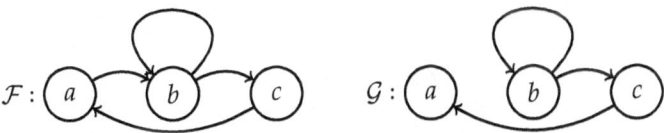

Figure 5.5: Non-Expansion Equivalent AFs

One possible scenario which makes the predicted different behaviour explicit is the following where $\mathcal{H} = (\{b,c,d\}, \{(b,d), (d,c)\})$. Observe that $\{\{a,d\}\} = \mathcal{E}_\sigma(\mathcal{F} \cup \mathcal{H}) \neq \{\emptyset\} = \mathcal{E}_\sigma(\mathcal{G} \cup \mathcal{H})$ for $\sigma \in \{ad, pr, id\}$.

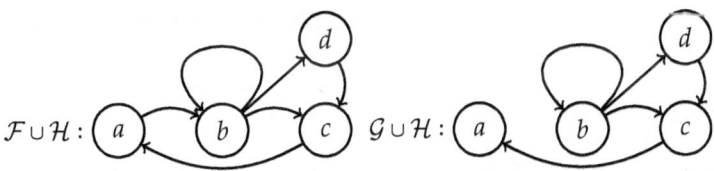

Figure 5.6: A Possible Scenario

5.2. Characterizing Strong Expansion Equivalence

Note that the already existing attack (a,b) in $\mathcal{F}$ becomes a *defending* attack of the newly added argument d in the augmented argumentation scenario $\mathcal{F} \cup \mathcal{H}$. This means, such attacks in fact play an important role with respect to further evaluation in case of arbitrary expansions. It is the main result of this section showing that AFs like $\mathcal{F}$ and $\mathcal{G}$ are strong expansion equivalent with respect to admissible, preferred and ideal semantics. This means, in particular, the attack (a,b) in $\mathcal{F}$ is redundant with respect to strong expansions and their evaluations.[3]

Here is the first novel kernel definition, the so-called *admissible-*-kernel* which (as we shall see) adequately describes strong expansion equivalence with respect to admissible, preferred and ideal semantics.

Definition 5.13. Given a an AF $\mathcal{F} = (A,R)$. We define the *admissible-*-kernel* of $\mathcal{F}$ as $\mathcal{F}^{k^*(ad)} = \left(A, R^{k^*(ad)}\right)$ where

$R^{k^*(ad)} = R \smallsetminus \{(a,b) \mid a \neq b, ((a,a) \in R \wedge \{(b,a),(b,b)\} \cap R \neq \emptyset) \vee$
$((b,b) \in R \wedge \forall c\, ((b,c) \in R \rightarrow \{(a,c),(c,a),(c,c),(c,b)\} \cap R \neq \emptyset))\}.$

The newly introduced kernel "forgets" an attack (a,b) if

1. a is self-attacking and at least one of the attacks (b,a) or (b,b) exists or

2. b is self-defeating and furthermore, for all arguments c which are attacked by b at least one of the following conditions holds:

    i) a attacks c,

    ii) c attacks a,

    iii) c attacks c or

    iv) c attacks b.

The first condition is exactly the same as in case of the ad-kernel (compare Definition 5.1). The motivation for the second disjunct is the following: At first observe that b cannot be an element of any conflict-free set. Thus, the attack (a,b) may only be relevant with respect to the defense of c since we are considering strong expansions. In the first three cases i), ii) and iii) this relevance becomes unimportant since $\{a,c\}$ does not even possess conflict-freeness. In case iv) the redundancy of (a,b) with respect to the defense of c is given by the fact that c already defends itself against b.

In the following we will prove that two AFs $\mathcal{F}$ and $\mathcal{G}$ possess the same ad-*-kernel if and only if they are strong expansion equivalent with respect to admissible, preferred and ideal semantics. At first we will show that any AF $\mathcal{F}$ and its ad-*-kernel possess the same extensions with respect to the aforementioned semantics.

Lemma 5.14. *For any AF $\mathcal{F}$ and $\sigma \in \{ad, pr, id\}$, $\mathcal{F} \equiv^\sigma \mathcal{F}^{k^*(ad)}$.*

---

[3]We invite and encourage the reader to try to show that this assertion does not hold.

Chapter 5. Notions of Equivalence and Replacement

Proof. At first we show that $\mathcal{F}$ and $\mathcal{F}^{k^*(ad)}$ contain the same conflict-free sets, i.e. $S \in cf(\mathcal{F})$ iff $S \in cf\left(\mathcal{F}^{k^*(ad)}\right)$. The if-direction is obvious because $R\left(\mathcal{F}^{k^*(ad)}\right) \subseteq R(\mathcal{F})$. It suffices to show that if $S \in cf\left(\mathcal{F}^{k^*(ad)}\right)$, then $S \in cf(\mathcal{F})$. Assume not, i.e. there are at least two arguments $a, b \in S$, such that $(a,b) \in R(\mathcal{F}) \setminus R\left(\mathcal{F}^{k^*(ad)}\right)$. Consequently, $(a,a) \in R(\mathcal{F}) \vee (b,b) \in R(\mathcal{F})$ has to hold. This contradicts the conflict-freeness of S in $\mathcal{F}^{k^*(ad)}$ because $\mathcal{F}^{k^*(ad)}$ and $\mathcal{F}$ share the same self-loops.

We now prove the result for $\sigma = ad$. We have to show that for each S conflict-free in $\mathcal{F}$ and $b \in S$, b is defended by S in $\mathcal{F}$ iff b is defended by S in $\mathcal{F}^{k^*(ad)}$. Hence, $\mathcal{F} \equiv^{ad} \mathcal{F}^{k^*(ad)}$ is implied. First, suppose b is defended by S in $\mathcal{F}$, i.e. for each $(a,b) \in R(\mathcal{F})$, $(S,a) \in R(\mathcal{F})$. Assume now b is not defended by S in $\mathcal{F}^{k^*(ad)}$, i.e. it exists $(a,b) \in R\left(\mathcal{F}^{k^*(ad)}\right)$, $(S,a) \notin R\left(\mathcal{F}^{k^*(ad)}\right)$. That means all counter-attacks $(c,a) \in R(\mathcal{F})$ have to be deleted. Since S is assumed to be conflict-free, $(c,c) \notin R(\mathcal{F})$ and hence, $(a,a) \in R(\mathcal{F})$ has to hold. If $c = b$, then $(a,b) \notin R\left(\mathcal{F}^{k^*(ad)}\right)$ because $(a,a) \in R(\mathcal{F})$ and $(b,a) \in R(\mathcal{F})$ was assumed. Let $c \neq b$. It follows $\{(b,c),(c,b),(b,b),(b,a)\} \cap R(\mathcal{F}) \neq \emptyset$. The first three attacks are impossible because conflict-freeness of S was assumed. Finally, if $(b,a) \in R(\mathcal{F})$, $(a,b) \notin R\left(\mathcal{F}^{k^*(ad)}\right)$ follows because $(a,a) \in R(\mathcal{F})$ was assumed.

Second, consider b is defended by S in $\mathcal{F}^{k^*(ad)}$ and b is not defended by S in $\mathcal{F}$, i.e. it exists $(a,b) \in R(\mathcal{F}) \setminus R\left(\mathcal{F}^{k^*(ad)}\right)$, $(S,a) \notin R(\mathcal{F})$. Since $(a,b) \notin R\left(\mathcal{F}^{k^*(ad)}\right)$, we deduce $(a,a) \in R(\mathcal{F})$ and $(b,b) \notin R(\mathcal{F})$ because conflict-freeness of S was assumed. Consequently, $(b,a) \in R(\mathcal{F})$ contradicting the assumption that b is not defended by S in $\mathcal{F}$. This concludes the proof for admissible semantics. Finally, applying Proposition 2.17, statement 1 the claim is verified for preferred and ideal semantics. □

The following lemma states that, if two AFs $\mathcal{F}$ and $\mathcal{G}$ possess equal ad-*-kernels, then the same applies to $\mathcal{F} \cup \mathcal{H}$ and $\mathcal{G} \cup \mathcal{H}$ where the latter AFs are strong expansions of the corresponding former ones.

Lemma 5.15. *If $\mathcal{F}^{k^*(ad)} = \mathcal{G}^{k^*(ad)}$, then $(\mathcal{F} \cup \mathcal{H})^{k^*(ad)} = (\mathcal{G} \cup \mathcal{H})^{k^*(ad)}$ for all AFs $\mathcal{H}$ which satisfy $\mathcal{F} \leq_S \mathcal{F} \cup \mathcal{H}$ and $\mathcal{G} \leq_S \mathcal{G} \cup \mathcal{H}$.*

Proof. First notice that the assumption $\mathcal{F}^{k^*(ad)} = \mathcal{G}^{k^*(ad)}$ implies $A(\mathcal{F}) = A\left(\mathcal{F}^{k^*(ad)}\right) = A\left(\mathcal{G}^{k^*(ad)}\right) = A(\mathcal{G})$. Given an AF $\mathcal{H}$, such that $\mathcal{F} \leq_S \mathcal{F} \cup \mathcal{H}$ and $\mathcal{G} \leq_S \mathcal{G} \cup \mathcal{H}$ is satisfied. Obviously, $(\mathcal{F} \cup \mathcal{H})^{k^*(ad)}$ and $(\mathcal{G} \cup \mathcal{H})^{k^*(ad)}$ share the same arguments. Hence, it suffices to show that $R\left((\mathcal{F} \cup \mathcal{H})^{k^*(ad)}\right) = R\left((\mathcal{G} \cup \mathcal{H})^{k^*(ad)}\right)$. Note that $\mathcal{F} = \mathcal{F} \cup \mathcal{H}$ if and only if $\mathcal{G} = \mathcal{G} \cup \mathcal{H}$. Hence, in case of equality we have nothing to show because $\mathcal{F}^{k^*(ad)} = \mathcal{G}^{k^*(ad)}$ guarantees $(\mathcal{F} \cup \mathcal{H})^{k^*(ad)} = (\mathcal{G} \cup \mathcal{H})^{k^*(ad)}$. This means, in the following we may assume that $\mathcal{F} \cup \mathcal{H}$ and $\mathcal{G} \cup \mathcal{H}$ are indeed strong expansions of $\mathcal{F}$ or $\mathcal{G}$. Consequently, $R(\mathcal{H}) \cap R(\mathcal{F}) = R(\mathcal{H}) \cap R(\mathcal{G}) = \emptyset$ can be assumed (compare Definition 2.13).

Let $(a,b) \in R\big((\mathcal{F} \cup \mathcal{H})^{k^*(ad)}\big)$. We will show $(a,b) \in R\big((\mathcal{G} \cup \mathcal{H})^{k^*(ad)}\big)$ by proof by cases (containedness of a and b in $A(\mathcal{F})$ or $A(\mathcal{H}) \setminus A(\mathcal{F})$). Since $\mathcal{F}^{k^*(ad)} = \mathcal{G}^{k^*(ad)}$ is assumed, it suffices to consider $a \neq b$ because the sharing of the same self-loops of $(\mathcal{F} \cup \mathcal{H})^{k^*(ad)}$ and $(\mathcal{G} \cup \mathcal{H})^{k^*(ad)}$ is already implied.
1st **case**: Let $a,b \in A(\mathcal{F})$. If $(a,b) \in R\big(\mathcal{F}^{k^*(ad)}\big)$, then $(a,b) \in R\big(\mathcal{G}^{k^*(ad)}\big)$ and $(a,b) \in R(\mathcal{G})$ follow. Furthermore $(a,b) \in R\big((\mathcal{G} \cup \mathcal{H})^{k^*(ad)}\big)$ since $\mathcal{G} \leq_S \mathcal{G} \cup \mathcal{H}$ was assumed, i.e. the AF $\mathcal{H}$ does not add relevant (with respect to the deletion of (a,b)) attacks. Assuming $(a,b) \notin R\big(\mathcal{F}^{k^*(ad)}\big)$ contradicts $(a,b) \in R\big((\mathcal{F} \cup \mathcal{H})^{k^*(ad)}\big)$ because the reason to remove an attack from $\mathcal{F} \cup \mathcal{H}$ remains untouched. 2nd **case**: Let $a,b \in A(\mathcal{H}) \setminus A(\mathcal{F})$. Assume $(a,b) \notin R\big((\mathcal{G} \cup \mathcal{H})^{k^*(ad)}\big)$. Hence, several reasons for removing have to be considered. The first possibility is $(a,a) \in R(\mathcal{H}) \wedge \{(b,b),(b,a)\} \cap R(\mathcal{H}) \neq \emptyset$ holds. This implies $(a,b) \notin R\big((\mathcal{F} \cup \mathcal{H})^{k^*(ad)}\big)$. The second one is $(b,b) \in R(\mathcal{H}) \wedge \forall c\, ((b,c) \in R(\mathcal{G} \cup \mathcal{H}) \rightarrow \{(a,c),(c,a),(c,c),(c,b)\} \cap R(\mathcal{G} \cup \mathcal{H}) \neq \emptyset)$ holds. If there is no c in $A(\mathcal{G})$ which is attacked by b we conclude $(a,b) \notin R\big((\mathcal{F} \cup \mathcal{H})^{k^*(ad)}\big)$ contradicting the assumption. So, consider $c \in A(\mathcal{G})$ and $(b,c) \in R(\mathcal{H})$. Consequently, $\{(a,c),(c,a),(c,c),(c,b)\} \cap R(\mathcal{G} \cup \mathcal{H}) \neq \emptyset$ has to hold. The attacks (c,a) and (c,b) are impossible since $\mathcal{G} \leq_S \mathcal{G} \cup \mathcal{H}$ was assumed. If $(a,c) \in R(\mathcal{G} \cup \mathcal{H})$, then $(a,c) \in R(\mathcal{H})$ and consequently $(a,c) \in R(\mathcal{F} \cup \mathcal{H})$ has to hold. If $(c,c) \in R(\mathcal{G} \cup \mathcal{H})$, then $(c,c) \in R(\mathcal{G})$ and $(c,c) \in R(\mathcal{F})$ (since $\mathcal{F}^{k^*(ad)} = \mathcal{G}^{k^*(ad)}$ was assumed), therefore $(c,c) \in R(\mathcal{F} \cup \mathcal{H})$. In all cases, $(a,b) \notin R\big((\mathcal{F} \cup \mathcal{H})^{k^*(ad)}\big)$. 3rd **case**: Let $a \in A(\mathcal{H}) \setminus A(\mathcal{F})$ and $b \in A(\mathcal{F})$. Assume $(a,b) \notin R\big((\mathcal{G} \cup \mathcal{H})^{k^*(ad)}\big)$. Again, several reasons for removing have to be considered. First consider $(a,a) \in R(\mathcal{H}) \wedge (b,b) \in R(\mathcal{G})$. We conclude $(b,b) \in R(\mathcal{F})$ because $\mathcal{F}^{k^*(ad)} = \mathcal{G}^{k^*(ad)}$ was assumed, thus $(a,a),(b,b) \in R(\mathcal{F} \cup \mathcal{H})$ holds which contradicts $(a,b) \in R\big((\mathcal{F} \cup \mathcal{H})^{k^*(ad)}\big)$. Note that $(a,a) \in R(\mathcal{H}) \wedge (b,a) \in R(\mathcal{G} \cup \mathcal{H})$ is impossible since $\mathcal{G} \leq_S \mathcal{G} \cup \mathcal{H}$ was assumed. Consider now $(b,b) \in R(\mathcal{G}) \wedge \forall c\, ((b,c) \in R(\mathcal{G} \cup \mathcal{H}) \rightarrow \{(a,c),(c,a),(c,c),(c,b)\} \cap R(\mathcal{G} \cup \mathcal{H}) \neq \emptyset)$. We observe $(b,b) \in R(\mathcal{F})$. Since $(a,b) \in R\big((\mathcal{F} \cup \mathcal{H})^{k^*(ad)}\big)$ was assumed there exists an argument $c \in A(\mathcal{F})$, such that $(b,c) \in R(\mathcal{F}) \wedge \{(a,c),(c,a),(c,c),(c,b)\} \cap R(\mathcal{F} \cup \mathcal{H}) = \emptyset$ holds. Thus, $\{(a,c),(c,a),(c,c)\} \cap R(\mathcal{G} \cup \mathcal{H}) = \emptyset$ holds. Remember that we assumed $\mathcal{F}^{k^*(ad)} = \mathcal{G}^{k^*(ad)}$. If $(b,c) \notin R(\mathcal{G})$, then (b,c) has to be deleted in $R\big(\mathcal{F}^{k^*(ad)}\big)$. But this is impossible since we already concluded $(c,c) \notin R(\mathcal{F}) \wedge (c,b) \notin R(\mathcal{F})$. If $(b,c) \in R(\mathcal{G})$, then $(c,b) \in R(\mathcal{G})$ has to hold since we assumed $(a,b) \notin R\big((\mathcal{G} \cup \mathcal{H})^{k^*(ad)}\big)$. Hence, (b,c) has to be deleted in $\mathcal{G}^{k^*(ad)}$ because $(b,b) \in R(\mathcal{G})$ was supposed. This contradicts $(b,c) \in \mathcal{F}^{k^*(ad)}$ concluding the proof. 4th **case**: Let $a \in A(\mathcal{F})$ and $b \in A(\mathcal{H}) \setminus A(\mathcal{F})$. Here we have nothing to show because the assumption $(a,b) \in R(\mathcal{F} \cup \mathcal{H})$ is impossible since $\mathcal{F} \leq_S \mathcal{F} \cup \mathcal{H}$ was supposed. $\square$

Now we are prepared to show that the syntactical equivalence of ad-*-

kernels characterizes strong expansion equivalence between two AFs $\mathcal{F}$ and $\mathcal{G}$ with respect to admissible, preferred and ideal semantics.

Theorem 5.16. *For any AFs $\mathcal{F}, \mathcal{G}$ and $\sigma \in \{ad, pr, id\}$:*

$$\mathcal{F}^{k^*(ad)} = \mathcal{G}^{k^*(ad)} \Leftrightarrow \mathcal{F} \equiv_S^\sigma \mathcal{G}.$$

Proof. Let $\mathcal{F}^{k^*(ad)} = \mathcal{G}^{k^*(ad)}$. Given an AF $\mathcal{H}$, such that $\mathcal{F} \leq_S \mathcal{F} \cup \mathcal{H}$ and $\mathcal{G} \leq_S \mathcal{G} \cup \mathcal{H}$. It suffices to show that $E \in \mathcal{E}_{ad}(\mathcal{F} \cup \mathcal{H})$ implies $E \in \mathcal{E}_{ad}(\mathcal{G} \cup \mathcal{H})$. Suppose $E \in \mathcal{E}_{ad}(\mathcal{F} \cup \mathcal{H})$. By Lemma 5.14, $E \in \mathcal{E}_{ad}\left((\mathcal{F} \cup \mathcal{H})^{k^*(ad)}\right)$ and applying Lemma 5.15, $E \in \mathcal{E}_{ad}\left((\mathcal{G} \cup \mathcal{H})^{k^*(ad)}\right)$. Finally, using Lemma 5.14, we derive $E \in \mathcal{E}_{ad}(\mathcal{G} \cup \mathcal{H})$ which concludes the if-direction for admissible semantics. Proposition 2.17 verifies the result for preferred and ideal semantics too, i.e. $\mathcal{F} \equiv_S^{pr} \mathcal{G}$ and $\mathcal{F} \equiv_S^{id} \mathcal{G}$.

We now show that $\mathcal{F}^{k^*(ad)} \neq \mathcal{G}^{k^*(ad)}$ implies $\mathcal{F} \not\equiv_S^\sigma \mathcal{G}$. **1st case:** Assume $A\left(\mathcal{F}^{k^*(ad)}\right) \neq A\left(\mathcal{G}^{k^*(ad)}\right)$. Hence, without loss of generality there exists an argument $a \in A(\mathcal{F}) \setminus A(\mathcal{G})$. We define $\mathcal{H} = ((A(\mathcal{F}) \cup A(\mathcal{G})) \setminus \{a\}, \emptyset)$. Consider the existence of a set E, such that $E \in \mathcal{E}_\sigma(\mathcal{F} \cup \mathcal{H})$ and $a \in E$. Consequently, $E \notin \mathcal{E}_\sigma(\mathcal{G} \cup \mathcal{H})$ holds. Assume now that for all extensions $E \in \mathcal{E}_\sigma(\mathcal{F} \cup \mathcal{H})$, $a \notin E$. We define $\mathcal{H}' = \mathcal{H} \cup (\{a\}, \emptyset)$. Hence, $\mathcal{F} \cup \mathcal{H} = \mathcal{F} \cup \mathcal{H}'$ and therefore, for all extensions $E \in \mathcal{E}_\sigma(\mathcal{F} \cup \mathcal{H}')$, $a \notin E$ holds. We observe $\{a\} \in \mathcal{E}_{ad}(\mathcal{G} \cup \mathcal{H}')$ and furthermore, for each $E \in \mathcal{E}_{pr}(\mathcal{G} \cup \mathcal{H}')$, $a \in E$ holds since a is unattacked in $\mathcal{G} \cup \mathcal{H}'$. This implies that a is contained in the unique ideal extension of $\mathcal{G} \cup \mathcal{H}'$. In all cases, $\mathcal{F} \not\equiv_S^\sigma \mathcal{G}$.

2nd case: Consider now $R\left(\mathcal{F}^{k^*(ad)}\right) \neq R\left(\mathcal{G}^{k^*(ad)}\right)$ and $A\left(\mathcal{F}^{k^*(ad)}\right) = A\left(\mathcal{G}^{k^*(ad)}\right) (= A(\mathcal{F}) = A(\mathcal{G}))$. Hence, without loss of generality there exists $a, b \in A(\mathcal{F})$, such that $(a,b) \in R\left(\mathcal{F}^{k^*(ad)}\right) \setminus R\left(\mathcal{G}^{k^*(ad)}\right)$. Let c be a new argument, i.e. $c \notin A(\mathcal{F})$. Furthermore we define

$$\mathcal{I} = \left(A(\mathcal{F}) \cup \{c\}, \{(c,c') \mid c' \in A(\mathcal{F}) \setminus \{a,b\}\}\right).$$

Case 2.1: Assume $a = b$. This means $(a,a) \in R\left(\mathcal{F}^{k^*(ad)}\right) \setminus R\left(\mathcal{G}^{k^*(ad)}\right)$ and consequently $(a,a) \in R(\mathcal{F}) \setminus R(\mathcal{G})$ by the definition of the ad-*-kernel (Definition 5.13). It can be checked (splitting results, Section 5.2.1) that $\{a,c\}$ is an admissible and the unique preferred extension of $\mathcal{G} \cup \mathcal{I}$. Hence, it follows that $\{a,c\}$ has to be the unique ideal extension of $\mathcal{G} \cup \mathcal{I}$ (Proposition 2.10). On the other hand, we have $\{a,c\} \notin \mathcal{E}_\sigma(\mathcal{F} \cup \mathcal{I})$ for $\sigma \subset \{ad, pr, id\}$ since $(a,a) \in R(\mathcal{F})$ was assumed. Thus from now on we assume that any self-loop is either contained in both $R\left(\mathcal{F}^{k^*(ad)}\right)$ and $R\left(\mathcal{G}^{k^*(ad)}\right)$ or in none of them.

Case 2.2: Let $a \neq b$, i.e. $(a,b) \in R\left(\mathcal{F}^{k^*(ad)}\right) \setminus R\left(\mathcal{G}^{k^*(ad)}\right)$ and $(a,b) \in R(\mathcal{F})$. Now we have to distinguish four cases for the presence or absence of attack (a,a) and (b,b). Keep in mind that $R(\mathcal{F}), R(\mathcal{G}), R\left(\mathcal{F}^{k^*(ad)}\right)$ and $R\left(\mathcal{G}^{k^*(ad)}\right)$ contain the same self-loops. **Case 2.2.1:** $(a,a), (b,b) \in R(\mathcal{F})$. This case is impossible because the definition of the ad-*-kernel enforce the deletion of

5.2. Characterizing Strong Expansion Equivalence

(a,b) in $R\left(\mathcal{F}^{k^*(ad)}\right)$. Case **2.2.2**: $(a,a),(b,b) \notin R(\mathcal{F})$. Note that $(a,b) \notin R(\mathcal{G})$ holds because a and b do not exhibit self-loops and $(a,b) \notin R\left(\mathcal{G}^{k^*(ad)}\right)$ was assumed. The attack (b,a) may or may not be an element of $R(\mathcal{F})$ or $R(\mathcal{G})$. The following results can be checked by using splitting results (compare Section 5.2.1). If $(b,a) \notin R(\mathcal{F})$, then $\{\{a,c\}\} = \mathcal{E}_\sigma(\mathcal{F} \cup \mathcal{I})$ for any $\sigma \in \{pr, id\}$. If not, i.e. $(b,a) \in R(\mathcal{F})$, then $\{\{a,c\},\{b,c\}\} = \mathcal{E}_{pr}(\mathcal{F} \cup \mathcal{I})$ and $\{\{c\}\} = \mathcal{E}_{id}(\mathcal{F} \cup \mathcal{I})$. On the other hand, if $(b,a) \notin R(\mathcal{G})$, then $\{\{a,b,c\}\} = \mathcal{E}_\sigma(\mathcal{G} \cup \mathcal{I})$ holds for any $\sigma \in \{pr, id\}$. If not, i.e. $(b,a) \in R(\mathcal{G})$ it follows $\{\{b,c\}\} = \mathcal{E}_\sigma(\mathcal{G} \cup \mathcal{I})$ for any $\sigma \in \{pr, id\}$. Thus, in all possible combinations we obtain different preferred and ideal extensions. Furthermore, different admissible extensions are implied (Proposition 2.17, statement 1). This means, we have shown that for any $\sigma \in \{ad, pr, id\}$, $\mathcal{F} \not\equiv_S^\sigma \mathcal{G}$ holds. Case **2.2.3**: $(a,a) \in R(\mathcal{F})$ and $(b,b) \notin R(\mathcal{F})$. First notice that $(b,a) \in R(\mathcal{F})$ cannot hold because $(a,a) \in R(\mathcal{F})$ would enforce the deletion of (a,b) in $R\left(\mathcal{F}^{k^*(ad)}\right)$ in contrast to the assumption. Using the standard construction we obtain $\{\{c\}\} = \mathcal{E}_\sigma(\mathcal{F} \cup \mathcal{I})$ for each $\sigma \in \{pr, id\}$. In the given self-loop constellation AF $\mathcal{G}$ may occur in three configurations with respect to the presence and absence of the attacks (a,b) and (b,a), namely: $(a,b),(b,a) \notin R(\mathcal{G})$ or $(a,b),(b,a) \in R(\mathcal{G})$ or $(a,b) \notin R(\mathcal{G})$ and $(b,a) \in R(\mathcal{G})$. Note that $(a,b) \in R(\mathcal{G})$ and $(b,a) \notin R(\mathcal{G})$ is impossible since $(a,b) \notin R\left(\mathcal{G}^{k^*(ad)}\right)$ was assumed. In all cases we obtain $\{\{b,c\}\} = \mathcal{E}_\sigma(\mathcal{G} \cup \mathcal{I})$ for each $\sigma \in \{pr, id\}$. By Proposition 2.17 we deduce $\mathcal{E}_{ad}(\mathcal{F} \cup \mathcal{I}) \neq \mathcal{E}_{ad}(\mathcal{G} \cup \mathcal{I})$. Altogether, we have shown that $\mathcal{F} \not\equiv_S^\sigma \mathcal{G}$ for each $\sigma \in \{ad, pr, id\}$. Case **2.2.4**: $(a,a) \notin R(\mathcal{F})$ and $(b,b) \in R(\mathcal{F})$. Since $(a,b) \in R\left(\mathcal{F}^{k^*(ad)}\right)$ is assumed, we deduce the existence of an argument $c \in A(\mathcal{F})$, such that $(b,c) \in R(\mathcal{F}) \land \{(a,c),(c,a),(c,c),(c,b)\} \cap R(\mathcal{F}) = \emptyset$ (compare Definition 5.13 of the ad-*-kernel). The following figures show the remaining two possibilities for AF $\mathcal{F}$. Note that we omit possible other arguments than a, b and c. This means, the AFs $\mathcal{F}_1$ and $\mathcal{F}_2$ as well as the subsequent AFs $\mathcal{G}_i$ are only representatives illustrating the relevant parts (consult Section 5.2.1).

So far we know $(a,a),(c,c) \notin R(\mathcal{G})$ and $(b,b) \in R(\mathcal{G})$. This means there are $2^6 = 64$ possibilities for the presence and absence of $(a,b),(b,a),(b,c),(c,b)$, (a,c) and (c,a) in $R(\mathcal{G})$. Note that some of them are impossible since $(a,b) \notin R\left(\mathcal{G}^{k^*(ad)}\right)$ was assumed. At first we modify the standard construction in the following way (d is a fresh argument):

$$\mathcal{I}' = \left(A(\mathcal{F}) \cup \{d\}, \{(d, c') \mid c' \in A(\mathcal{F}) \setminus \{a, b, c\}\}\right).$$

The following extensions can be checked by applying splitting results (compare Section 5.2.1). It can be easily seen that for each $\sigma \in \{pr, id\}$,

Chapter 5. Notions of Equivalence and Replacement

$\mathcal{E}_\sigma(\mathcal{F}_1 \cup \mathcal{I}') = \mathcal{E}_\sigma(\mathcal{F}_2 \cup \mathcal{I}') = \{\{a,c,d\}\}$ holds. If $(a,c) \in R(\mathcal{G})$ or $(c,a) \in R(\mathcal{G})$, then for each $\sigma \in \{ad, pr, id\}$, $\{a,c,d\} \notin \mathcal{E}_\sigma(\mathcal{G} \cup \mathcal{H}')$ holds since $\{a,c,d\}$ is not conflict-free. Hence, without loss of generality we may assume $(a,c),(c,a) \notin R(\mathcal{G})$. Thus, $2^4 = 16$ possibilities with respect to the presence or absence of $(a,b),(b,a),(b,c)$ and (c,b) remain. For clarity, we will present all possibilities.

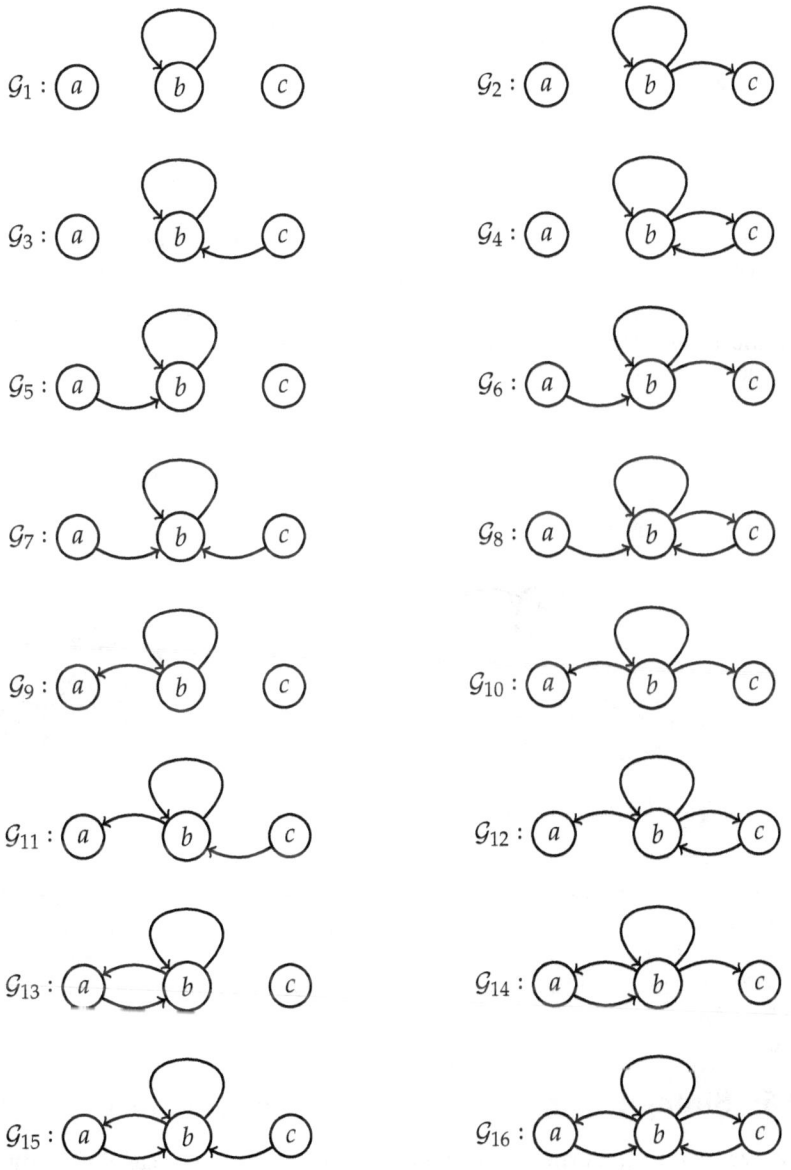

First note that $\mathcal{G}_6$ (= $\mathcal{F}_2$) and $\mathcal{G}_{14}$ (= $\mathcal{F}_1$) are impossible since $(a,b) \notin R(\mathcal{G}^{k^*(ad)})$ was assumed. Furthermore, the cases $\mathcal{G}_2$, $\mathcal{G}_9$ and $\mathcal{G}_{10}$ can be checked by considering the union with AF $\mathcal{I}'$. For each $\sigma \in \{pr, id\}$, $\{\{a,d\}\} =$

5.2. Characterizing Strong Expansion Equivalence

$\mathcal{E}_\sigma(\mathcal{G}_2 \cup \mathcal{I}')$, $\{\{c,d\}\} = \mathcal{E}_\sigma(\mathcal{G}_9 \cup \mathcal{I}')$ and $\{\{d\}\} = \mathcal{E}_\sigma(\mathcal{G}_{10} \cup \mathcal{I}')$. For all other cases we define a slightly different version of $\mathcal{I}'$, namely

$$\mathcal{I}'' = \left(A(\mathcal{F}) \cup \{d\}, \{(d,c') \mid c' \in A(\mathcal{F}) \smallsetminus \{b,c\}\} \right).$$

At first we have to check the extensions of $\mathcal{F}_1 \cup \mathcal{I}''$ and $\mathcal{F}_2 \cup \mathcal{I}''$. It turns out that for any $\sigma \in \{pr, id\}$, $\{\{d\}\} = \mathcal{E}_\sigma(\mathcal{F}_1 \cup \mathcal{I}'') = \mathcal{E}_\sigma(\mathcal{F}_2 \cup \mathcal{I}'')$ holds. On the other hand we have $\{\{c,d\}\} = \mathcal{E}_\sigma(\mathcal{G}_i \cup \mathcal{I}'')$ for each $\sigma \in \{pr, id\}$ and every $i \in \{1,3,4,5,7,8,11,12,13,15,16\}$. Remember that different preferred extensions imply different admissible extensions (Proposition 2.17, statement 1). This means, finally, we have shown that for each $\sigma \in \{ad, pr, id\}$, $\mathcal{F} \not\equiv_S^\sigma \mathcal{G}$ holds. □

Let us consider again Example 5.12 from the beginning of this section.

Example 5.17 (Example 5.12 continued). According to Theorem 5.16 we have now formally proven that $\mathcal{F} \equiv_S^\sigma \mathcal{G}$ for $\sigma \in \{ad, pr, id\}$ since both possess the same ad-*-kernels, namely $\mathcal{F}^{k^*(ad)} = \mathcal{G}^{k^*(ad)} = \mathcal{G}$. In consideration of Theorem 5.11 the interested reader may ask for an example showing that $\mathcal{F}$ and $\mathcal{G}$ are not strong expansion equivalent with respect to semi-stable and eager semantics. Here is a counter-example for both semantics. We define

$$\mathcal{H} = (\{b,c,d,e\}, \{(d,b),(d,c),(d,e),(e,c),(e,d)\}).$$

The graph representation of $\mathcal{F} \cup \mathcal{H}$ and $\mathcal{G} \cup \mathcal{H}$ is as follows.

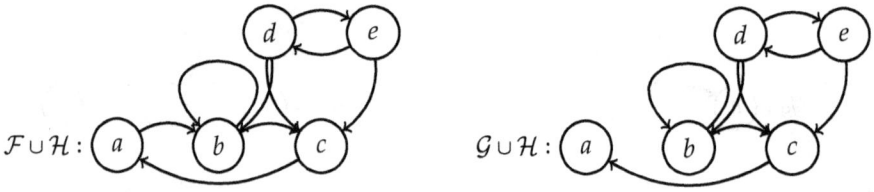

Figure 5.7: Counter-example for Semi-stable and Eager Semantics

The AFs $\mathcal{F} \cup \mathcal{H}$ and $\mathcal{G} \cup \mathcal{H}$ possess different semi-stable and eager extensions. In fact, $\{a,d\}$ and $\{a,e\}$ are semi-stable extensions in $\mathcal{F} \cup \mathcal{H}$ and furthermore, $\{a\}$ is not admissible. Consequently, $\mathcal{E}_{eg}(\mathcal{F} \cup \mathcal{H}) = \{\emptyset\}$. In case of $\mathcal{G} \cup \mathcal{H}$ we observe that $\{a,d\}$ is the unique semi-stable extension and thus, $\mathcal{E}_{eg}(\mathcal{G} \cup \mathcal{H}) = \{\{a,d\}\}$. This means, $\mathcal{F} \not\equiv_S^\sigma \mathcal{G}$ for $\sigma \in \{ss, eg\}$.

5.2.5 Strong Expansion Equivalence for Grounded Semantics

Now we turn to the grounded semantics. Similarly to the case of admissible, preferred and ideal semantics we will see that strong expansion equivalence between two AFs is not sufficient for their expansion equivalence with respect to grounded semantics. We therefore introduce a novel kernel, the so-called grounded-*-kernel which is defined as follows.

Definition 5.18. Given a an AF $\mathcal{F} = (A, R)$. We define the *grounded-*-kernel* of $\mathcal{F}$ as $\mathcal{F}^{k^*(gr)} = \left(A, R^{k^*(gr)}\right)$ where

$$R^{k^*(gr)} = R \setminus \{(a,b) \mid a \neq b, ((b,b) \in R \wedge \{(a,a),(b,a)\} \cap R \neq \emptyset) \vee \\ ((b,b) \in R \wedge \forall c \, ((b,c) \in R \rightarrow \{(a,c),(c,a),(c,c)\} \cap R \neq \emptyset))\}.$$

The newly introduced kernel "forgets" an attack (a, b) if

1. b is self-attacking and at least one of the attacks (a, a) or (b, a) exists or

2. b is self-defeating and furthermore, for all arguments c which are attacked by b at least one of the following conditions holds:

   i) a attacks c,

   ii) c attacks a or

   iii) c attacks c.

As explained in Section 5.2.4 a distinguishing feature of strong expansions in contrast to arbitrary expansions is that an old argument will never become a defender of a newly introduced and attacked argument. This means, there is more potential for irrelevant attacks which is reflected by the definition above.

The first disjunct captures attacks which are even redundant with respect to arbitrary expansions (compare Definition 5.1 and Theorem 5.5). Similar to the ad-*-kernel (see Definition 5.13) the second disjunct allows the deletions of an attack (a, b) if b is self-attacking and for all c's which are attacked by b we have $\{a, c\}$ is conflicting encoded by i), ii) and iii). In these cases the potential defense of c by a becomes irrelevant since conflict-freeness is violated. In contrast to admissible, preferred and ideal semantics the fourth possibility, namely the presence of the attack (c, b), i.e. c defends itself against b does not justify a deletion of (a, b). This can be easily seen by considering the original definition of the grounded semantics introduced in [Dung, 1995]. The grounded extension of an AF $\mathcal{F} = (A, R)$ is alternatively given as the least fixpoint of the so-called *characteristic function* $\Gamma_\mathcal{F} : 2^A \rightarrow 2^A$, where $\Gamma_\mathcal{F}(S) = \{a \in A \mid a \text{ is defended by } S \text{ in } \mathcal{F}\}$. In case of finite AFs, this least fixpoint can be achieved by applying iteratively $\Gamma_\mathcal{F}$ on the empty set. Furthermore, $\Gamma_\mathcal{F}$ can be shown to be monotonic [Dung, 1995]. This means, the fourth possibility is excluded because the defense of c against b by a may be essential for c being an element of the grounded extension, although c defends itself against b.

The following example shows that we actually need a further kernel definition since the grounded extensions of an AF $\mathcal{F}$ and $\mathcal{F}^{k^*(ad)}$ are not necessarily the same.

Example 5.19. Observe that $\mathcal{F}^{k^*(ad)} = \mathcal{G}$ and $\mathcal{F}^{k^*(gr)} = \mathcal{H}$. Hence, $\mathcal{F} \equiv_S^\sigma \mathcal{G}$ for any $\sigma \in \{ad, pr, id\}$ (Theorem 5.16). Furthermore, $\{\emptyset\} = \mathcal{E}_{gr}(\mathcal{F}) \neq \mathcal{E}_{gr}(\mathcal{G}) = \{\{a, c\}\}$ which proves $\mathcal{F} \not\equiv_S^{gr} \mathcal{G}$. Note that the empty set is also the unique grounded extension of $\mathcal{H}$. We even claim that $\mathcal{F} \equiv_S^{gr} \mathcal{H}$ which will be a consequence of Theorem 5.22.

5.2. Characterizing Strong Expansion Equivalence

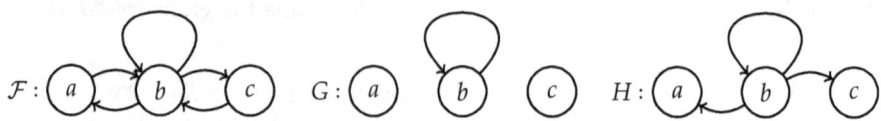

Figure 5.8: Grounded-*- vs. Admissible-*-Kernel

Analogously to the subsection before we will proceed with two technical lemmata paving the way for the main theorem.

Lemma 5.20. *For any AF $\mathcal{F}$, $\mathcal{F} \equiv^{gr} \mathcal{F}^{k^*(gr)}$.*

Proof. It suffices to show that for all $i \geq 1$, $\Gamma^i_{\mathcal{F}}(\emptyset) = \Gamma^i_{\mathcal{F}^{k^*(gr)}}(\emptyset)$ holds. We will prove this by induction.

First, we show that the sets of unattacked arguments coincide, i.e. $\Gamma^1_{\mathcal{F}}(\emptyset) = \Gamma^1_{\mathcal{F}^{k^*(gr)}}(\emptyset)$. Furthermore, $\Gamma^1_{\mathcal{F}}(\emptyset) \subseteq \Gamma^1_{\mathcal{F}^{k^*(gr)}}(\emptyset)$ is obvious since $R\left(\mathcal{F}^{k^*(gr)}\right) \subseteq R(\mathcal{F})$ holds. Given $a \in \Gamma^1_{\mathcal{F}^{k^*(gr)}}(\emptyset)$, then $(a,a) \notin R\left(\mathcal{F}^{k^*(gr)}\right)$ and therefore $(a,a) \notin R(\mathcal{F})$. Assuming that a is attacked in $\mathcal{F}$, i.e. there is an argument b, such that $(b,a) \in R(\mathcal{F})$ yields to $(b,a) \in R\left(\mathcal{F}^{k^*(gr)}\right)$ (compare Definition 5.18). This contradicts the assumption that a is unattacked in $\mathcal{F}^{k^*(gr)}$.

Suppose now that for all $i < k$, $\Gamma^i_{\mathcal{F}}(\emptyset) = \Gamma^i_{\mathcal{F}^{k^*(gr)}}(\emptyset)$ holds. We will show that $\Gamma^k_{\mathcal{F}}(\emptyset) = \Gamma^k_{\mathcal{F}^{k^*(gr)}}(\emptyset)$ is implied. ($\subseteq$) Assume $a \in \Gamma^k_{\mathcal{F}}(\emptyset)$ and $a \notin \Gamma^k_{\mathcal{F}^{k^*(gr)}}(\emptyset)$. Hence there is an attack $(b,a) \in R\left(\mathcal{F}^{k^*(gr)}\right)$, such that b is not attacked by $\Gamma^{k-1}_{\mathcal{F}^{k^*(gr)}}(\emptyset)$ in $\mathcal{F}^{k^*(gr)}$. Since $a \in \Gamma^k_{\mathcal{F}}(\emptyset)$ was assumed it follows that there is at least one argument $c \in \Gamma^{k-1}_{\mathcal{F}}(\emptyset)$ ($\subseteq \Gamma^k_{\mathcal{F}}(\emptyset)$), such that $(c,b) \in R(\mathcal{F})$ holds (note that $b \notin \Gamma^{k-1}_{\mathcal{F}}(\emptyset)$ is implied). Consequently, all these attacks have to be deleted in $\mathcal{F}^{k^*(gr)}$. We have to consider several reasons for deletion. First notice that $(b,b) \in R(\mathcal{F})$ (thus $(b,b) \in R\left(\mathcal{F}^{k^*(gr)}\right)$) has to hold. Furthermore none of the attacks $(a,a), (a,c), (c,a) \in R(\mathcal{F})$ are possible since $(a,c) \in \Gamma^k_{\mathcal{F}}(\emptyset)$ has to be conflict-free. Hence, all arguments $c \in \Gamma^{k-1}_{\mathcal{F}}(\emptyset)$ with the property $(c,b) \in R(\mathcal{F})$ has to be counterattacked by b itself, i.e. $(b,c) \in R(\mathcal{F})$ (compare Definition 5.18). Note that all these (b,c)'s survive in $R\left(\mathcal{F}^{k^*(gr)}\right)$ because $c \in \Gamma^k_{\mathcal{F}}(\emptyset)$ guarantees $(c,c) \notin R(\mathcal{F})$. By inductive hypothesis we get $c \in \Gamma^{k-1}_{\mathcal{F}^{k^*(gr)}}(\emptyset)$ and finally with $(b,c) \in R\left(\mathcal{F}^{k^*(gr)}\right)$ and the observation that all counter-attacks to b are deleted we contradict the admissibility of $\Gamma^{k-1}_{\mathcal{F}^{k^*(gr)}}(\emptyset)$ in $\mathcal{F}^{k^*(gr)}$. ($\supseteq$) Given $a \in \Gamma^k_{\mathcal{F}^{k^*(gr)}}(\emptyset)$, i.e. a is defended by $\Gamma^{k-1}_{\mathcal{F}^{k^*(gr)}}(\emptyset)$ in $\mathcal{F}^{k^*(gr)}$. Furthermore $(a,a) \notin R\left(\mathcal{F}^{k^*(gr)}\right)$ (thus $(a,a) \notin R(\mathcal{F})$) holds since $\Gamma^k_{\mathcal{F}^{k^*(gr)}}(\emptyset)$ is conflict-free in $\mathcal{F}^{k^*(gr)}$. This means, $(b,a) \in R\left(\mathcal{F}^{k^*(gr)}\right)$ iff $(b,a) \in R(\mathcal{F})$. Hence, using $\Gamma^{k-1}_{\mathcal{F}^{k^*(gr)}}(\emptyset) = \Gamma^{k-1}_{\mathcal{F}}(\emptyset)$ (inductive hypothesis) and the observation above we deduce that a is defended by $\Gamma^{k-1}_{\mathcal{F}}(\emptyset)$ in $\mathcal{F}$. Thus, $a \in \Gamma^k_{\mathcal{F}}(\emptyset)$. □

Chapter 5. Notions of Equivalence and Replacement

Lemma 5.21. *If* $\mathcal{F}^{k^*}(gr) = \mathcal{G}^{k^*}(gr)$, *then* $(\mathcal{F} \cup \mathcal{H})^{k^*}(gr) = (\mathcal{G} \cup \mathcal{H})^{k^*}(gr)$ *for all AFs* $\mathcal{H}$ *which satisfy* $\mathcal{F} \leq_S \mathcal{F} \cup \mathcal{H}$ *and* $\mathcal{G} \leq_S \mathcal{G} \cup \mathcal{H}$.

Proof. Assume $\mathcal{F}^{k^*}(gr) = \mathcal{G}^{k^*}(gr)$. Consequently, $A(\mathcal{F}) = A(\mathcal{F}^{k^*}(gr)) = A(\mathcal{G}^{k^*}(gr)) = A(\mathcal{G})$ holds. Consider now an AF $\mathcal{H}$ satisfying the specified properties (strong expansion or equality). Note that in case of equality there is nothing to show since $\mathcal{F} = \mathcal{F} \cup \mathcal{H}$ implies $\mathcal{G} = \mathcal{G} \cup \mathcal{H}$ and vice versa and hence, $(\mathcal{F} \cup \mathcal{H})^{k^*}(gr) = (\mathcal{G} \cup \mathcal{H})^{k^*}(gr)$ is implied. From now on we may suppose that $\mathcal{F} \cup \mathcal{H}$ and $\mathcal{G} \cup \mathcal{H}$ are indeed strong expansions of $\mathcal{F}$ or $\mathcal{G}$. Thus, $R(\mathcal{H}) \cap R(\mathcal{F}) = \emptyset$ and $R(\mathcal{H}) \cap R(\mathcal{G}) = \emptyset$ can be assumed (compare Definition 2.13). Let $(a,b) \in R\left((\mathcal{F} \cup \mathcal{H})^{k^*}(gr)\right)$, therefore $(a,b) \in R(\mathcal{F} \cup \mathcal{H})$. We will show $(a,b) \in R\left((\mathcal{G} \cup \mathcal{H})^{k^*}(gr)\right)$ by proof by cases (containedness of a and b in $A(\mathcal{F})$ or $A(\mathcal{H}) \setminus A(\mathcal{F})$). Again we suppose $a \neq b$ for all cases (containedness of self-loops is obvious).

1st **case:** Let $a, b \in A(\mathcal{F})$. If $(a,b) \in R(\mathcal{F}^{k^*}(gr))$, then $(a,b) \in R(\mathcal{G}^{k^*}(gr))$ and $(a,b) \in R(\mathcal{G})$ follow. Furthermore $(a,b) \in R\left((\mathcal{G} \cup \mathcal{H})^{k^*}(gr)\right)$ is implied because $\mathcal{G} \cup \mathcal{H}$ was assumed to be a strong expansion of $\mathcal{G}$ and so, no relevant attacks are added. The assumption $(a,b) \notin R(\mathcal{F}^{k^*}(gr))$ contradicts $(a,b) \in R\left((\mathcal{F} \cup \mathcal{H})^{k^*}(gr)\right)$ because the reason to remove an attack remains untouched in $\mathcal{F} \cup \mathcal{H}$. 2nd **case:** Let $a, b \in A(\mathcal{H}) \setminus A(\mathcal{F})$. Hence, $(a,b) \in R(\mathcal{G} \cup \mathcal{H})$ is implied. Assume now $(a,b) \notin R\left((\mathcal{G} \cup \mathcal{H})^{k^*}(gr)\right)$. This means, several reasons for removing have to be considered. Observe that $(b,b) \in R(\mathcal{H})$ has to hold. If (a,a) or (b,a) are contained in $R(\mathcal{H})$ we deduce $(a,b) \notin R\left((\mathcal{F} \cup \mathcal{H})^{k^*}(gr)\right)$ in contrast to the assumption. Assume now that $\forall c\, ((b,c) \in R(\mathcal{G} \cup \mathcal{H}) \to \{(a,c),(c,a),(c,c)\} \cap R(\mathcal{G} \cup \mathcal{H}) \neq \emptyset)$ holds. If there is no c in $A(\mathcal{G})$ which is attacked by b we conclude $(a,b) \notin R\left((\mathcal{F} \cup \mathcal{H})^{k^*}(gr)\right)$. So, consider $c \in A(\mathcal{G})$ and $(b,c) \in R(\mathcal{H})$. We obtain $\{(a,c),(c,a),(c,c)\} \cap R(\mathcal{G} \cup \mathcal{H}) \neq \emptyset$. The attack $(c,a) \in R(\mathcal{G} \cup \mathcal{H})$ is impossible since $\mathcal{G} \leq_S \mathcal{G} \cup \mathcal{H}$ was assumed. If $(a,c) \in R(\mathcal{G} \cup \mathcal{H})$, then $(a,c) \in R(\mathcal{H})$ and consequently $(a,c) \in R(\mathcal{F} \cup \mathcal{H})$ has to hold. If $(c,c) \in R(\mathcal{G} \cup \mathcal{H})$, then $(c,c) \in R(\mathcal{G})$ and $(c,c) \in R(\mathcal{F})$ (since $\mathcal{F}^{k^*}(gr) = \mathcal{G}^{k^*}(gr)$ was assumed), therefore $(c,c) \in R(\mathcal{F} \cup \mathcal{H})$. In all cases we get $(a,b) \notin R\left((\mathcal{F} \cup \mathcal{H})^{k^*}(gr)\right)$ in contrast to the assumption. 3rd **case:** Let $a \in A(\mathcal{H}) \setminus A(\mathcal{F})$ and $b \in A(\mathcal{F})$. Hence, $(a,b) \in R(\mathcal{G} \cup \mathcal{H})$ is implied. Assume now $(a,b) \notin R\left((\mathcal{G} \cup \mathcal{H})^{k^*}(gr)\right)$. Again, several reasons for removing have to be considered. First, notice that $(b,b) \in R(\mathcal{G})$ (thus $(b,b) \in R(\mathcal{F})$) has to hold. If $(a,a) \in R(\mathcal{H})$ holds we deduce $(a,a),(b,b) \in R(\mathcal{F} \cup \mathcal{H})$ which contradicts $(a,b) \in R\left((\mathcal{F} \cup \mathcal{H})^{k^*}(gr)\right)$. Note that $(b,a) \in R(\mathcal{G} \cup \mathcal{H})$ is just impossible since $\mathcal{G} \leq_S \mathcal{G} \cup \mathcal{H}$ was assumed. Assume now that $\forall c\, ((b,c) \in R(\mathcal{G} \cup \mathcal{H}) \to \{(a,c),(c,a),(c,c)\} \cap R(\mathcal{G} \cup \mathcal{H}) \neq \emptyset)$ holds. Since $(a,b) \in R\left((\mathcal{F} \cup \mathcal{H})^{k^*}(gr)\right)$ was assumed there exists an argument $c \in A(\mathcal{F})$, such that $(b,c) \in R(\mathcal{F}) \wedge \{(a,c),(c,a),(c,c)\} \cap R(\mathcal{F} \cup \mathcal{H}) = \emptyset$. We observe that

5.2. Characterizing Strong Expansion Equivalence

$\{(a,c),(c,a),(c,c)\} \cap R(\mathcal{G} \cup \mathcal{H}) = \emptyset$ is implied and hence, if $(b,c) \in R(\mathcal{G})$, then $(a,b) \in R\left((\mathcal{G} \cup \mathcal{H})^{k^*(gr)}\right)$ follows contradicting the assumption. Remember that $\mathcal{F}^{k^*(gr)} = \mathcal{G}^{k^*(gr)}$ has to hold. Hence, if $(b,c) \notin R(\mathcal{G})$, then (b,c) has to be deleted in $R\left(\mathcal{F}^{k^*(gr)}\right)$. This is impossible because $(c,c) \notin R(\mathcal{F} \cup \mathcal{H})$ (thus $(c,c) \notin R(\mathcal{F})$) is already shown. 4th **case:** Let $a \in A(\mathcal{F})$ and $b \in A(\mathcal{H}) \setminus A(\mathcal{F})$. This case is impossible because $(a,b) \in R(\mathcal{F} \cup \mathcal{H})$ cannot hold if $\mathcal{F} \leq_S \mathcal{F} \cup \mathcal{H}$ is fulfilled. □

With the help of the two lemmata above we will prove now that syntactical equivalence of gr-*-kernels of two AFs characterizes their strong expansion equivalence with respect to grounded semantics.

Theorem 5.22. *For any AFs $\mathcal{F}, \mathcal{G}$:*

$$\mathcal{F}^{k^*(gr)} = \mathcal{G}^{k^*(gr)} \Leftrightarrow \mathcal{F} \equiv_S^{gr} \mathcal{G}.$$

Proof. The if-direction, namely $\mathcal{F}^{k^*(gr)} = \mathcal{G}^{k^*(gr)} \Rightarrow \mathcal{F} \equiv_S^{gr} \mathcal{G}$ follows by applying Lemmata 5.20 and 5.21 (similarly to Theorem 5.16). We will show the only-if-direction by proving the contrapositive, i.e. $\mathcal{F}^{k^*(gr)} \neq \mathcal{G}^{k^*(gr)} \Rightarrow \mathcal{F} \not\equiv_S^{gr} \mathcal{G}$.

1st case: Assume $A\left(\mathcal{F}^{k^*(gr)}\right) \neq A\left(\mathcal{G}^{k^*(gr)}\right)$. Hence, without loss of generality there exists an argument $a \in A(\mathcal{F}) \setminus A(\mathcal{G})$. We define $\mathcal{H} = ((A(\mathcal{F}) \cup A(\mathcal{G})) \setminus \{a\}, \emptyset)$. Let E be the unique grounded extension of $\mathcal{F} \cup \mathcal{H}$. If $a \in E$, $E \notin \mathcal{E}_{gr}(\mathcal{G} \cup \mathcal{H})$ follows. Consider now $a \notin E$. We define $\mathcal{H}' = \mathcal{H} \cup (\{a\}, \emptyset)$. Hence, $\mathcal{F} \cup \mathcal{H} = \mathcal{F} \cup \mathcal{H}'$ and therefore, E is the unique grounded extension of $\mathcal{F} \cup \mathcal{H}'$. Furthermore we observe that a is unattacked in $\mathcal{G} \cup \mathcal{H}'$ and so, a is contained in the unique grounded extension E' of $\mathcal{G} \cup \mathcal{H}'$. Hence, $\mathcal{F} \not\equiv_S^{gr} \mathcal{G}$ follows.

2nd case: Consider $R\left(\mathcal{F}^{k^*(gr)}\right) \neq R\left(\mathcal{G}^{k^*(gr)}\right)$ and $A\left(\mathcal{F}^{k^*(gr)}\right) = A\left(\mathcal{G}^{k^*(gr)}\right)$ $(= A(\mathcal{F}) = A(\mathcal{G}))$. Hence, without loss of generality there exists $a,b \in A(\mathcal{F})$, such that $(a,b) \in R\left(\mathcal{F}^{k^*(gr)}\right) \setminus R\left(\mathcal{G}^{k^*(gr)}\right)$. Let c be a new argument, i.e. $c \notin A(\mathcal{F})$. Furthermore we define

$$\mathcal{I} = \left(A(\mathcal{F}) \cup \{c\}, \{(c,c') \mid c' \in A(\mathcal{F}) \setminus \{a,b\}\}\right).$$

Case 2.1: Let $a = b$. This means $(a,a) \in R\left(\mathcal{F}^{k^*(gr)}\right) \setminus R\left(\mathcal{G}^{k^*(gr)}\right)$ and consequently $(a,a) \in R(\mathcal{F}) \setminus R(\mathcal{G})$ by Definition 5.18 of the gr-*-kernel. It is easy to see (splitting results, Section 5.2.1) that $\{\{c\}\} = \mathcal{E}_{gr}(\mathcal{F} \cup \mathcal{I}) \neq \mathcal{E}_{gr}(\mathcal{G} \cup \mathcal{I}) = \{\{a,c\}\}$ holds. From now on we suppose that any self-loop is either contained in both $R\left(\mathcal{F}^{k^*(gr)}\right)$ and $R\left(\mathcal{G}^{k^*(gr)}\right)$ or in none of them.

Case 2.2: Consider now $a \neq b$, i.e. $(a,b) \in R\left(\mathcal{F}^{k^*(gr)}\right) \setminus R\left(\mathcal{G}^{k^*(gr)}\right)$ and $(a,b) \in R(\mathcal{F})$. We have to distinguish four cases for the presence or absence of attack (a,a) and (b,b). Keep in mind that $R(\mathcal{F}), R(\mathcal{G}), R\left(\mathcal{F}^{k^*(gr)}\right)$ and $R\left(\mathcal{G}^{k^*(gr)}\right)$ contain the same self-loops. **Case 2.2.1:** $(a,a),(b,b) \in R(\mathcal{F})$.

This case is impossible because $(a,b) \in R\left(\mathcal{F}^{k^*(gr)}\right)$ cannot hold (Definition 5.18). Case **2.2.2**: $(a,a),(b,b) \notin R(\mathcal{F})$. Note that $(a,b) \notin R(\mathcal{G})$ holds because $(b,b) \notin R(\mathcal{G})$ and $(a,b) \notin R\left(\mathcal{G}^{k^*(gr)}\right)$ was assumed. The attack (b,a) may or may not be an element of $R(\mathcal{F})$ or $R(\mathcal{G})$. If $(b,a) \notin R(\mathcal{F})$, $\{\{a,c\}\} = \mathcal{E}_{gr}(\mathcal{F} \cup \mathcal{I})$ follows. If not, i.e. $(b,a) \in R(\mathcal{F})$, then $\{\{c\}\} = \mathcal{E}_{gr}(\mathcal{F} \cup \mathcal{I})$ holds. Furthermore, if $(b,a) \notin R(\mathcal{G})$ we deduce $\{\{a,b,c\}\} = \mathcal{E}_{gr}(\mathcal{G} \cup \mathcal{I})$ and if not, i.e. $(b,a) \in R(\mathcal{G})$ it follows $\{\{b,c\}\} = \mathcal{E}_{gr}(\mathcal{G} \cup \mathcal{I})$. Thus, in all possible combinations we obtain different grounded extensions, i.e. $\mathcal{F} \not\equiv_S^{gr} \mathcal{G}$. **2.2.3**: $(a,a) \in R(\mathcal{F})$ and $(b,b) \notin R(\mathcal{F})$. Again, it is impossible that $(a,b) \in R(\mathcal{G})$ holds since $(b,b) \notin R(\mathcal{G})$ and $(a,b) \notin R\left(\mathcal{G}^{k^*(gr)}\right)$ was assumed. The attack (b,a) may or may not be an element of $R(\mathcal{F})$ and $R(\mathcal{G})$. Either way, $\{\{c\}\} = \mathcal{E}_{gr}(\mathcal{F} \cup \mathcal{I}) \neq \mathcal{E}_{gr}(\mathcal{G} \cup \mathcal{I}) = \{\{b,c\}\}$ follows. Hence, $\mathcal{F} \not\equiv_S^{gr} \mathcal{G}$. **2.2.4**: $(a,a) \notin R(\mathcal{F})$ and $(b,b) \in R(\mathcal{F})$. Since $(a,b) \in R\left(\mathcal{F}^{k^*(gr)}\right)$ is assumed, we deduce $(b,a) \notin R(\mathcal{F})$ and furthermore the existence of an argument $c \in A(\mathcal{F}) : (b,c) \in R(\mathcal{F}) \wedge \{(a,c),(c,a),(c,c)\} \cap R(\mathcal{F}) = \emptyset$ (compare Definition 5.18). The following figures show the remaining two possibilities for AF $\mathcal{F}$. Note that we omit possible other arguments than a, b and c.

Up to now we know $(a,a),(c,c) \notin R(\mathcal{G})$ and $(b,b) \in R(\mathcal{G})$. Hence, there are $2^6 = 64$ possibilities for the presence and absence of $(a,b),(b,a),(b,c),(c,b)$, (a,c) and (c,a) in $R(\mathcal{G})$. We will show that some of them are impossible since $(a,b) \notin R\left(\mathcal{G}^{k^*(gr)}\right)$ was assumed. Again, we use the slightly different version of the standard construction $\mathcal{I}$, namely

$$\mathcal{I}' = \left(A(\mathcal{F}) \cup \{d\}, \{(d,c') \,|\, c' \in A(\mathcal{F}) \setminus \{a,b,c\}\}\right).$$

It can be checked that $\mathcal{E}_{gr}(\mathcal{F}_1 \cup \mathcal{I}') = \mathcal{E}_{gr}(\mathcal{F}_2 \cup \mathcal{I}') = \{\{a,c,d\}\}$. If $(a,c) \in R(\mathcal{G})$ or $(c,a) \in R(\mathcal{G})$, then $\{\{a,c,d\}\} \neq \mathcal{E}_{gr}(\mathcal{G} \cup \mathcal{I}')$ because a grounded extension has to be conflict-free. From now on we assume $(a,c),(c,a) \notin R(\mathcal{G})$. This means, $2^4 = 16$ possibilities with respect to the presence or absence of $(a,b),(b,a),(b,c)$ and (c,b) remain. These sixteen remaining possibilities are listed in Theorem 5.16. $\mathcal{G}_6$ (= $\mathcal{F}_2$) and $\mathcal{G}_8$ (= $\mathcal{F}_1$) are impossible since $(a,b) \notin R\left(\mathcal{G}^{k^*(gr)}\right)$ was assumed. The cases $\mathcal{G}_2, \mathcal{G}_4, \mathcal{G}_9, \mathcal{G}_{10}, \mathcal{G}_{12}, \mathcal{G}_{13}, \mathcal{G}_{14}$ and $\mathcal{G}_{16}$ can be checked by considering the union with AF $\mathcal{I}'$. For every $i \in \{2,4,9,10,12,13,14,16\}$, $\{\{a,c,d\}\} \neq \mathcal{E}_{gr}(\mathcal{G}_i \cup \mathcal{I}')$ holds. For all other cases we use

$$\mathcal{I}'' = \left(A(\mathcal{F}) \cup \{d\}, \{(d,c') \,|\, c' \in A(\mathcal{F}) \setminus \{b,c\}\}\right).$$

Combining $\mathcal{F}_1$ and $\mathcal{F}_2$ with $\mathcal{I}''$ we get $\{\{d\}\} = \mathcal{E}_{gr}(\mathcal{F}_1 \cup \mathcal{I}'') = \mathcal{E}_{gr}(\mathcal{F}_2 \cup \mathcal{I}'')$. Furthermore we have $\{\{c,d\}\} = \mathcal{E}_{gr}(\mathcal{G}_i \cup \mathcal{I}'')$ for every $i \in \{1,3,5,7,11,15\}$. Hence, $\mathcal{F} \not\equiv_S^{gr} \mathcal{G}$ concluding the proof. □

5.2. Characterizing Strong Expansion Equivalence

Finally, we will give a counter-example showing that strong expansion equivalence is not sufficient for expansion equivalence with respect to grounded semantics as stated at the very beginning of this section.

Example 5.23. The AFs $\mathcal{F}$ and $\mathcal{G}$ are strong expansion equivalent since they possess equal gr-*-kernels, namely $\mathcal{F}^{k^*(gr)} = \mathcal{G}^{k^*(gr)} = \mathcal{G}$ (Theorem 5.22).

Figure 5.9: Strong Expansion Equivalent AFs

Furthermore, they are not expansion equivalent with respect to grounded semantics which can be demonstrated by the following expansions $\mathcal{F} \cup \mathcal{H}$ of $\mathcal{F}$ and $\mathcal{G} \cup \mathcal{H}$ of $\mathcal{G}$, where $\mathcal{H} = (\{b,d\}, \{(b,d)\})$.

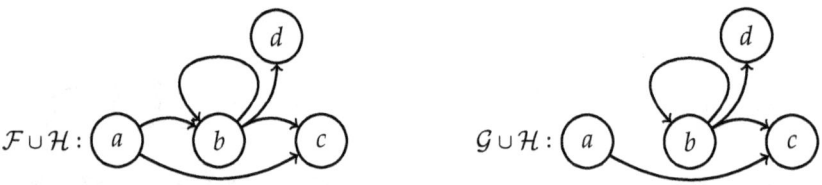

Figure 5.10: Non-coincidence of Expansion and Strong Expansion Equivalence

One may easily identify different grounded extensions for $\mathcal{F} \cup \mathcal{H}$ and $\mathcal{G} \cup \mathcal{H}$, namely $\{a,d\}$ or $\{a\}$, respectively.

5.2.6 Strong Expansion Equivalence for Complete Semantics

We now turn to complete semantics. Remember that any preferred extension is a complete one and furthermore, any complete extension is admissible (compare Proposition 2.7). Moreover, the ad-kernel serves as a uniform characterization for expansion equivalence with respect to complete and admissible semantics (see Theorems 5.5, 5.7). Similarly, we have already shown that the ad-*-kernel characterizes both semantics with respect to strong expansion equivalence (Theorem 5.16). In the light of these relations it is quite surprising or at least interesting that neither in case of arbitrary expansions (Theorem 5.5) nor in case of strong expansions (as we will see) complete semantics agree with preferred or admissible semantics.

We now introduce a further novel kernel definition, the so-called *complete-*-*kernel* which characterizes strong expansion equivalence with respect to complete semantics. Here is the formal definition.

Chapter 5. Notions of Equivalence and Replacement

Definition 5.24. Given an AF $\mathcal{F} = (A, R)$. We define the *complete-*-kernel* of $\mathcal{F}$ as $\mathcal{F}^{k^*(co)} = \left(A, R^{k^*(co)}\right)$ where

$R^{k^*(co)} = R \smallsetminus \{(a,b) \mid a \neq b, ((a,a),(b,b) \in R) \vee$
$((b,b) \in R \wedge (b,a) \notin R \wedge \forall c\ ((b,c) \in R \rightarrow \{(a,c),(c,a),(c,c)\} \cap R \neq \emptyset))\}.$

The newly introduced kernel "forgets" an attack (a, b) if

1. a and b are self-attacking or

2. b is self-defeating, b does not attack a and furthermore, for all arguments c which are attacked by b at least one of the following conditions holds:

    i) a attacks c,

    ii) c attacks a or

    iii) c attacks c.

The first disjunct describes attacks which are even redundant with respect to arbitrary expansions (compare co-kernel, Definition 5.1). The additional part, namely the second disjunct of the co-*-kernel is very similar to the gr-*-kernel (Definition 5.18). The difference is that the deletion of an attack (a,b) requires the additional precondition that b does not attack a. This is due to the fact that the attack (a,b) may be crucial for the acceptance of the argument a if (b,a) is established. Roughly speaking, the argument a may justify its acceptance itself in contrast to grounded semantics where the reason for being a member of the unique grounded extension has to come from the outside, i.e. former accepted arguments have to defend a. Consider the following example.

Example 5.25. The AFs $\mathcal{F}$ and $\mathcal{G}\left(= \mathcal{F}^{k^*(gr)}\right)$ are strong expansion equivalent with respect to grounded semantics (Theorem 5.22). In particular, the deletion of (a,b) is irrelevant with respect to the grounded extensions of $\mathcal{F}$ and $\mathcal{G}$. Observe that in case of complete semantics (a,b) is essential since $\{a\}$ is no longer complete in $\mathcal{G}$.

Figure 5.11: Grounded-*- vs. Complete-*-Kernel

We proceed with some useful properties of the newly introduced kernel. The following lemma shows that any AF $\mathcal{F}$ and its co-*-kernel possess the same complete extensions.

5.2. Characterizing Strong Expansion Equivalence

Lemma 5.26. *For any AF $\mathcal{F}$, $\mathcal{F} \equiv^{co} \mathcal{F}^{k^*(co)}$.*

Proof. The first step is to show that $\mathcal{F}$ and $\mathcal{F}^{k^*(co)}$ contain the same conflict-free sets, i.e. $S \in cf(\mathcal{F})$ iff $S \in cf(\mathcal{F}^{k^*(co)})$. The if-direction is obvious because $R(\mathcal{F}^{k^*(co)}) \subseteq R(\mathcal{F})$ holds (compare Definition 5.24). Assume now $S \in cf(\mathcal{F}^{k^*(co)})$ and $S \notin cf(\mathcal{F})$. Consequently, there are two arguments a and b in S with the property $(a,b) \in R(\mathcal{F}) \smallsetminus R(\mathcal{F}^{k^*(co)})$. In any case, $(b,b) \in R(\mathcal{F})$ has to hold. The same applies to $R(\mathcal{F}^{k^*(co)})$ which contradicts the assumption $S \in cf(\mathcal{F}^{k^*(co)})$.

We now prove that $E \in \mathcal{E}_{co}(\mathcal{F})$ implies $E \in \mathcal{E}_{co}(\mathcal{F}^{k^*(co)})$. At first we will show that E is admissible in $\mathcal{F}^{k^*(co)}$. Assume $E \in \mathcal{E}_{co}(\mathcal{F})$ and E does not defend all its elements in $\mathcal{F}^{k^*(co)}$. This means, there is an argument $a \in E$ and an argument $b \notin E$ (conflict-freeness) such that $(b,a) \in R(\mathcal{F}^{k^*(co)})$ and $(E,b) \notin R(\mathcal{F}^{k^*(co)})$. Since $R(\mathcal{F}^{k^*(co)}) \subseteq R(\mathcal{F})$ and $\mathcal{E}_{co}(\mathcal{F}) \subseteq \mathcal{E}_{ad}(\mathcal{F})$ hold, we deduce the existence of an argument $c \in E$, such that $(c,b) \in R(\mathcal{F}) \smallsetminus R(\mathcal{F}^{k^*(co)})$. There are two possibilities for the deletion of (c,b) in $R(\mathcal{F}^{k^*(co)})$. First, $(c,c),(b,b) \in R(\mathcal{F})$ and second, $(b,b) \in R(\mathcal{F})$, $(b,c) \notin R(\mathcal{F})$ and at least $\{(a,c),(c,a),(a,a)\} \cap R(\mathcal{F}) \neq \emptyset$. Due to the conflict-freeness of E in $\mathcal{F}$ and the membership of a and c in E both options fail. Assume now $E \in \mathcal{E}_{co}(\mathcal{F})$ but E does not contain all defended elements in $\mathcal{F}^{k^*(co)}$. Hence, there is an argument $a \notin E$, such that for all arguments c with $(c,a) \in R(\mathcal{F}^{k^*(co)})$, $(E,c) \in R(\mathcal{F}^{k^*(co)})$. Since E is assumed to be complete in $\mathcal{F}$ and $a \notin E$ we deduce the existence of an argument c with the property $(c,a) \in R(\mathcal{F})$ and $(E,c) \notin R(\mathcal{F})$. Combining both conclusions we get $(c,a) \in R(\mathcal{F}) \smallsetminus R(\mathcal{F}^{k^*(co)})$. In any case, $(a,a) \in R(\mathcal{F})$ and thus, $(a,a) \in R(\mathcal{F}^{k^*(co)})$. Since a is defended by E in $\mathcal{F}^{k^*(co)}$, $(E,a) \in R(\mathcal{F}^{k^*(co)})$ has to hold. Finally, $(E,E) \in R(\mathcal{F}^{k^*(co)})$ follows contradicting the conflict-freeness of E in $\mathcal{F}^{k^*(co)}$.

We now prove that $E \in \mathcal{E}_{co}(\mathcal{F}^{k^*(co)})$ implies $E \in \mathcal{E}_{co}(\mathcal{F})$. First of all, we show the admissibility of E in $\mathcal{F}$. Given $E \in \mathcal{E}_{co}(\mathcal{F}^{k^*(co)})$, we assume the existence of an argument $a \in E$ and an argument $b \notin E$ (conflict-freeness), such that $(b,a) \in R(\mathcal{F})$ and $(E,b) \notin R(\mathcal{F})$ holds. Due to the relations $R(\mathcal{F}^{k^*(co)}) \subseteq R(\mathcal{F})$ and $\mathcal{E}_{co}(\mathcal{F}^{k^*(co)}) \subseteq \mathcal{E}_{ad}(\mathcal{F}^{k^*(co)})$, $(b,a) \in R(\mathcal{F}) \smallsetminus R(\mathcal{F}^{k^*(co)})$ follows. Consequently, $(a,a) \in R(\mathcal{F})$ has to hold contradicting the conflict-freeness of E in $\mathcal{F}$. Assume now that E does not contain all defended elements in $\mathcal{F}$, i.e. it exists an argument $a \notin E$, such that for all arguments c with the property $(c,a) \in R(\mathcal{F}),(E,c) \in R(\mathcal{F})$ holds. Since E is assumed to be complete in $\mathcal{F}^{k^*(co)}$ and $a \notin E$ holds, we deduce the existence of an argument c, such that $(c,a) \in R(\mathcal{F}^{k^*(co)})$ and $(E,c) \notin R(\mathcal{F}^{k^*(co)})$. Altogether, $(c,a) \in R(\mathcal{F}^{k^*(co)})$

and $(E,c) \in R(\mathcal{F}) \setminus R\left(\mathcal{F}^{k^*(co)}\right)$. Let d be the argument in E which attacks c, i.e. $(d,c) \in R(\mathcal{F}) \setminus R\left(\mathcal{F}^{k^*(co)}\right)$. We observe that $(d,d) \in R(\mathcal{F})$ is impossible because $E \in cf(\mathcal{F})$ is assumed. Hence, $(c,c) \in R(\mathcal{F})$, $(c,d) \notin R(\mathcal{F})$ and $\{(a,a),(a,d),(d,a)\} \cap R(\mathcal{F}) \neq \emptyset$ follows. The cases $(a,a) \in R(\mathcal{F})$ and $(d,a) \in R(\mathcal{F})$ contradict the conflict-freeness of E in $\mathcal{F}$ because a is assumed to be defended by E in $\mathcal{F}$. In case of $(a,d) \in R(\mathcal{F})$ we use the already shown admissibility of E in $\mathcal{F}$ to infer $(E,a) \in R(\mathcal{F})$. Again, we get a contradiction to the conflict-freeness of E in $\mathcal{F}$ if we apply that a is defended by E in $\mathcal{F}$. □

The next lemma proves the robustness of the co-*-kernel. That means, if two AFs $\mathcal{F}$ and $\mathcal{G}$ possess the same co-*-kernel, then the same applies for any compositions $\mathcal{F} \cup \mathcal{H}$ and $\mathcal{G} \cup \mathcal{H}$ under the condition that the latter are strong expansions of their initial frameworks $\mathcal{F}$ and $\mathcal{G}$, respectively.

Lemma 5.27. *If* $\mathcal{F}^{k^*(co)} = \mathcal{G}^{k^*(co)}$, *then* $(\mathcal{F} \cup \mathcal{H})^{k^*(co)} = (\mathcal{G} \cup \mathcal{H})^{k^*(co)}$ *for all AFs* $\mathcal{H}$ *which satisfy* $\mathcal{F} \leq_S \mathcal{F} \cup \mathcal{H}$ *and* $\mathcal{G} \leq_S \mathcal{G} \cup \mathcal{H}$.

Proof. First notice that the assumption $\mathcal{F}^{k^*(co)} = \mathcal{G}^{k^*(co)}$ implies $A(\mathcal{F}) = A\left(\mathcal{F}^{k^*(co)}\right) = A\left(\mathcal{G}^{k^*(co)}\right) = A(\mathcal{G})$. Consider now an AF $\mathcal{H}$ satisfying the specified properties (strong expansion or equality). If $\mathcal{F} = \mathcal{F} \cup \mathcal{H}$, then $\mathcal{G} = \mathcal{G} \cup \mathcal{H}$ is implied and vice versa. Consequently, in this case it is nothing to show because $(\mathcal{F} \cup \mathcal{H})^{k^*(co)} = (\mathcal{G} \cup \mathcal{H})^{k^*(co)}$ follows immediately. Without loss of generality we may assume that $\mathcal{F} \cup \mathcal{H}$ and $\mathcal{G} \cup \mathcal{H}$ are indeed strong expansions of $\mathcal{F}$ or $\mathcal{G}$. Thus, $R(\mathcal{H}) \cap R(\mathcal{F}) = \emptyset$ and $R(\mathcal{H}) \cap R(\mathcal{G}) = \emptyset$ can be assumed. Let $(a,b) \in R\left((\mathcal{F} \cup \mathcal{H})^{k^*(co)}\right)$, therefore $(a,b) \in R(\mathcal{F} \cup \mathcal{H})$. We will show $(a,b) \in R\left((\mathcal{G} \cup \mathcal{H})^{k^*(co)}\right)$ by proof by cases (containedness of a and b in $A(\mathcal{F})$ or $A(\mathcal{H}) \setminus A(\mathcal{F})$). For all cases we will suppose $a \neq b$ since the self-loop case is obvious.

1st **case:** Let $a,b \in A(\mathcal{F})$. Assuming $(a,b) \in R\left(\mathcal{F}^{k^*(co)}\right)$ implies $(a,b) \in R\left(\mathcal{G}^{k^*(co)}\right)$ and therefore $(a,b) \in R(\mathcal{G})$. Consequently $(a,b) \in R\left((\mathcal{G} \cup \mathcal{H})^{k^*(co)}\right)$ holds since $\mathcal{G} \cup \mathcal{H}$ was assumed to be a strong expansion of $\mathcal{G}$ and so, no relevant attacks are added. The assumption $(a,b) \notin R\left(\mathcal{F}^{k^*(co)}\right)$ contradicts $(a,b) \in R\left((\mathcal{F} \cup \mathcal{H})^{k^*(co)}\right)$ because the reason to remove an attack remains untouched in $\mathcal{F} \cup \mathcal{H}$. 2nd **case:** Let $a,b \in A(\mathcal{H}) \setminus A(\mathcal{F})$. Thus, $(a,b) \in R(\mathcal{G} \cup \mathcal{H})$ is implied. Suppose now $(a,b) \notin R\left((\mathcal{G} \cup \mathcal{H})^{k^*(co)}\right)$. This means, several reasons for removing have to be checked. The assumption $(a,a),(b,b) \in R(\mathcal{H})$ is inconsistent with $(a,b) \in R\left((\mathcal{F} \cup \mathcal{H})^{k^*(co)}\right)$. Thus, $(b,b) \in R(\mathcal{H})$, $(b,a) \notin R(\mathcal{H})$ and $\forall c \left((b,c) \in R(\mathcal{G} \cup \mathcal{H}) \to \{(a,c),(c,a),(c,c)\} \cap R(\mathcal{G} \cup \mathcal{H}) \neq \emptyset\right)$ has to hold. If there is no c in $A(\mathcal{G})$ which is attacked by b we deduce $(a,b) \notin R\left((\mathcal{F} \cup \mathcal{H})^{k^*(co)}\right)$. Thus, consider an argument $c \in A(\mathcal{G})$ with the property $(b,c) \in R(\mathcal{H})$. Hence, $\{(a,c),(c,a),(c,c)\} \cap R(\mathcal{G} \cup \mathcal{H}) \neq \emptyset$ has to hold. In the first case, namely $(a,c) \in R(\mathcal{G} \cup \mathcal{H})$, $(a,c) \in R(\mathcal{H})$ and consequently $(a,c) \in R(\mathcal{F} \cup \mathcal{H})$ follows. The second case, i.e. $(c,a) \in R(\mathcal{G} \cup \mathcal{H})$, is just impossible since $\mathcal{G} \cup \mathcal{H}$ was assumed to be a strong expansion of $\mathcal{G}$. If $(c,c) \in R(\mathcal{G} \cup \mathcal{H})$, then $(c,c) \in R(\mathcal{G})$ and $(c,c) \in R(\mathcal{F})$ (since $\mathcal{F}^{k^*(co)} = \mathcal{G}^{k^*(co)}$ was assumed), therefore $(c,c) \in R(\mathcal{F} \cup \mathcal{H})$. In all cases we deduce $(a,b) \notin R\left((\mathcal{F} \cup \mathcal{H})^{k^*(co)}\right)$ contradicting the assumption. 3rd **case:** Let $a \in A(\mathcal{H}) \setminus A(\mathcal{F})$ and $b \in A(\mathcal{F})$. Consequently, $(a,b) \in R(\mathcal{G} \cup \mathcal{H})$ holds. Assume now $(a,b) \notin R\left((\mathcal{G} \cup \mathcal{H})^{k^*(co)}\right)$. Again, several reasons for removing have to be considered. We observe that $(b,b) \in R(\mathcal{G})$ (thus $(b,b) \in R(\mathcal{F})$) has to hold. If $(a,a) \in R(\mathcal{H})$ holds we deduce $(a,a), (b,b) \in R(\mathcal{F} \cup \mathcal{H})$ contrary to $(a,b) \in R\left((\mathcal{F} \cup \mathcal{H})^{k^*(co)}\right)$. Furthermore, we observe $(b,a) \notin R(\mathcal{F} \cup \mathcal{H}), R(\mathcal{G} \cup \mathcal{H})$ because $\mathcal{F} \cup \mathcal{H}$ and $\mathcal{G} \cup \mathcal{H}$ are assumed to be strong expansions of $\mathcal{F}$ and $\mathcal{G}$ respectively. Together with the assumption $(a,b) \in R\left((\mathcal{F} \cup \mathcal{H})^{k^*(co)}\right)$ we deduce the existence of an argument $c \in A(\mathcal{F})$, such that $(b,c) \in R(\mathcal{F}) \wedge \{(a,c), (c,a), (c,c)\} \cap R(\mathcal{F} \cup \mathcal{H}) = \emptyset$. Hence, $(c,c) \notin R(\mathcal{F})$ (therefore $(c,c) \notin R(\mathcal{G})$) and $(a,c), (c,a) \notin R(\mathcal{H})$ is implied. Consequently, if $(b,c) \in R(\mathcal{G})$, then $(a,b) \in R\left((\mathcal{G} \cup \mathcal{H})^{k^*(co)}\right)$ contradicting the assumption. Remember that $\mathcal{F}^{k^*(co)} = \mathcal{G}^{k^*(co)}$ has to hold. Hence, if $(b,c) \notin R(\mathcal{G})$, then (b,c) has to be deleted in $R\left(\mathcal{F}^{k^*(co)}\right)$. This is impossible because $(c,c) \notin R(\mathcal{F})$ is already shown. 4th **case:** Let $a \in A(\mathcal{F})$ and $b \in A(\mathcal{H}) \setminus A(\mathcal{F})$. Here is nothing to show because $(a,b) \in R(\mathcal{F} \cup \mathcal{H})$ cannot hold if $\mathcal{F} \leq_S \mathcal{F} \cup \mathcal{H}$ is fulfilled. $\square$

Now we are prepared to show the main theorem for the case of complete semantics. The notion of co-*-kernel is suitable to describe strong expansions equivalence with respect to complete semantics.

Theorem 5.28. *For any AFs $\mathcal{F}, \mathcal{G}$:*

$$\mathcal{F}^{k^*(co)} = \mathcal{G}^{k^*(co)} \Leftrightarrow \mathcal{F} \equiv_S^{co} \mathcal{G}.$$

Proof. The first direction, namely $\mathcal{F}^{k^*(co)} = \mathcal{G}^{k^*(co)} \Rightarrow \mathcal{F} \equiv_S^{co} \mathcal{G}$ can be shown by applying Lemmata 5.26 and 5.27 (similarly to Theorem 5.16). We will prove the only-if-direction by showing the contrapositive, i.e. $\mathcal{F}^{k^*(co)} \neq \mathcal{G}^{k^*(co)} \Rightarrow \mathcal{F} \not\equiv_S^{co} \mathcal{G}$.

1st **case:** Suppose $A\left(\mathcal{F}^{k^*(co)}\right) \neq A\left(\mathcal{G}^{k^*(co)}\right)$. Thus, without loss of generality there exists an argument $a \in A(\mathcal{F}) \setminus A(\mathcal{G})$. We define $\mathcal{H} = ((A(\mathcal{F}) \cup A(\mathcal{G})) \setminus \{a\}, \emptyset)$. Consider the existence of an extension $E \in \mathcal{E}_{co}(\mathcal{F} \cup \mathcal{H})$, such that $a \in E$ holds. Consequently, $E \notin \mathcal{E}_{co}(\mathcal{G} \cup \mathcal{H})$ and therefore $\mathcal{F} \not\equiv_S^{co} \mathcal{G}$. Hence, we may assume that for all extensions $E \in \mathcal{E}_{co}(\mathcal{F} \cup \mathcal{H})$, $a \notin E$ holds. We define $\mathcal{H}' = \mathcal{H} \cup (\{a\}, \emptyset)$. Observe that a is unattacked in $\mathcal{H}'$. Since $\mathcal{G} \cup \mathcal{H}' = \mathcal{H}'$ we deduce that for any extension $E \in \mathcal{E}_{co}(\mathcal{G} \cup \mathcal{H}')$, $a \in E$ holds. Remember that the existence of a complete extension is guaranteed. Finally, since $\mathcal{E}_{co}(\mathcal{F} \cup \mathcal{H}) = \mathcal{E}_{co}(\mathcal{F} \cup \mathcal{H}')$ obviously holds we are done.

2nd **case:** Assume $R\left(\mathcal{F}^{k^*(co)}\right) \neq R\left(\mathcal{G}^{k^*(co)}\right)$ and $A\left(\mathcal{F}^{k^*(co)}\right) = A\left(\mathcal{G}^{k^*(co)}\right)$ $(= A(\mathcal{F}) = A(\mathcal{G}))$. Thus, without loss of generality there exist some arguments $a, b \in A(\mathcal{F})$ with the property $(a,b) \in R\left(\mathcal{F}^{k^*(co)}\right) \setminus R\left(\mathcal{G}^{k^*(co)}\right)$. Let c be

Chapter 5. Notions of Equivalence and Replacement

a fresh argument, i.e. $c \notin A(\mathcal{F})$. Furthermore we define

$$\mathcal{I} = \left(A(\mathcal{F}) \cup \{c\}, \{(c,c') \mid c' \in A(\mathcal{F}) \setminus \{a,b\}\}\right).$$

Case 2.1: Let $a = b$ (self-loop case). Hence, $(a,a) \in R\left(\mathcal{F}^{k^*(co)}\right) \setminus R\left(\mathcal{G}^{k^*(co)}\right)$ and therefore $(a,a) \in R(\mathcal{F}) \setminus R(\mathcal{G})$ follows. We obtain $\{\{c\}\} = \mathcal{E}_{gr}(\mathcal{F} \cup \mathcal{I}) \neq \mathcal{E}_{gr}(\mathcal{G} \cup \mathcal{I}) = \{\{a,c\}\}$. From now on we suppose that any self-loop is either contained in both $R\left(\mathcal{F}^{k^*(co)}\right)$ and $R\left(\mathcal{G}^{k^*(co)}\right)$ or in none of them.

Case 2.2: Assume $a \neq b$. This means, $(a,b) \in R\left(\mathcal{F}^{k^*(co)}\right) \setminus R\left(\mathcal{G}^{k^*(co)}\right)$ and $(a,b) \in R(\mathcal{F})$. We will distinguish four cases for the presence or absence of the self-loops (a,a) and (b,b). Remember that $R(\mathcal{F}), R(\mathcal{G}), R\left(\mathcal{F}^{k^*(co)}\right)$ and $R\left(\mathcal{G}^{k^*(co)}\right)$ contain the same self-loops. **Case 2.2.1:** $(a,a),(b,b) \in R(\mathcal{F})$. This case contradicts the assumption because $(a,b) \in R\left(\mathcal{F}^{k^*(co)}\right)$ cannot be fulfilled (compare co-*-kernel, Definition 5.24). **Case 2.2.2:** $(a,a),(b,b) \notin R(\mathcal{F})$. Observe that $(a,b) \notin R(\mathcal{G})$ holds because $(b,b) \notin R(\mathcal{G})$ and $(a,b) \notin R\left(\mathcal{G}^{k^*(co)}\right)$ was assumed. The attack (b,a) may or may not be an element of $R(\mathcal{F})$ or $R(\mathcal{G})$. In any case, $\{a,c\} \in \mathcal{E}_{co}(\mathcal{F} \cup \mathcal{I})$. This can be checked by applying splitting results (compare Section 5.2.1). In the following we will leave this comment out. If $(b,a) \notin R(\mathcal{G})$, then $\{\{a,b,c\}\} = \mathcal{E}_{co}(\mathcal{G} \cup \mathcal{I})$ follows and if not, we deduce $\{\{b,c\}\} = \mathcal{E}_{co}(\mathcal{G} \cup \mathcal{I})$. Thus, $\mathcal{F} \not\equiv_S^{co} \mathcal{G}$ is shown. **Case 2.2.3:** $(a,a) \in R(\mathcal{F})$ and $(b,b) \notin R(\mathcal{F})$. Again, it is impossible that $(a,b) \in R(\mathcal{G})$ holds since $(b,b) \notin R(\mathcal{G})$ and $(a,b) \notin R\left(\mathcal{G}^{k^*(co)}\right)$ was assumed. The attack (b,a) may be contained in $R(\mathcal{F}), R(\mathcal{G})$ or not. In any case, $\{c\} \in \mathcal{E}_{co}(\mathcal{F} \cup \mathcal{I})$ and $\mathcal{E}_{co}(\mathcal{G} \cup \mathcal{I}) = \{\{b,c\}\}$ holds. Hence, $\mathcal{F} \not\equiv_S^{co} \mathcal{G}$. **Case 2.2.4:** $(a,a) \notin R(\mathcal{F})$ and $(b,b) \in R(\mathcal{F})$. Since $(a,b) \in R\left(\mathcal{F}^{k^*(gr)}\right)$ is assumed, we have to consider two sub-cases: First, $(b,a) \in R(\mathcal{F})$ and second, $(b,a) \notin R(\mathcal{F}) \wedge \exists c \in A(\mathcal{F}) : (b,c) \in R(\mathcal{F}) \wedge \{(a,c),(c,a),(c,c)\} \cap R(\mathcal{F}) = \emptyset$ (compare co-*-kernel, Definition 5.24). If $(b,a) \in R(\mathcal{F})$, then $\{\{c\},\{c,a\}\} = \mathcal{E}_{co}(\mathcal{F} \cup \mathcal{I})$ follows. Since $(a,b) \notin \mathcal{G}^{k^*(co)}$ is assumed, we deduce that if $(a,b) \in R(\mathcal{G})$, then $(b,a) \notin R(\mathcal{G})$ has to hold. In this case we obtain $\{\{c,a\}\} = \mathcal{E}_{co}(\mathcal{G} \cup \mathcal{I})$. Let $(a,b) \notin R(\mathcal{G})$. Hence, (b,a) may or may not be an element of $R(\mathcal{G})$. If (b,a) is contained in $R(\mathcal{G})$, $\{\{c\}\} = \mathcal{E}_{co}(\mathcal{G} \cup \mathcal{I})$ and if not, $\{\{a,c\}\} = \mathcal{E}_{co}(\mathcal{G} \cup \mathcal{I})$. Altogether, we have shown that in the first sub-case $\mathcal{F} \not\equiv_S^{co} \mathcal{G}$ is implied. Consider now $(b,a) \notin R(\mathcal{F}) \wedge \exists c \in A(\mathcal{F}) : (b,c) \in R(\mathcal{F}) \wedge \{(a,c),(c,a),(c,c)\} \cap R(\mathcal{F}) = \emptyset$. Just like in case of grounded semantics, two possibilities for AF $\mathcal{F}$ remain. Again, we omit possible other arguments than a, b, and c.

So far we know $(a,a),(c,c) \notin R(\mathcal{G})$ and $(b,b) \in R(\mathcal{G})$. Thus, there are $2^6 = 64$ combinations with respect to the presence and absence of $(a,b),(b,a),(b,c)$,

103

5.2. Characterizing Strong Expansion Equivalence

$(c,b), (a,c)$ and (c,a) in $R(\mathcal{G})$. Let d be a fresh argument. We define

$$\mathcal{I}' = \left(A(\mathcal{F}) \cup \{d\}, \{(d,c') \mid c' \in A(\mathcal{F}) \setminus \{a,b,c\}\}\right).$$

Observe that $\mathcal{E}_{co}(\mathcal{F}_1 \cup \mathcal{I}') = \mathcal{E}_{co}(\mathcal{F}_2 \cup \mathcal{I}') = \{\{a,c,d\}\}$ holds. If $(a,c) \in R(\mathcal{G})$ or $(c,a) \in R(\mathcal{G})$, then $\{a,c,d\} \notin \mathcal{E}_{co}(\mathcal{G} \cup \mathcal{I}')$ because complete extensions are conflict-free. Hence, we may assume $(a,c), (c,a) \notin R(\mathcal{G})$. This means, $2^4 = 16$ possibilities with respect to the presence or absence of $(a,b), (b,a), (b,c)$ and (c,b) remain. These sixteen AFs are already listed in Theorem 5.16. $\mathcal{G}_6$ ($= \mathcal{F}_2$) and $\mathcal{G}_8$ ($= \mathcal{F}_1$) as well as $\mathcal{G}_{13}, \mathcal{G}_{14}, \mathcal{G}_{15}$ and $\mathcal{G}_{16}$ are impossible because $(a,b) \notin R\left(\mathcal{G}^{k^*(co)}\right)$ was assumed. Remember that different grounded extensions imply different complete extensions (Proposition 2.17, statement 2). Thus, the counter-examples presented in the proof of Theorem 5.22 also serve to show that the remaining possibilities and $\mathcal{F}_1$ respectively $\mathcal{F}_2$ are not strong expansion equivalent with respect to the grounded semantics.

For the sake of completeness we present the counter-examples again. The cases $\mathcal{G}_2, \mathcal{G}_4, \mathcal{G}_9, \mathcal{G}_{10}$ and $\mathcal{G}_{12}$ can be checked by considering the union with AF $\mathcal{I}'$. For every $i \in \{2,4,9,10,12\}$, $\{a,c,d\} \notin \mathcal{E}_{co}(\mathcal{G}_i \cup \mathcal{I}')$ holds. The other cases can be proven by taken the union with

$$\mathcal{I}'' = \left(A(\mathcal{F}) \cup \{d\}, \{(d,c') \mid c' \in A(\mathcal{F}) \setminus \{b,c\}\}\right).$$

Combining $\mathcal{F}_1$ and $\mathcal{F}_2$ with $\mathcal{I}''$ we get $\{d\} \in \mathcal{E}_{co}(\mathcal{F}_1 \cup \mathcal{I}''), \mathcal{E}_{co}(\mathcal{F}_2 \cup \mathcal{I}'')$. Furthermore we have $\{\{c,d\}\} = \mathcal{E}_{co}(\mathcal{G}_i \cup \mathcal{I}'')$ for every $i \in \{1,3,5,7,11\}$. Hence, $\mathcal{F} \not\equiv_S^{co} \mathcal{G}$ concluding the proof. $\square$

We finish this section by giving an example showing that strong expansion equivalence and expansion equivalence with respect to complete semantics do not coincide.

Example 5.29. On the one hand, the AFs $\mathcal{F}$ and $\mathcal{G}$ are strong expansion equivalent with respect to complete semantics because $\mathcal{F}^{k^*(co)} = \mathcal{G}^{k^*(co)} = \mathcal{G}$ (Theorem 5.28). On the other hand, they are not equivalent with respect to arbitrary expansions which can be made explicit by conjoining them with $\mathcal{H} = (\{a,b\}, \{(b,a)\})$. We have $\{a\} \in \mathcal{E}_{co}(\mathcal{F} \cup \mathcal{H})$ and $\{a\} \notin \mathcal{E}_{co}(\mathcal{G} \cup \mathcal{H})$.

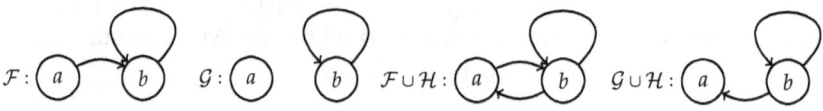

Figure 5.12: Non-coincidence of Expansion and Strong Expansion Equivalence

5.2.7 Strong Expansion Equivalence for Naive Semantics

Finally, in this last subsection we will prove that strong expansion equivalence with respect to naive semantics can be characterized by the following two conditions: First, possessing the same naive extensions and second, sharing the same arguments. This means, in contrast to the other semantics considered in this book the characterization of strong expansion equivalence in case of naive semantics is not purely syntactical. We mention that in [Gaggl and Woltran, 2012] it was already shown that expansion and local expansion equivalence are characterizable by the two properties listed above. This means, one consequence of the theorem we will show below is that in case of naive semantics expansion, normal expansion, strong expansion and local expansion equivalence coincide.

Theorem 5.30. *For any AFs $\mathcal{F}, \mathcal{G}$:*

$$\mathcal{E}_{na}(\mathcal{F}) = \mathcal{E}_{na}(\mathcal{G}) \text{ and } A(\mathcal{F}) = A(\mathcal{G}) \Leftrightarrow \mathcal{F} \equiv_S^{na} \mathcal{G}.$$

Proof. It suffices to prove that $\mathcal{F} \equiv_S^{na} \mathcal{G}$ implies $\mathcal{E}_{na}(\mathcal{F}) = \mathcal{E}_{na}(\mathcal{G})$ and $A(\mathcal{F}) = A(\mathcal{G})$. The converse direction is given by Theorem 5.13 [Gaggl and Woltran, 2012] and Proposition 2.18. First, assuming $\mathcal{E}_{na}(\mathcal{F}) \neq \mathcal{E}_{na}(\mathcal{G})$ immediately entails $\mathcal{F} \not\equiv_S^{na} \mathcal{G}$ (Proposition 2.18). Consider now $A(\mathcal{F}) \neq A(\mathcal{G})$ and $\mathcal{E}_{na}(\mathcal{F}) = \mathcal{E}_{na}(\mathcal{G})$. Without loss of generality there exists an argument $a \in A(\mathcal{F}) \setminus A(\mathcal{G})$. Obviously, for any $E \in \mathcal{E}_{na}(\mathcal{G})$, $a \notin E$. Consider $\mathcal{H} = (\{a\}, \emptyset)$. We have $a \in E'$, for any $E' \in \mathcal{E}_{na}(\mathcal{G} \cup \mathcal{H})$. Furthermore, $E' \notin \mathcal{E}_{na}(\mathcal{F} \cup \mathcal{H})$ since $\mathcal{F} \cup \mathcal{H} = \mathcal{F}$. Observe that $\mathcal{G} \cup \mathcal{H}$ is a strong expansion of $\mathcal{G}$ because we added an isolated argument. Consequently, $\mathcal{F} \not\equiv_S^{na} \mathcal{G}$. □

5.3 Characterizing Normal Expansion Equivalence

The main aim of this section is the characterization of normal expansion equivalence with respect to all semantics considered in this book. Remember that any arbitrary expansion can be split into a normal and local part (compare Definition 2.13). So one natural conjecture is that normal and local expansion equivalence jointly imply expansion equivalence. Using the results presented in this section we will not only verify the addressed conjecture but even a significantly stronger result. In fact, the main and quite surprisingly result for the considered semantics can be briefly and concisely presented in the following "equality":

normal expansion equivalence = expansion equivalence

This means, if two AFs $\mathcal{F}$ and $\mathcal{G}$ are proven to be normal expansion equivalent, then the requirement that $\mathcal{F}$ and $\mathcal{G}$ are equivalent when conjoined with any further framework $\mathcal{H}$ is fulfilled too. This is quite surprising since the class of normal expansions is obviously a proper subset of the class of arbitrary expansions. In other words, if different implicit information of two AFs $\mathcal{F}$ and $\mathcal{G}$ is made explicit by conjoining them with an AF $\mathcal{H}$ which adds further attacks between former arguments, then there exists an AF $\mathcal{H}'$ showing this difference without changing the former attack-relations of $\mathcal{F}$ and $\mathcal{G}$. Consider the following example.

5.3. Characterizing Normal Expansion Equivalence

Example 5.31 (Example 5.29 continued). In Figure 5.12 we showed that $\mathcal{F}$ and $\mathcal{G}$ are not expansion equivalent with respect to complete semantics. This property can be shown by conjoining them with $\mathcal{H}' = (\{a,b,c,d\}, \{(b,d),(c,a),(d,c)\}$ which do not add further attacks between the old arguments a and b. Note that $\{a,d\} \in \mathcal{E}_{co}(\mathcal{F} \cup \mathcal{H}')$ and $\{a,d\} \notin \mathcal{E}_{co}(\mathcal{G} \cup \mathcal{H}')$.

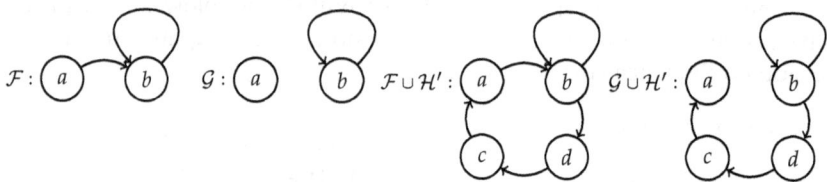

Figure 5.13: Unchanged Attack-relation

5.3.1 Normal Expansion Equivalence for Stable and Stage Semantics

At first we consider stable and stage semantics. In Section 5.2.2 we have already shown that the notion of *stb*-kernel not only characterizes expansion equivalence but also strong expansion equivalence with respect to stable and stage semantics. In consideration that the class of normal expansions lie inbetween (with respect to subset-relation) the classes of arbitrary and strong expansions the following theorems follow immediately.

Theorem 5.32. *For any AFs $\mathcal{F}, \mathcal{G}$,*

$$\mathcal{F}^{k(stb)} = \mathcal{G}^{k(stb)} \Leftrightarrow \mathcal{F} \equiv_N^{stb} \mathcal{G} \Leftrightarrow \mathcal{F} \equiv_N^{stg} \mathcal{G}.$$

Proof. Combine Theorems 5.5, 5.8 and 5.10 and Proposition 2.18. □

5.3.2 Normal Expansion Equivalence for Semi-Stable, Eager, Admissible, Preferred and Ideal Semantics

In [Oikarinen and Woltran, 2011] it was shown that the *ad*-kernel serves as a uniform characterization for expansion equivalence with respect to semi-stable, eager, admissible, preferred and ideal semantics. The following theorem proves that this result carries over to normal expansion equivalence. Remember that we have already proven that a similar uniform behaviour of the five mentioned semantics with respect to strong expansion equivalence is not given (compare Sections 5.2.3, 5.2.4).

Theorem 5.33. *For any AFs $\mathcal{F}, \mathcal{G}$ and $\sigma \in \{ss, eg, ad, pr, id\}$,*

$$\mathcal{F}^{k(ad)} = \mathcal{G}^{k(ad)} \Leftrightarrow \mathcal{F} \equiv_N^{\sigma} \mathcal{G}.$$

Proof. In case of semi-stable and eager semantics the assertion follows by combining Theorems 5.5, 5.7, 5.11 and Proposition 2.18.

Let us turn to admissible, preferred and ideal semantics. Note that Theorems 5.5, 5.7 and Proposition 2.18 imply $\mathcal{F}^{k(ad)} = \mathcal{G}^{k(ad)} \Rightarrow \mathcal{F} \equiv_N^\sigma \mathcal{G}$ for any $\sigma \in \{ad, pr, id\}$. This means, it suffices to show that $\mathcal{F}^{k(ad)} \neq \mathcal{G}^{k(ad)} \Rightarrow \mathcal{F} \not\equiv_N^\sigma \mathcal{G}$. Almost all cases are already proven in Theorem 5.11 (compare remarks at the beginning of the proof). We will prove now the remaining case 2.2.4.

The case **2.2.4** is based on the following assumptions: $A\left(\mathcal{F}^{k(ad)}\right) = A\left(\mathcal{G}^{k(ad)}\right)$ and $R\left(\mathcal{F}^{k(ad)}\right) \neq R\left(\mathcal{G}^{k(ad)}\right)$. Hence, there are two arguments $a, b \in A(\mathcal{F})$, such that $(a, b) \in R\left(\mathcal{F}^{k(ad)}\right) \smallsetminus R\left(\mathcal{G}^{k(ad)}\right)$. Furthermore, we assume $a \neq b$ and b is self-defeating, i.e. $(a, a) \notin R(\mathcal{F})$ and $(b, b) \in R(\mathcal{F})$. Using the following AF $\mathcal{J}$ we will prove that the AFs $\mathcal{F}$ and $\mathcal{G}$ are not normal expansion equivalent with respect to admissible, preferred and ideal semantics. Let c be a fresh argument and $B = A(\mathcal{F}) \smallsetminus \{a, b\}$, then

$$\mathcal{J} = \left(A(\mathcal{F}) \cup \{c\}, \{(c, c') \mid c' \in B\} \cup \{(b, c)\}\right).$$

The following figure illustrates $\mathcal{F} \cup \mathcal{J}$ and $\mathcal{G} \cup \mathcal{J}$. Note that (b, a) may or not be in $R(\mathcal{F})$ or $R(\mathcal{G})$ (indicated by dashed arrows). Furthermore, $(a, b) \notin R(\mathcal{G})$ since $(a, a) \notin R(\mathcal{G})$ and $(a, b) \notin R\left(\mathcal{G}^{k(ad)}\right)$ was assumed. For reasons of clarity we left out possible attacks between the arguments in B and $\{a, b\}$.

Whether (b, a) is in $R(\mathcal{F})$ or not we obtain $\mathcal{E}_{pr}(\mathcal{F} \cup \mathcal{J}) = \{\{a, c\}\}$. If $(b, a) \in R(\mathcal{G})$ we observe that apart from the empty set no other set is admissible in $\mathcal{G} \cup \mathcal{J}$. Hence, $\mathcal{E}_{pr}(\mathcal{G} \cup \mathcal{J}) = \{\emptyset\}$. Consider now $(b, a) \notin R(\mathcal{G})$. The preferred extension of $\mathcal{G} \cup \mathcal{J}$ depends on whether a defends itself in $\mathcal{G}$ (and so in $\mathcal{G} \cup \mathcal{J}$) or not. If so, we have $\mathcal{E}_{pr}(\mathcal{G} \cup \mathcal{J}) = \{\{a\}\}$. If not, we get $\mathcal{E}_{pr}(\mathcal{G} \cup \mathcal{J}) = \{\emptyset\}$. In all cases the preferred extension of $\mathcal{G} \cup \mathcal{J}$ is unique and differs from $\{a, c\}$. Thus, the same holds for admissible (Proposition 2.17, statement 1) and ideal semantics (Proposition 2.10). Hence, $\mathcal{F} \not\equiv_N^\sigma \mathcal{G}$ for $\sigma \in \{ad, pr, id\}$ concluding the proof. □

5.3.3 Normal Expansion Equivalence for Grounded Semantics

Expansion equivalence with respect to the very cautious grounded semantics can be captured by the gr-kernel which "identifies" attacks (a, b) as redundant if both a and b are self-attacking or there is an attack (b, a) and b is self-defeating. The latter condition is a unique feature of the grounded semantics and reflects its subset-minimality among the complete extensions. The following theorem shows that the gr-kernel is even suitable to characterize normal expansion equivalence with respect to grounded semantics.

5.3. Characterizing Normal Expansion Equivalence

Theorem 5.34. *For any AFs $\mathcal{F}$ and $\mathcal{G}$,*

$$\mathcal{F}^{k(gr)} = \mathcal{G}^{k(gr)} \Leftrightarrow \mathcal{F} \equiv_N^{gr} \mathcal{G}.$$

Proof. The if-direction, namely $\mathcal{F}^{k(gr)} = \mathcal{G}^{k(gr)} \Rightarrow \mathcal{F} \equiv_N^{gr} \mathcal{G}$ is a consequence of the Theorems 5.5 and 5.6 as well as Proposition 2.18. Hence, it suffices to show the only-if-direction, i.e. $\mathcal{F} \equiv_N^{gr} \mathcal{G} \Rightarrow \mathcal{F}^{k(gr)} = \mathcal{G}^{k(gr)}$. We will prove the contrapositive.

Suppose $\mathcal{F}^{k(gr)} \neq \mathcal{G}^{k(gr)}$. We skip the consideration of different arguments, i.e. **1st case**: $A(\mathcal{F}^{k(gr)}) \neq A(\mathcal{G}^{k(gr)})$ as well as the occurrence of different self-loops, i.e. case **2.1**: $(a,b) \in R(\mathcal{F}^{k(gr)}) \setminus R(\mathcal{G}^{k(gr)})$ where $a = b$ holds since the proofs of them are exactly the same as in Theorem 5.22. In the following we assume $A(\mathcal{F}^{k(gr)}) = A(\mathcal{G}^{k(gr)})$ and any self-loop is either contained in both $R(\mathcal{F}^{k(gr)})$ and $R(\mathcal{G}^{k(gr)})$ or none of them.

Case 2.2: Consider $(a,b) \in R(\mathcal{F}^{k(gr)}) \setminus R(\mathcal{G}^{k(gr)})$ with $a \neq b$. Note that $(a,b) \in R(\mathcal{F})$ is implied. Consequently, at least one of the following two statements has to hold: $(b,b) \notin R(\mathcal{F})$; $(a,a) \notin R(\mathcal{F})$ and $(b,a) \notin R(\mathcal{F})$. We will use the standard construction $\mathcal{I}$ to prove some cases (c is a fresh argument).

$$\mathcal{I} = (A(\mathcal{F}) \cup \{c\}, \{(c,c') \mid c' \in A(\mathcal{F}) \setminus \{a,b\}\}).$$

Case 2.2.1: Let $(b,b) \notin R(\mathcal{F})$. Consequently, $(b,b) \notin R(\mathcal{G})$ and hence, $(a,b) \notin R(\mathcal{G})$ since $(a,b) \notin R(\mathcal{G}^{k(gr)})$ holds. The following extensions can be obtained by applying splitting results (compare Section 5.2.1). If $(b,a) \in R(\mathcal{F})$ or $(a,a) \in R(\mathcal{F})$, then $\mathcal{E}_{gr}(\mathcal{F} \cup \mathcal{I}) = \{\{c\}\}$ holds. If not, we get $\mathcal{E}_{gr}(\mathcal{F} \cup \mathcal{I}) = \{\{a,c\}\}$. On the other hand, if $(b,a) \in R(\mathcal{G})$ or $(a,a) \in R(\mathcal{G})$ holds, we obtain $\mathcal{E}_{gr}(\mathcal{G} \cup \mathcal{I}) = \{\{b,c\}\}$. If not, we conclude $\mathcal{E}_{gr}(\mathcal{G} \cup \mathcal{I}) = \{\{a,b,c\}\}$. This means, for all possible combinations $\mathcal{E}_{gr}(\mathcal{F} \cup \mathcal{I}) \neq \mathcal{E}_{gr}(\mathcal{G} \cup \mathcal{I})$ follows. **Case 2.2.2**: Let $(b,b) \in R(\mathcal{F})$ and furthermore, $(a,a) \notin R(\mathcal{F})$ and $(b,a) \notin R(\mathcal{F})$. Note that the attacks in $\mathcal{F}$ with respect to the arguments a and b are uniquely determined. The union of $\mathcal{F}$ and $\mathcal{I}$ yields $\{a,c\}$ as the unique grounded extension. The AF $\mathcal{G}$ may occur in three configurations (remember that $(a,a) \notin R(\mathcal{G})$ and $(b,b) \in R(\mathcal{G})$ is already assumed), namely i) $(a,b),(b,a) \in R(\mathcal{G})$, ii) $(a,b) \notin R(\mathcal{G})$, $(b,a) \in R(\mathcal{G})$ and iii) $(a,b),(b,a) \notin R(\mathcal{G})$. In the first two cases $\mathcal{E}_{gr}(\mathcal{G} \cup \mathcal{H}) = \{\{c\}\}$ is implied. The third possibility establishes the grounded extension $\{a,c\}$ too. Hence, we have to find another AF $\mathcal{K}$, such that $\mathcal{E}_{gr}(\mathcal{F} \cup \mathcal{K}) \neq \mathcal{E}_{gr}(\mathcal{G} \cup \mathcal{K})$ is implied. Let c and d be fresh arguments and $B = A(\mathcal{F}) \setminus \{a,b\}$. We define

$$\mathcal{K} = (A(\mathcal{F}) \cup \{c,d\}, \{(b,d)\} \cup \{(c,c') \mid c' \in B\}).$$

The following figure illustrates $\mathcal{F} \cup \mathcal{K}$ and $\mathcal{G} \cup \mathcal{K}$. We left out possible attacks between the arguments in B and $\{a,b\}$ since they can be "ignored" in case of evaluating the AFs with respect to grounded semantics.

Chapter 5. Notions of Equivalence and Replacement

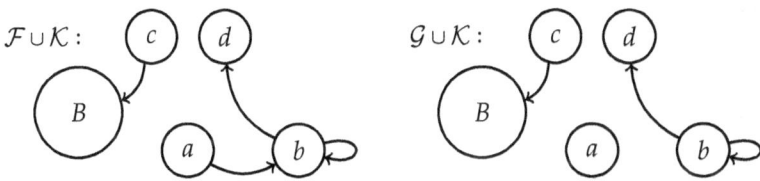

The grounded extensions of $\mathcal{F} \cup \mathcal{K}$ and $\mathcal{G} \cup \mathcal{K}$ differ, namely $\{\{a,c,d\}\} = \mathcal{E}_{gr}(\mathcal{F} \cup \mathcal{K}) \neq \mathcal{E}_{gr}(\mathcal{G} \cup \mathcal{K}) = \{\{a,c\}\}$. This can be seen as follows: The argument c is unattacked in $\mathcal{F} \cup \mathcal{K}$ and $\mathcal{G} \cup \mathcal{K}$. Furthermore, in both AFs a is defended by $\{c\}$ and hence, $\{a,c\}$ has to be a subset of the grounded extension in both AFs. We observe that d has to belong to the grounded extension of $\mathcal{F} \cup \mathcal{K}$ since it is defended by $\{a,c\}$. This does not apply to $\mathcal{G} \cup \mathcal{K}$. Consequently, $\mathcal{F} \not\equiv_N^{gr} \mathcal{G}$ concluding the proof. □

5.3.4 Normal Expansion Equivalence for Complete Semantics

We now turn to complete semantics. Similarly to the other semantics considered in this book we will show that a different semantical behaviour of two AFs with respect to arbitrary expansions and complete semantics is sufficient for not being normal expansion equivalent with respect to complete semantics. This claim is illustrated in Example 5.31 from the beginning of this section and will be a consequence of the following theorem showing that the co-kernel adequately describes normal expansion equivalence with respect to complete semantics.

Theorem 5.35. *For any AFs $\mathcal{F}$ and $\mathcal{G}$,*

$$\mathcal{F}^{k(co)} = \mathcal{G}^{k(co)} \Leftrightarrow \mathcal{F} \equiv_N^{co} \mathcal{G}.$$

Proof. The if-direction, namely $\mathcal{F}^{k(co)} = \mathcal{G}^{k(co)} \Rightarrow \mathcal{F} \equiv_N^{co} \mathcal{G}$ can be obtained by combining Theorem 5.5 and Proposition 2.18. Hence, it suffices to show that $\mathcal{F} \equiv_N^{co} \mathcal{G} \Rightarrow \mathcal{F}^{k(co)} = \mathcal{G}^{k(co)}$ holds. By Theorem 5.6 the latter implication is equivalent to $\mathcal{F} \equiv_N^{co} \mathcal{G} \Rightarrow \mathcal{F}^{k(ad)} = \mathcal{G}^{k(ad)}$ and $\mathcal{F}^{k(gr)} = \mathcal{G}^{k(gr)}$. Using Theorems 5.33 and 5.34 we may replace the kernel-equalities by normal expansion equivalence with respect to preferred or complete semantics. Thus, we obtain the implication we have to prove, $\mathcal{F} \equiv_N^{co} \mathcal{G} \Rightarrow \mathcal{F} \equiv_N^{pr} \mathcal{G}$ and $\mathcal{F} \equiv_N^{gr} \mathcal{G}$. If $\mathcal{F} \not\equiv_N^{pr} \mathcal{G}$ (or $\mathcal{F} \not\equiv_N^{gr} \mathcal{G}$), then there exists an AF $\mathcal{H}$ with the property $\mathcal{F} \leq^N \mathcal{F} \cup \mathcal{H}$ and $\mathcal{G} \leq^N \mathcal{G} \cup \mathcal{H}$, such that $\mathcal{F} \cup \mathcal{H} \not\equiv^{pr} \mathcal{G} \cup \mathcal{H}$ (or $\mathcal{F} \cup \mathcal{H} \not\equiv^{gr} \mathcal{G} \cup \mathcal{H}$). Hence, in both cases $\mathcal{F} \cup \mathcal{H} \not\equiv^{co} \mathcal{G} \cup \mathcal{H}$ by applying the contrapositive of Proposition 2.17, statement 2. Consequently, $\mathcal{F} \not\equiv_N^{co} \mathcal{G}$ is shown concluding the proof. □

5.3.5 Normal Expansion Equivalence for Naive Semantics

Finally, we consider naive semantics. As already stated in Section 5.3.5 normal expansion equivalence with respect to naive semantics can be characterized by the following two conditions: First, possessing the same naive extensions and second, sharing the same arguments.

5.4. Characterizing Local and Weak Expansion Equivalence

Theorem 5.36. *For any AFs $\mathcal{F}$, $\mathcal{G}$:*

$$\mathcal{E}_{na}(\mathcal{F}) = \mathcal{E}_{na}(\mathcal{G}) \text{ and } A(\mathcal{F}) = A(\mathcal{G}) \Leftrightarrow \mathcal{F} \equiv_N^{na} \mathcal{G}.$$

Proof. Combine Theorem 5.13 in [Gaggl and Woltran, 2012] as well as Theorem 5.30 and Proposition 2.18 of this book. □

5.4 Characterizing Local and Weak Expansion Equivalence

In this section we establish characterization theorems for weak expansion equivalence in case of stable and preferred semantics, arguably the most important semantics for Dung frameworks. The characterization will not be purely syntactical as we will see. Furthermore we provide an alternative characterization for local equivalence with respect to stage semantics in terms of a kernel. Remember that stage and stable semantics behave in the same manner with respect to the characterization of expansion, normal expansion and strong expansion equivalence. This means, all mentioned equivalence notions are captured by the *stb*-kernel (compare Theorems 5.5, 5.8, 5.10 and 5.32). In case of local expansion equivalence we observe a certain difference. Consider therefore the following example.

Example 5.37. The AFs $\mathcal{F}$ and $\mathcal{G}$ are local expansion equivalent with respect to stable semantics. This can be seen by checking that $\mathcal{E}_{stb}(\mathcal{F} \cup \mathcal{H}) = \mathcal{E}_{stb}(\mathcal{G} \cup \mathcal{H})$ for any $\mathcal{H}$, such that $A(\mathcal{H}) \subseteq \{a,b,c\}$.[4]

Figure 5.14: *Non-coincidence of Stable and Stage Semantics*

Obviously, $\mathcal{F}$ and $\mathcal{G}$ are not local expansion equivalent with respect to stage semantics because $\{\{b\}\} = \mathcal{E}_{stg}(\mathcal{F}) \neq \mathcal{E}_{stg}(\mathcal{G}) = \{\emptyset\}$. Let us consider now two further AFs $\mathcal{F}'$ and $\mathcal{G}'$ slightly different to $\mathcal{F}$ and $\mathcal{G}$.

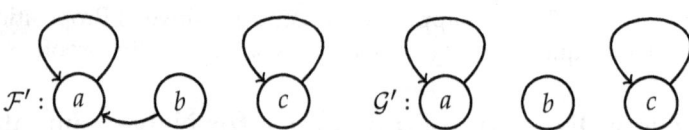

Figure 5.15: *Local Expansion Equivalent AFs*

---

[4]More precisely, local expansion equivalence is due to the fact that $\mathcal{E}_{stb}(\mathcal{F}) = \mathcal{E}_{stb}(\mathcal{G}) = \emptyset$, $A(\mathcal{F}) \setminus A(\mathcal{G}) = \{b\}$, $\{(b,b),(b,c)\} \cap R(\mathcal{F} \cup \mathcal{G}) = \emptyset$ and $(c,c) \in R(\mathcal{F} \cup \mathcal{G})$ (compare [Oikarinen and Woltran, 2011, Theorem 8]).

We state that $\mathcal{F}'$ and $\mathcal{G}'$ are not local expansion equivalent with respect to stable semantics. One possible scenario which makes the predicted different behaviour explicit is the following where $\mathcal{H} = (\{b,c\}, \{(b,c)\})$. Obviously, $\{\{b\}\} = \mathcal{E}_{stb}(\mathcal{F}' \cup \mathcal{H}) \neq \mathcal{E}_{stb}(\mathcal{G}' \cup \mathcal{H}) = \emptyset$. It is one main result of this section showing that AFs like $\mathcal{F}'$ and $\mathcal{G}'$ are local expansion equivalent with respect to stage semantics because they possess identical *stage-*-kernels*.[5]

5.4.1 Weak Expansion Equivalence for Stable Semantics

Weak expansions are normal expansions which add weak arguments only, i.e. the added arguments never attack the previous ones and furthermore the former attack relation remain unchanged. We already mentioned that weak expansions naturally occur in case of Value Based AFs [Bench-Capon, 2003]. Former arguments may be arguments which advance higher values than the further arguments. Consequently, the new arguments cannot attack the former. The following theorem characterizes weak expansion equivalence for stable semantics.

Theorem 5.38. *For any AFs $\mathcal{F}$ and $\mathcal{G}$ we have:*

$\mathcal{F} \equiv_W^{stb} \mathcal{G} \Leftrightarrow$ *i)* $A(\mathcal{F}) = A(\mathcal{G})$ *and* $\mathcal{E}_{stb}(\mathcal{F}) = \mathcal{E}_{stb}(\mathcal{G})$ *or ii)* $\mathcal{E}_{stb}(\mathcal{F}) = \mathcal{E}_{stb}(\mathcal{G}) = \emptyset$.

Proof. ($\Leftarrow$) Given an AF $\mathcal{H}$ such that $\mathcal{F} \leq_W \mathcal{F} \cup \mathcal{H}$ and $\mathcal{G} \leq_W \mathcal{G} \cup \mathcal{H}$ holds. We show that $E \in \mathcal{E}_{stb}(\mathcal{F} \cup \mathcal{H})$ implies $E \in \mathcal{E}_{stb}(\mathcal{G} \cup \mathcal{H})$ (and vice versa). 1st case: Let us assume $A(\mathcal{F}) = A(\mathcal{G})$ and $\mathcal{E}_{stb}(\mathcal{F}) = \mathcal{E}_{stb}(\mathcal{G})$. Consequently, if $\mathcal{F} = \mathcal{F} \cup \mathcal{H}$, then $\mathcal{G} = \mathcal{G} \cup \mathcal{H}$ is implied because $\mathcal{F}$ and $\mathcal{G}$ share the same arguments. Thus, $\mathcal{E}_{stb}(\mathcal{F} \cup \mathcal{H}) = \mathcal{E}_{stb}(\mathcal{G} \cup \mathcal{H})$ has to hold. From now on we may assume that $\mathcal{F} \cup \mathcal{H}$ and $\mathcal{G} \cup \mathcal{H}$ are indeed weak expansions of $\mathcal{F}$ or $\mathcal{G}$. In consideration of Proposition 4.2 we are able to apply splitting results. This means, $(\mathcal{F}, A_2, R_3)$ and $(\mathcal{G}, A_2, R_3)$ are splittings of $\mathcal{F} \cup \mathcal{H}$ or $\mathcal{G} \cup \mathcal{H}$, whereas $A_2 = (A(\mathcal{H}) \setminus A(\mathcal{F}), R(\mathcal{H}) \cap (A(\mathcal{H}) \setminus A(\mathcal{F}) \times A(\mathcal{H}) \setminus A(\mathcal{F})))$ and $R_3 = R(\mathcal{H}) \cap (A(\mathcal{F}) \times A(\mathcal{H}) \setminus A(\mathcal{F}))$. Assume $E \in \mathcal{E}_{stb}(\mathcal{F} \cup \mathcal{H})$. Thus, statement 2 of the splitting theorem 4.10 implies 1. $E \cap A(\mathcal{F}) \in \mathcal{E}_{stb}(\mathcal{F})$ and 2. $E \cap (A(\mathcal{H}) \setminus A(\mathcal{F})) \in \mathcal{E}_{stb}\left(mod_{U_{E \cap A(\mathcal{F})}, R_3}\left(A_2^{E \cap A(\mathcal{F}), R_3}\right)\right)$. Thus, $E \cap A(\mathcal{G}) \in \mathcal{E}_{stb}(\mathcal{G})$ since $A(\mathcal{F}) = A(\mathcal{G})$ and $\mathcal{E}_{stb}(\mathcal{F}) = \mathcal{E}_{stb}(\mathcal{G})$ was assumed. Furthermore, applying Proposition 4.9, statement 1 we deduce $E \cap (A(\mathcal{H}) \setminus A(\mathcal{F})) \in \mathcal{E}_{stb}\left(A_2^{E \cap A(\mathcal{F}), R_3}\right)$ since $E \cap A(\mathcal{F})$ is a stable extension of $\mathcal{F}$. Moreover, using that $A(\mathcal{F})$ equals $A(\mathcal{G})$ we derive $E \cap (A(\mathcal{H}) \setminus A(\mathcal{G})) \in \mathcal{E}_{stb}\left(A_2^{E \cap A(\mathcal{G}), R_3}\right)$. Again, by statement 1, Proposition 4.9 it follows $E \cap (A(\mathcal{H}) \setminus A(\mathcal{G})) \in \mathcal{E}_{stb}\left(mod_{U_{E \cap A(\mathcal{G})}, R_3}\left(A_2^{E \cap A(\mathcal{G}), R_3}\right)\right)$ because $E \cap A(\mathcal{G}) \in \mathcal{E}_{stb}(\mathcal{G})$ is already shown. Finally, statement 1, splitting theorem 4.10 justifies $(E \cap A(\mathcal{G})) \cup (E \cap (A(\mathcal{H}) \setminus A(\mathcal{G}))) = E \in \mathcal{E}_{stb}(\mathcal{F} \cup \mathcal{H})$. We omit the other direction, i.e. $E \in \mathcal{E}_{stb}(\mathcal{G} \cup \mathcal{H})$ implies $E \in \mathcal{E}_{stb}(\mathcal{F} \cup \mathcal{H})$, since it can be shown in a similar way.

---

[5]Note that a characterization theorem for local expansion equivalence with respect to stage semantics is already given in [Gaggl and Woltran, 2011, Theorem 6]. In this book we will present a kernel-based characterization. We want to mention that both results were found independently.

5.4. Characterizing Local and Weak Expansion Equivalence

2nd case: Suppose $\mathcal{E}_{stb}(\mathcal{F}) = \mathcal{E}_{stb}(\mathcal{G}) = \emptyset$. Thus, $\mathcal{E}_{stb}(\mathcal{F} \cup \mathcal{H}) = \mathcal{E}_{stb}(\mathcal{G} \cup \mathcal{H}) = \emptyset$ because assuming $E \in \mathcal{E}_{stb}(\mathcal{F} \cup \mathcal{H})$ yields to $E \cap A(\mathcal{F}) \in \mathcal{E}_{stb}(\mathcal{F})$ (splitting theorem 4.10, statement 1) contradicting $\mathcal{E}_{st}(\mathcal{F}) = \emptyset$. Hence, $E \in \mathcal{E}_{stb}(\mathcal{F} \cup \mathcal{H})$ implies $E \in \mathcal{E}_{stb}(\mathcal{G} \cup \mathcal{H})$. The converse direction can be shown in a similar way.

($\Rightarrow$) We show the contrapositive, i.e. $(\mathcal{E}_{stb}(\mathcal{F}) \neq \mathcal{E}_{stb}(\mathcal{G}) \vee A(\mathcal{F}) \neq A(\mathcal{G})) \wedge (\mathcal{E}_{stb}(\mathcal{F}) \neq \mathcal{E}_{stb}(\mathcal{G}) \vee \mathcal{E}_{stb}(\mathcal{G}) \neq \emptyset)$ implies the existence of an AF $\mathcal{H}$, such that $\mathcal{F} \leq_W \mathcal{F} \cup \mathcal{H}, \mathcal{G} \leq_W \mathcal{G} \cup \mathcal{H}$ and $\mathcal{E}_{stb}(\mathcal{F} \cup \mathcal{H}) \neq \mathcal{E}_{stb}(\mathcal{G} \cup \mathcal{H})$. First, assume $\mathcal{E}_{stb}(\mathcal{F}) \neq \mathcal{E}_{stb}(\mathcal{G})$. Without loss of generality we assume the existence of a set E, such that $E \in \mathcal{E}_{stb}(\mathcal{F}) \wedge E \notin \mathcal{E}_{stb}(\mathcal{G})$. Let a be a fresh argument and consider $\mathcal{H} = (\{a\}, \emptyset)$. We immediately derive that $E \cup \{a\}$ is a stable extension of $\mathcal{F} \cup \mathcal{H}$. Furthermore it is impossible that $E \cup \{a\} \in \mathcal{E}_{stb}(\mathcal{G} \cup \mathcal{H})$ since $E \notin \mathcal{E}_{stb}(\mathcal{G})$ was assumed (statement 2, splitting theorem 4.10). Second, let $\mathcal{E}_{stb}(\mathcal{F}) = \mathcal{E}_{stb}(\mathcal{G}) \wedge \mathcal{E}_{stb}(\mathcal{G}) \neq \emptyset \wedge A(\mathcal{F}) \neq A(\mathcal{G})$. Without loss of generality let $a \in A(\mathcal{F}) \setminus A(\mathcal{G})$. Observe that for all stable extensions E of $\mathcal{F}$ and $\mathcal{G}$, $a \notin E$ since $\mathcal{E}_{stb}(\mathcal{G}) \neq \emptyset$ was supposed. Assume $E \in \mathcal{E}_{st}(\mathcal{G})$ and define $\mathcal{H} = (\{a\}, \emptyset)$. Hence, $E' = E \cup \{a\} \in \mathcal{E}_{stb}(\mathcal{G} \cup \mathcal{H})$ (statement 1, splitting theorem 4.10). On the other hand, $E' \notin \mathcal{E}_{stb}(\mathcal{F} \cup \mathcal{H})$ because of $\mathcal{F} = \mathcal{F} \cup \mathcal{H}$ and $a \in E'$ concluding the proof. $\square$

We close this section with an example.

Example 5.39. Consider the following AFs $\mathcal{F}$ and $\mathcal{G}$.

Figure 5.16: Weak Expansion Equivalent AFs

Obviously, $A(\mathcal{F}) = A(\mathcal{G})$ and $\mathcal{E}_{stb}(\mathcal{F}) = \mathcal{E}_{stb}(\mathcal{G}) = \{\{a_1\}\}$. Thus, applying the characterization theorem 5.38 we obtain $\mathcal{F} \equiv_W^{stb} \mathcal{G}$.

5.4.2 Weak Expansion Equivalence for Preferred Semantics

How does the characterization look if we turn to the more relaxed notion of preferred semantics? It turns out that the characterization is very similar but not identical to stable semantics, namely: two AFs are weak expansion equivalent with respect to preferred semantics if and only if they share the same arguments, possess the same preferred extensions and furthermore, for any extension E the set of arguments which are not in the extension without being refuted has to coincide in both AFs.

Theorem 5.40. *For any two AFs $\mathcal{F}, \mathcal{G}$ we have: $\mathcal{F} \equiv_W^{pr} \mathcal{G}$ iff $A(F) = A(G)$, $\mathcal{E}_{pr}(\mathcal{F}) = \mathcal{E}_{pr}(\mathcal{G})$ and for each $E \in \mathcal{E}_{pr}(\mathcal{F}) : U_E^{\mathcal{F}} = U_E^{\mathcal{G}}$ where $U_E^{\mathcal{A}} = \{a \in A(\mathcal{A}) \mid a \notin E \wedge (E, a) \notin R(\mathcal{A})\}$.*

Proof. ($\Leftarrow$) Given an AF $\mathcal{H}$, such that $\mathcal{F} \leq_W \mathcal{F} \cup \mathcal{H}$ and $\mathcal{G} \leq_W \mathcal{G} \cup \mathcal{H}$. We have to show that $\mathcal{E}_{pr}(\mathcal{F} \cup \mathcal{H}) = \mathcal{E}_{pr}(\mathcal{G} \cup \mathcal{H})$. If $\mathcal{F} = \mathcal{F} \cup \mathcal{H}$, then $\mathcal{G} = \mathcal{G} \cup \mathcal{H}$ since $A(\mathcal{F}) = A(\mathcal{G})$ is assumed. In consideration of $\mathcal{E}_{pr}(\mathcal{F}) = \mathcal{E}_{pr}(\mathcal{G})$ the assertion follows. Assume now that $\mathcal{F} \neq \mathcal{F} \cup \mathcal{H}$. Using splitting theorem 4.10 one may easily show that $E \in \mathcal{E}_{pr}(\mathcal{F} \cup \mathcal{H})$ implies $E \in \mathcal{E}_{pr}(\mathcal{G} \cup \mathcal{H})$ and vice versa (compare Theorem 5.38).

($\Rightarrow$) We will show the contrapositive, i.e. if $A(F) \neq A(G)$ or $\mathcal{E}_{pr}(\mathcal{F}) \neq \mathcal{E}_{pr}(\mathcal{G})$ or there exists an $E \in \mathcal{E}_{pr}(\mathcal{F})$, such that $U_E^{\mathcal{F}} \neq U_E^{\mathcal{G}}$, then $\mathcal{F} \not\equiv_W^{pr} \mathcal{G}$. Consider $E \in \mathcal{E}_{pr}(\mathcal{F})$ and $E \notin \mathcal{E}_{pr}(\mathcal{G})$. Consequently, $E \cup \{d\} \in \mathcal{E}_{pr}(\mathcal{F} \cup \mathcal{H})$ and $E \cup \{d\} \notin \mathcal{E}_{pr}(\mathcal{G} \cup \mathcal{H})$ where $\mathcal{H} = (\{d\}, \emptyset)$ and d is a fresh argument, i.e. $d \notin A(\mathcal{F}) \cup A(\mathcal{G})$. Assume now $A(F) \neq A(G)$ and $\mathcal{E}_{pr}(\mathcal{F}) = \mathcal{E}_{pr}(\mathcal{G})$. W.l.o.g. let $a \in A(F) \setminus A(G)$. Consequently, there is no preferred extension E, such that $a \in E$. If $\mathcal{H} = (\{a\}, \emptyset)$, then $\mathcal{F} \cup \mathcal{H} = \mathcal{F}$ and thus, there is no $E \in \mathcal{E}_{pr}(\mathcal{F} \cup \mathcal{H})$, such that $a \in E$. On the other hand, since a is unattacked in $\mathcal{G} \cup \mathcal{H}$ we deduce that a is contained in the grounded extension of $\mathcal{G} \cup \mathcal{H}$. Thus, $\mathcal{E}_{pr}(\mathcal{F} \cup \mathcal{H}) \neq \mathcal{E}_{pr}(\mathcal{G} \cup \mathcal{H})$ is shown. Finally, we consider $A(F) = A(G)$ and $\mathcal{E}_{pr}(\mathcal{F}) = \mathcal{E}_{pr}(\mathcal{G})$ but there exists an $E \in \mathcal{E}_{pr}(\mathcal{F})$, such that $U_E^{\mathcal{F}} \neq U_E^{\mathcal{G}}$. W.l.o.g. let $a \in U_E^{\mathcal{G}} \setminus U_E^{\mathcal{F}}$. This means, $a \notin E$, $(E, a) \notin R(\mathcal{G})$ and $(E, a) \in R(\mathcal{F})$, i.e. a is attacked by E in $\mathcal{F}$. Consider now $\mathcal{H} = (\{a, b\}, \{(a, b)\})$ where b is a fresh argument. One can easily see that $E \cup \{b\} \in \mathcal{E}_{pr}(\mathcal{F} \cup \mathcal{H})$ (b is defended by E) but $E \cup \{b\} \notin \mathcal{E}_{pr}(\mathcal{G} \cup \mathcal{H})$ (b is not defended by E). Altogether, $\mathcal{F} \not\equiv_W^{pr} \mathcal{G}$ is shown. □

5.4.3 Local Expansion Equivalence for Stage Semantics

We already know that expansion equivalence implies local expansion equivalence for any semantics σ (compare Proposition 2.18). Furthermore, there are certain semantics where local expansion equivalence is even sufficient for expansion equivalence. This is, for example, the case if we consider semi-stable semantics [Oikarinen and Woltran, 2011, Theorem 8]. In case of stage semantics an analogous result does not hold, i.e., for this semantics, local expansion equivalence is a properly weaker concept than expansion equivalence. We will see that the characterization of local expansion equivalence with respect to stage semantics can be done by comparing a newly introduced *stage-*-kernel* which is defined as follows.

Definition 5.41. Given a an AF $\mathcal{F} = (A, R)$. We define the *stage-*-kernel* of $\mathcal{F}$ as $\mathcal{F}^{k^*(stg)} = (A, R^{k^*(stg)})$ where

$$R^{k^*(stg)} = R \setminus \{(a, b) \mid a \neq b, (a, a) \in R \vee \forall c \, (c \neq a \rightarrow (c, c) \in R)\}.$$

The first disjunct allows the deletion of an attack (a, b) if a is self-attacking. These attacks are even redundant with respect to arbitrary expansions (compare [Gaggl and Woltran, 2012, Theorem 5.11]). The second disjunct reflects the intuition that an attack (a, b) becomes irrelevant with respect to local expansions and the evaluation given by stage semantics if all arguments different from a are self-attacking. Given the latter scenario we observe that only two sets may be a stage extension, namely $\emptyset$ or $\{a\}$. In particular, the conflict-freeness of $\{a\}$ is sufficient for being the unique stage extension.

5.4. Characterizing Local and Weak Expansion Equivalence

At first we will prove two technical lemmata paving the way for the main theorem showing that the syntactical equivalence of stage-*-kernels characterizes local expansion equivalence between two AFs with respect to stage semantics.

Lemma 5.42. *If* $\mathcal{F}^{k^*(stg)} = \mathcal{G}^{k^*(stg)}$, *then* $(\mathcal{F} \cup \mathcal{H})^{k^*(stg)} = (\mathcal{G} \cup \mathcal{H})^{k^*(stg)}$ *for any AF* $\mathcal{H}$, *such that* $A(\mathcal{H}) \subseteq A(\mathcal{F} \cup \mathcal{G})$.

Proof. First notice that the assumption $\mathcal{F}^{k^*(stg)} = \mathcal{G}^{k^*(stg)}$ implies that $R(\mathcal{F})$ and $R(\mathcal{G})$ contain the same self-loops. Furthermore, $A(\mathcal{F}) = A(\mathcal{G})$ and thus, $A(\mathcal{F}) = A\left((\mathcal{F} \cup \mathcal{H})^{k^*(stg)}\right) = A\left((\mathcal{G} \cup \mathcal{H})^{k^*(stg)}\right)$. Hence, it suffices to show that $R\left((\mathcal{F} \cup \mathcal{H})^{k^*(stg)}\right) = R\left((\mathcal{G} \cup \mathcal{H})^{k^*(stg)}\right)$. Assume $(a,b) \in R\left((\mathcal{F} \cup \mathcal{H})^{k^*(stg)}\right) \setminus R\left((\mathcal{G} \cup \mathcal{H})^{k^*(ad)}\right)$. It suffices to consider $a \neq b$ since the sharing of the same self-loops of $(\mathcal{F} \cup \mathcal{H})^{k^*(stg)}$ and $(\mathcal{G} \cup \mathcal{H})^{k^*(stg)}$ is implied. We deduce $(a,b) \in R(\mathcal{F} \cup \mathcal{H})$ and furthermore, $(a,a) \notin R(\mathcal{F}), R(\mathcal{H})$ and thus, $(a,a) \notin R(\mathcal{G})$. Additionally, there exists an argument c, such that $c \neq a$ and $(c,c) \notin R(\mathcal{F} \cup \mathcal{H})$. Thus, $(c,c) \notin R(\mathcal{F}), R(\mathcal{H})$ and consequently, $(c,c) \notin R(\mathcal{G})$. We have to consider two cases. If $(a,b) \in R(\mathcal{H})$, then $(a,b) \in R(\mathcal{G} \cup \mathcal{H})$. If $(a,b) \in R(\mathcal{F})$, then $(a,b) \in R(\mathcal{G})$ because $\mathcal{F}^{k^*(stg)} = \mathcal{G}^{k^*(stg)}$ is assumed and $(a,b) \in R\left(\mathcal{F}^{k^*(stg)}\right)$ can be derived since neither (a,a) nor (c,c) are included in $R(\mathcal{F})$. In both cases, $(a,b) \in R(\mathcal{G} \cup \mathcal{H})$. Furthermore, $(a,b) \in R\left((\mathcal{G} \cup \mathcal{H})^{k^*(stg)}\right)$ (in contrast to the assumption) since neither (a,a) nor (c,c) are included in $R(\mathcal{G} \cup \mathcal{H})$. □

We have shown that the notion of an stage-*-kernel is robust with respect to local expansions. The next technical lemma shows that evaluation given by stage semantics is insensitive with respect to transitions to the associated stage-*-kernel.

Lemma 5.43. *For any AF* $\mathcal{F}$, $\mathcal{F} \equiv^{stg} \mathcal{F}^{k^*(stg)}$.

Proof. First we show that $\mathcal{F}$ and $\mathcal{F}^{k^*(stg)}$ contain the same conflict-free sets, i.e. $S \in cf(\mathcal{F})$ iff $S \in cf\left(\mathcal{F}^{k^*(stg)}\right)$. The if-direction is given by $R\left(\mathcal{F}^{k^*(stg)}\right) \subseteq R(\mathcal{F})$. It suffices to show that if $S \in cf\left(\mathcal{F}^{k^*(stg)}\right)$, then $S \in cf(\mathcal{F})$. Assume not, i.e. there are two arguments $a,b \in S$, such that $(a,b) \in R(\mathcal{F}) \setminus R\left(\mathcal{F}^{k^*(ad)}\right)$. without loss of generality $a \neq b$. Consequently, $(a,a) \in R(\mathcal{F})$ or at least $(b,b) \in R(\mathcal{F})$ has to hold. This contradicts the conflict-freeness of S in $\mathcal{F}^{k^*(stg)}$ because $\mathcal{F}^{k^*(stg)}$ and $\mathcal{F}$ share the same self-loops.

We show now that $\mathcal{E}_{stg}(\mathcal{F}) = \mathcal{E}_{stg}\left(\mathcal{F}^{k^*(stg)}\right)$. Assume $E \in \mathcal{E}_{stg}(\mathcal{F}) \setminus \mathcal{E}_{stg}\left(\mathcal{F}^{k^*(stg)}\right)$. Hence, $E \in cf(\mathcal{F})$ and thus, $E \in cf\left(\mathcal{F}^{k^*(stg)}\right)$. Furthermore, there exists a conflict-free set E', such that $R^+_{\mathcal{F}^{k^*(stg)}}(E) \subset R^+_{\mathcal{F}^{k^*(stg)}}(E')$. We deduce $E' \in cf(\mathcal{F})$ and $E' \not\subseteq E$. Since $R\left(\mathcal{F}^{k^*(stg)}\right) \subseteq R(\mathcal{F})$ and $E \in \mathcal{E}_{stg}(\mathcal{F}) \setminus \mathcal{E}_{stg}\left(\mathcal{F}^{k^*(stg)}\right)$ we deduce the existence of at least two arguments a and b, such that $a \in E$, $b \notin E$ and $(a,b) \in R(\mathcal{F}) \setminus R\left(\mathcal{F}^{k^*(stg)}\right)$ because E cannot maintain the same range in $\mathcal{F}^{k^*(stg)}$. Observe that $(a,a) \in R(\mathcal{F})$ yields a contradiction since $E \in cf(\mathcal{F})$ is already deduced. On the other hand, $(c,c) \in R(\mathcal{F})$ for any $c \neq a$ implies that $E = \{a\}$ and $E' \subseteq E$ in contradiction to $E' \not\subseteq E$.

Assume now $E \in \mathcal{E}_{stg}\left(\mathcal{F}^{k^*(stg)}\right) \setminus \mathcal{E}_{stg}(\mathcal{F})$. Thus, E is conflict-free in $\mathcal{F}^{k^*(stg)}$ and $\mathcal{F}$. We deduce the existence of a conflict-free set E', such that $R_\mathcal{F}^+(E) \subset R_\mathcal{F}^+(E')$. Hence, $E' \in cf\left(\mathcal{F}^{k^*(stg)}\right)$ and furthermore, $E \not\subseteq E'$ because $E \in \mathcal{E}_{stg}\left(\mathcal{F}^{k^*(stg)}\right)$ is assumed. Since $R\left(\mathcal{F}^{k^*(stg)}\right) \subseteq R(\mathcal{F})$ and $E \in \mathcal{E}_{stg}\left(\mathcal{F}^{k^*(stg)}\right) \setminus \mathcal{E}_{stg}(\mathcal{F})$ we deduce the existence of at least two arguments a and b, such that $a \in E'$, $b \notin E'$ and $(a,b) \in R(\mathcal{F}) \setminus R\left(\mathcal{F}^{k^*(stg)}\right)$ since E' has to reduce its range in $\mathcal{F}^{k^*(stg)}$. Again, $(a,a) \in R(\mathcal{F})$ yields a contradiction because $E' \in cf(\mathcal{F})$ is already shown. Furthermore, if $(c,c) \in R(\mathcal{F})$ for any $c \neq a$ we deduce $E' = \{a\}$ and $E \subseteq E'$ in contrast to $E \not\subseteq E'$. □

We are now prepared to prove the main theorem of this section.

Theorem 5.44. *For any AFs $\mathcal{F}$ and $\mathcal{G}$,*

$$\mathcal{F}^{k^*(stg)} = \mathcal{G}^{k^*(stg)} \Leftrightarrow \mathcal{F} \equiv_L^{stg} \mathcal{G}.$$

Proof. Let $\mathcal{F}^{k^*(stg)} = \mathcal{G}^{k^*(stg)}$ and given an AF $\mathcal{H}$, such that $A(\mathcal{H}) \subseteq A(\mathcal{F} \cup \mathcal{G})$. It suffices to show that $E \in \mathcal{E}_{stg}(\mathcal{F} \cup \mathcal{H})$ if, and only if $E \in \mathcal{E}_{stg}(\mathcal{G} \cup \mathcal{H})$. Suppose $E \in \mathcal{E}_{stg}(\mathcal{F} \cup \mathcal{H})$. By Lemma 5.43, $E \in \mathcal{E}_{stg}\left((\mathcal{F} \cup \mathcal{H})^{k^*(stg)}\right)$ and applying Lemma 5.42, $E \in \mathcal{E}_{stg}\left((\mathcal{G} \cup \mathcal{H})^{k^*(stg)}\right)$. Finally, using Lemma 5.43, we derive $E \in \mathcal{E}_{stg}(\mathcal{G} \cup \mathcal{H})$ concluding this case. Showing that $E \in \mathcal{E}_{stg}(\mathcal{G} \cup \mathcal{H})$ implies $E \in \mathcal{E}_{stg}(\mathcal{F} \cup \mathcal{H})$ can be done in a similar way. Consequently, $\mathcal{F} \equiv_L^{stg} \mathcal{G}$ is shown.

Assume now $\mathcal{F}^{k^*(stg)} \neq \mathcal{G}^{k^*(stg)}$. We will show that $\mathcal{F} \not\equiv_L^{stg} \mathcal{G}$ is implied. Without loss of generality we may assume $A(\mathcal{F}) = A(\mathcal{G})$ and $(a,a) \in R(\mathcal{F}) \Leftrightarrow (a,a) \in R(\mathcal{G})$ (compare [Gaggl and Woltran, 2012, Lemmata 5.3, 5.4]). Consider $a \neq b$ and $(a,b) \in R\left(\mathcal{F}^{k^*(stg)}\right) \setminus R\left(\mathcal{G}^{k^*(stg)}\right)$. It follows that $(a,b) \in R(\mathcal{F})$, $(a,a) \notin R(\mathcal{F})$ and consequently, $(a,a) \notin R(\mathcal{G})$. Now we have to distinguish two cases with respect to the presence or absence of the self-loop (b,b).

1st **case:** Assume $(b,b) \notin R(\mathcal{F})$. Thus, $(b,b) \notin R(\mathcal{G})$ and consequently, $(a,b) \notin R(\mathcal{G})$. Note the attack (b,a) may or may not be in $R(\mathcal{F})$ or $R(\mathcal{G})$. We define

$$\mathcal{K} = (A(\mathcal{F}), \{(a,c),(b,c),(c,c) \mid c \in A(\mathcal{F}) \setminus \{a,b\}\}).$$

In any case, $\{a\} \in \mathcal{E}_{st}(\mathcal{J} \cup \mathcal{K})$ and consequently, $\{a\} \in \mathcal{E}_{stg}(\mathcal{F} \cup \mathcal{K})$. On the other hand, we state $\{a\} \notin \mathcal{E}_{stg}(\mathcal{G} \cup \mathcal{K})$ because $\{\{a,b\}\} = \mathcal{E}_{stg}(\mathcal{G} \cup \mathcal{I})$ if $(b,a) \notin R(\mathcal{G})$ and $\{\{b\}\} = \mathcal{E}_{stg}(\mathcal{G} \cup \mathcal{K})$ if $(b,a) \in R(\mathcal{G})$.

2nd **case:** Consider $(b,b) \in R(\mathcal{F})$. Thus, $(b,b) \in R(\mathcal{G})$ and furthermore, there exists an argument c, such that $(c,c) \notin R(\mathcal{F})$. Consequently, $(c,c) \notin R(\mathcal{G})$ and therefore, $(a,b) \notin R(\mathcal{G})$. Unfortunately, there are $2^5 = 32$ possibilities with respect to the presence or absence of $(b,a),(b,c),(c,b),(a,c)$ and (c,a) in $R(\mathcal{F})$ and $R(\mathcal{G})$. Consequently, there are $2^{10} = 1024$ combination possibilities

5.4. Characterizing Local and Weak Expansion Equivalence

of $\mathcal{F}$ and $\mathcal{G}$ with respect to to the aforementioned attacks. Fortunately, we do not have to consider every single possibility because we have already shown that if two AFs possess the same stg-*-kernel, then local expansion equivalence of them is implied. Since $(b,b) \in R(\mathcal{F})$ and consequently, $(b,b) \in R(\mathcal{G})$ is assumed we may omit the consideration of (b,a) and (b,c) in $R(\mathcal{F})$ and $R(\mathcal{G})$. Thus, for both AFs $2^3 = 8$ possibilities with respect to the presence or absence of $(c,b), (a,c)$ and (c,a) remain. For clarity, we will present all possibilities.

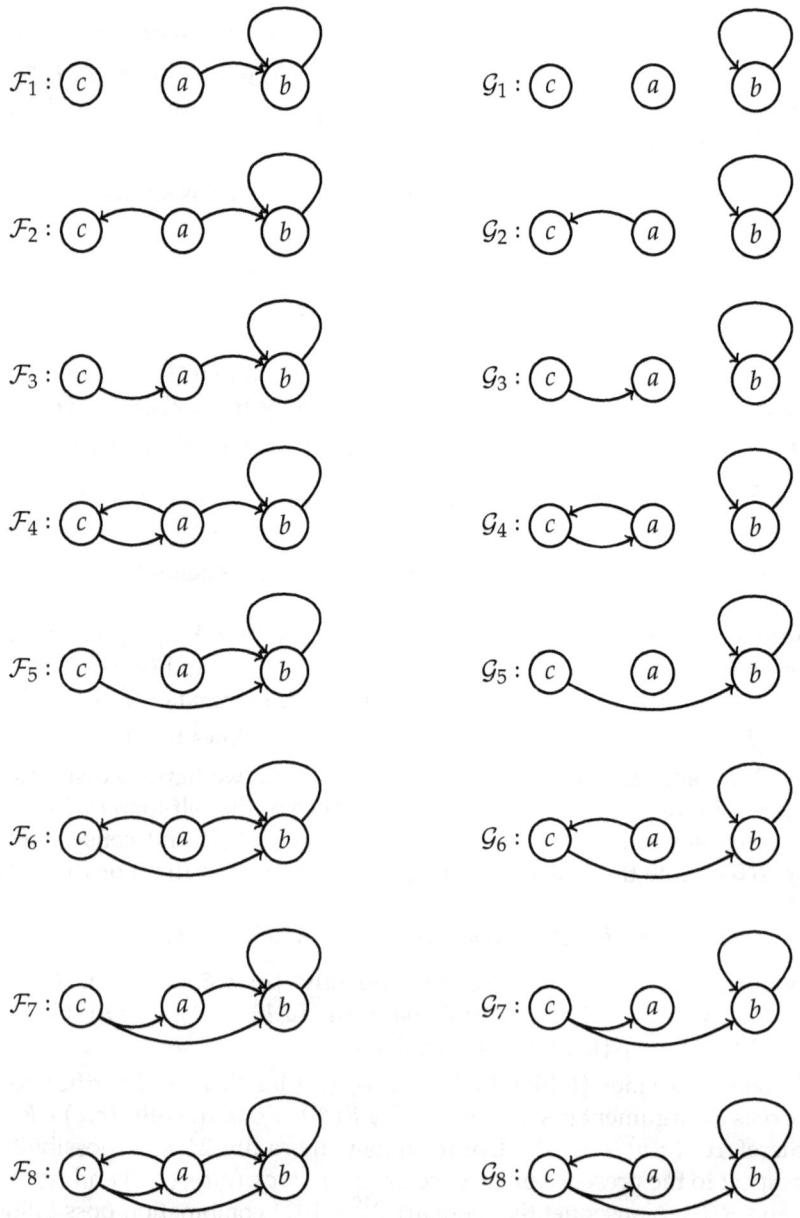

Chapter 5. Notions of Equivalence and Replacement

In order to prove that any pair of AFs $(\mathcal{F}_i, \mathcal{G}_j)$ for $1 \leq i,j \leq 8$ can be distinguished by a local expansion we present the following table. An empty cell $(\mathcal{F}_i, \mathcal{G}_j)$ in the table means that $\mathcal{E}_{stg}(\mathcal{F}_i \cup \mathcal{L}) \neq \mathcal{E}_{stg}(\mathcal{G}_j \cup \mathcal{L})$ where

$$\mathcal{L} = (A(\mathcal{F}), \{(a,d), (c,d), (d,d) \mid d \in A(\mathcal{F}) \setminus \{a,b,c\}\}).$$

Furthermore, an entry "(a,b)" means that $\mathcal{E}_{stg}(\mathcal{F}_i \cup \mathcal{L}') \neq \mathcal{E}_{stg}(\mathcal{G}_j \cup \mathcal{L}')$ is fulfilled if we consider $\mathcal{L}' = \mathcal{L} \cup (\{a,b\}, \{(a,b)\})$. For instance, $\mathcal{E}_{stg}(\mathcal{F}_8 \cup \mathcal{L}') = \{\{a\}, \{c\}\} \neq \{\{a\}\} = \mathcal{E}_{stg}(\mathcal{G}_4 \cup \mathcal{L}')$

	$\mathcal{G}_1$	$\mathcal{G}_2$	$\mathcal{G}_3$	$\mathcal{G}_4$	$\mathcal{G}_5$	$\mathcal{G}_6$	$\mathcal{G}_7$	$\mathcal{G}_8$
$\mathcal{F}_1$	(c,a)				(c,a)			
$\mathcal{F}_2$		(a,c) (c,a)		(a,c) (c,a)		(a,c) (c,a)		
$\mathcal{F}_3$		(a,c) (c,a)		(a,c) (c,a)		(a,c) (c,a)		
$\mathcal{F}_4$		(a,c) (c,a)		(a,c) (c,a)		(a,c) (c,a)		
$\mathcal{F}_5$	(a,c) (c,b)				(a,c) (c,b)			
$\mathcal{F}_6$		(c,b)						
$\mathcal{F}_7$			(a,b)				(a,c)	(a,c)
$\mathcal{F}_8$			(a,b)			(a,b)		

Finally, we have shown that $\mathcal{F} \not\equiv_L^{stg} \mathcal{G}$ concluding the proof. □

We want to conclude this section by providing an example showing that equivalence classes with respect to local expansion equivalence and stage semantics may be very huge sets in the presence of self-loops.

Example 5.45. The following figure represents 64 different AFs (any combination of dashed arrows is suitable). All of them are local expansion equivalent with respect to stage semantics.

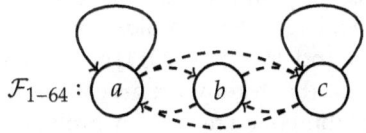

Figure 5.17: A Huge Equivalence Class

5.5 Summary of Results and Implications

5.5.1 Overview: "Strength" of Kernels

In our results the notions of a kernel played a crucial role. Indeed, kernels are interesting from several perspectives: First, they allow to decide the corresponding notion of equivalence by a simple check for topological (i.e. syntactical) equality. Moreover, all kernels we have obtained so far can be efficiently constructed from a given argumentation framework.

In this subsection we want to provide a quick overview of the considered kernels as well as their potential with respect to characterizing equivalence notions. In the following we recall the resulting attack-relation of the σ-kernel or σ-*-kernel of an AF $\mathcal{F} = (A, R)$. Remember that the considered kernels do not change the initial set of arguments, i.e. $\mathcal{F}^{k(\sigma)} = (A, R^{k(\sigma)})$ or $\mathcal{F}^{k^*(\sigma)} = (A, R^{k^*(\sigma)})$, respectively.

1. $R^{k(stb)} = R \setminus \{(a,b) \mid a \neq b, (a,a) \in R\}$,

2. $R^{k^*(stg)} = R \setminus \{(a,b) \mid a \neq b, (a,a) \in R \vee \forall c\, (c \neq a \rightarrow (c,c) \in R)\}$.

3. $R^{k(ad)} = R \setminus \{(a,b) \mid a \neq b, (a,a) \in R, \{(b,a),(b,b)\} \cap R \neq \emptyset\}$,

4. $R^{k^*(ad)} = R \setminus \{(a,b) \mid a \neq b, ((a,a) \in R \wedge \{(b,a),(b,b)\} \cap R \neq \emptyset) \vee ((b,b) \in R \wedge \forall c\, ((b,c) \in R \rightarrow \{(a,c),(c,a),(c,c),(c,b)\} \cap R \neq \emptyset))\}$,

5. $R^{k(gr)} = R \setminus \{(a,b) \mid a \neq b, (b,b) \in R, \{(a,a),(b,a)\} \cap R \neq \emptyset\}$,

6. $R^{k^*(gr)} = R \setminus \{(a,b) \mid a \neq b, ((b,b) \in R \wedge \{(a,a),(b,a)\} \cap R \neq \emptyset) \vee ((b,b) \in R \wedge \forall c\, ((b,c) \in R \rightarrow \{(a,c),(c,a),(c,c)\} \cap R \neq \emptyset))\}$,

7. $R^{k(co)} = R \setminus \{(a,b) \mid a \neq b, (a,a),(b,b) \in R\}$,

8. $R^{k^*(co)} = R \setminus \{(a,b) \mid a \neq b, ((a,a),(b,b) \in R) \vee ((b,b) \in R \wedge (b,a) \notin R \wedge \forall c\, ((b,c) \in R \rightarrow \{(a,c),(c,a),(c,c)\} \cap R \neq \emptyset))\}$.

The following table provides a comprehensive overview of the potential of the above mentioned kernels. For the sake of completeness we also mentioned local expansion equivalence (second line) since Oikarinen and Woltran have shown that the *ad*-kernel even characterizes local expansion equivalence with respect to semi-stable, eager, admissible, preferred and ideal semantics (Theorem 8 in [Oikarinen and Woltran, 2011]). The entry "$k(\sigma)$" in line X and column τ indicates that the σ-kernel characterizes $\equiv_X^\tau$. The entry "$[m]_n$" indicates two facts: First, the characterization problem is already solved in [m] (Theorem n), but there is no kernel provided so far and second, none of the considered kernels serve as a characterization. Hereby, we use [1] as a

Chapter 5. Notions of Equivalence and Replacement

shorthand for [Oikarinen and Woltran, 2011] as well as [2] for [Gaggl and Woltran, 2012]. A red-highlighted entry indicates a new result.

	stg	stb	ss	eg	ad	pr	id	gr	co	na
L	$k^*(stg)$	$[1]_9$	$k(ad)$	$k(ad)$	$k(ad)$	$k(ad)$	$k(ad)$	$[1]_{10}$	$[1]_{11}$	$[2]_{5.13}$
E	$k(stb)$	$k(stb)$	$k(ad)$	$k(ad)$	$k(ad)$	$k(ad)$	$k(ad)$	$k(gr)$	$k(co)$	$[2]_{5.13}$
N	$k(stb)$	$k(stb)$	$k(ad)$	$k(ad)$	$k(ad)$	$k(ad)$	$k(ad)$	$k(gr)$	$k(co)$	$[2]_{5.13}$
S	$k(stb)$	$k(stb)$	$k(ad)$	$k(ad)$	$k^*(ad)$	$k^*(ad)$	$k^*(ad)$	$k^*(gr)$	$k^*(co)$	$[2]_{5.13}$

Figure 5.18: The Whole Landscape of Characterizations

Figure 5.18 shows the entire collections of characterizations for local expansion, expansion, normal expansion and strong expansion equivalence with respect to the ten semantics studied in this theses. The fact that different notions of equivalence might or might not coincide is interesting from a conceptual point of view. To illustrate this let us have a look at normal and strong expansion equivalence. Recall that normal expansions add new arguments and possibly new attacks which involve at least one of the fresh arguments, while strong expansions (a subclass of normal expansions) restrict the possible attacks between the new arguments and the old ones to a single direction. In dynamic settings, both concepts can be justified in the sense that new arguments might be raised but this will not influence the relation between already existing arguments. For strong expansions, only strong arguments will be raised, i.e. arguments which cannot be attacked by existing ones. The corresponding equivalence notions now check whether two AFs are "equally robust" to such new arguments, and indeed, normal expansion equivalence always implies strong expansion equivalence but the other direction is only true for some of the semantics, namely stage, stable, semi-stable, eager and naive semantics. One interpretation is that when two AFs are not normal expansion equivalent, then this can be made explicit by only posing strong arguments (not attacked by existing ones), while for the other semantics this is not the case. For this particular example, it seems that the notion of admissibility which is more "explicit" in the admissible, preferred, ideal, grounded and complete semantics is responsible for the fact that frameworks might be strong expansion equivalent but not normal expansion equivalent.

5.5. Summary of Results and Implications

5.5.2 Relations Between Different Notions of Equivalence

In Section 2.1.5 we considered preliminary relations between several notions of equivalence which hold for **any** semantics (compare Figure 2.9). Using the characterization theorems presented in this book we may provide a more fine-grained picture for the considered semantics.

We will present the results in one single theorem. For a better understanding we provide arrowed diagrams just like in Figure 2.9. The obtained relations hardly need a proof since they are simply combinations of former theorems. For this reason we only list the involved statements instead of providing full proofs.

Theorem 5.46. *For any AFs $\mathcal{F}$ and $\mathcal{G}$,*

1. $\mathcal{F} \equiv_E^{stg} \mathcal{G} \Leftrightarrow \mathcal{F} \equiv_N^{stg} \mathcal{G} \Leftrightarrow \mathcal{F} \equiv_S^{stg} \mathcal{G} \Rightarrow \mathcal{F} \equiv_L^{stg} \mathcal{G}, \mathcal{F} \equiv_W^{stg} \mathcal{G} \Rightarrow \mathcal{F} \equiv^{stg} \mathcal{G}$

2. $\mathcal{F} \equiv_E^{stb} \mathcal{G} \Leftrightarrow \mathcal{F} \equiv_N^{stb} \mathcal{G} \Leftrightarrow \mathcal{F} \equiv_S^{stb} \mathcal{G} \Rightarrow \mathcal{F} \equiv_L^{stb} \mathcal{G} \Rightarrow \mathcal{F} \equiv_W^{stb} \mathcal{G} \Rightarrow \mathcal{F} \equiv^{stb} \mathcal{G}$

3. $\mathcal{F} \equiv_E^{ss} \mathcal{G} \Leftrightarrow \mathcal{F} \equiv_N^{ss} \mathcal{G} \Leftrightarrow \mathcal{F} \equiv_S^{ss} \mathcal{G} \Leftrightarrow \mathcal{F} \equiv_L^{ss} \mathcal{G} \Rightarrow \mathcal{F} \equiv_W^{ss} \mathcal{G} \Rightarrow \mathcal{F} \equiv^{ss} \mathcal{G}$

4. $\mathcal{F} \equiv_E^{eg} \mathcal{G} \Leftrightarrow \mathcal{F} \equiv_N^{eg} \mathcal{G} \Leftrightarrow \mathcal{F} \equiv_S^{eg} \mathcal{G} \Leftrightarrow \mathcal{F} \equiv_L^{eg} \mathcal{G} \Rightarrow \mathcal{F} \equiv_W^{eg} \mathcal{G} \Rightarrow \mathcal{F} \equiv^{eg} \mathcal{G}$

5. $\mathcal{F} \equiv_E^{ad} \mathcal{G} \Leftrightarrow \mathcal{F} \equiv_N^{ad} \mathcal{G} \Leftrightarrow \mathcal{F} \equiv_L^{ad} \mathcal{G} \Rightarrow \mathcal{F} \equiv_S^{ad} \mathcal{G}, \mathcal{F} \equiv_W^{ad} \mathcal{G} \Rightarrow \mathcal{F} \equiv^{ad} \mathcal{G}$

6. $\mathcal{F} \equiv_E^{pr} \mathcal{G} \Leftrightarrow \mathcal{F} \equiv_N^{pr} \mathcal{G} \Leftrightarrow \mathcal{F} \equiv_L^{pr} \mathcal{G} \Rightarrow \mathcal{F} \equiv_S^{pr} \mathcal{G}, \mathcal{F} \equiv_W^{pr} \mathcal{G} \Rightarrow \mathcal{F} \equiv^{pr} \mathcal{G}$

7. $\mathcal{F} \equiv_E^{id} \mathcal{G} \Leftrightarrow \mathcal{F} \equiv_N^{id} \mathcal{G} \Leftrightarrow \mathcal{F} \equiv_L^{id} \mathcal{G} \Rightarrow \mathcal{F} \equiv_S^{id} \mathcal{G}, \mathcal{F} \equiv_W^{id} \mathcal{G} \Rightarrow \mathcal{F} \equiv^{id} \mathcal{G}$

8. $\mathcal{F} \equiv_E^{gr} \mathcal{G} \Leftrightarrow \mathcal{F} \equiv_N^{gr} \mathcal{G} \Rightarrow \mathcal{F} \equiv_L^{gr} \mathcal{G}, \mathcal{F} \equiv_S^{gr} \mathcal{G}, \mathcal{F} \equiv_W^{gr} \mathcal{G} \Rightarrow \mathcal{F} \equiv^{gr} \mathcal{G}$

9. $\mathcal{F} \equiv_E^{co} \mathcal{G} \Leftrightarrow \mathcal{F} \equiv_N^{co} \mathcal{G} \Rightarrow \mathcal{F} \equiv_L^{co} \mathcal{G}, \mathcal{F} \equiv_S^{co} \mathcal{G}, \mathcal{F} \equiv_W^{co} \mathcal{G} \Rightarrow \mathcal{F} \equiv^{co} \mathcal{G}$

10. $\mathcal{F} \equiv_E^{na} \mathcal{G} \Leftrightarrow \mathcal{F} \equiv_N^{na} \mathcal{G} \Leftrightarrow \mathcal{F} \equiv_S^{na} \mathcal{G} \Leftrightarrow \mathcal{F} \equiv_L^{na} \mathcal{G} \Rightarrow \mathcal{F} \equiv_W^{na} \mathcal{G} \Rightarrow \mathcal{F} \equiv^{na} \mathcal{G}$

Proof. We only list the involved statements.
ad 1.) Combine Proposition 2.18, Theorems 5.5, 5.8, 5.10, 5.32, 5.44
ad 2.) Combine Proposition 2.18, Theorems 5.5, 5.10, 5.32, 5.38 and Theorem 9 in [Oikarinen and Woltran, 2011]
ad 3.-4.) Combine Proposition 2.18, Theorems 5.5, 5.7, 5.11, 5.33 and Theorem 8 in [Oikarinen and Woltran, 2011]
ad 5.-7.) Combine Proposition 2.18, Theorems 5.5, 5.7, 5.16, 5.33, 5.40 and Theorem 8 in [Oikarinen and Woltran, 2011]

Chapter 5. Notions of Equivalence and Replacement

ad 8.) Combine Proposition 2.18, Theorems 5.5, 5.7, 5.22, 5.34
ad 9.) Combine Proposition 2.18, Theorems 5.5, 5.7, 5.28, 5.35
ad 10.) Combine Proposition 2.18, Theorems 5.36, 5.30 and Theorem 5.13 in [Gaggl and Woltran, 2012] □

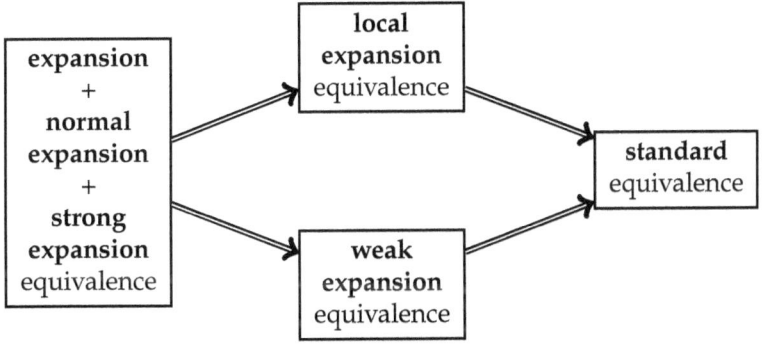

Figure 5.19: Relations for Stage Semantics

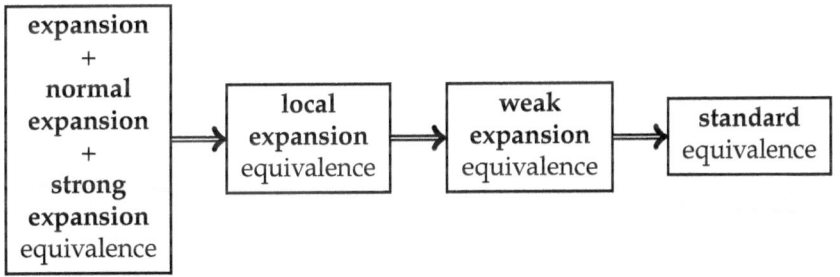

Figure 5.20: Relations for Stable Semantics

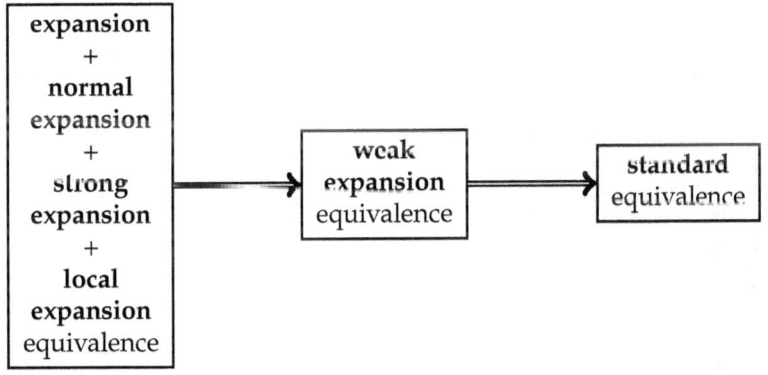

Figure 5.21: Relations for Semi-stable and Eager Semantics

5.5. Summary of Results and Implications

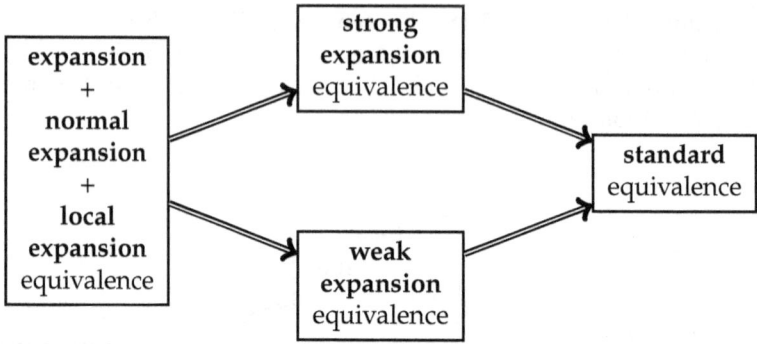

Figure 5.22: Relations for Admissible, Preferred and Ideal Semantics

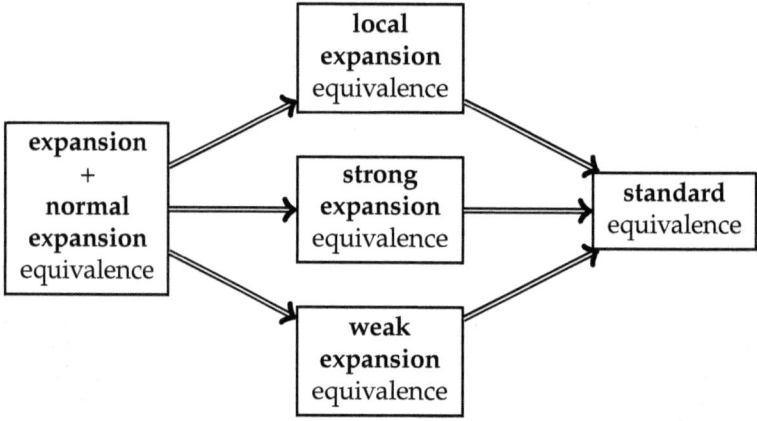

Figure 5.23: Relations for Grounded and Complete Semantics

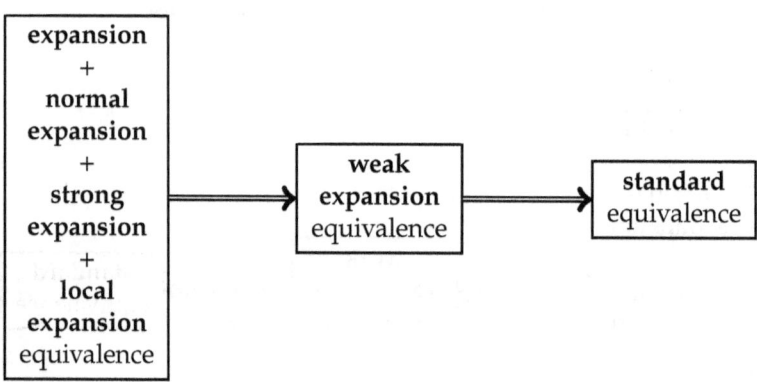

Figure 5.24: Relations for Naive Semantics

In our point of view, the most remarkable relations are those of stable, semi-stable, eager and naive semantics since their corresponding equivalence relations are totally ordered with respect to subset-relation. Bearing in mind that strong, weak and local expansions are completely different concepts the containedness or coincidence of their corresponding equivalence relations is unexpected.

5.5.3 The Role of Self-loop-free AFs

If we take a closer look at the definitions of the σ-kernel or σ-*-kernel of an AF $\mathcal{F}$ we observe that in case of self-loop-free AFs nothing changes, i.e. $\mathcal{F}$ and its corresponding kernel are identical. Consequently, any equivalence notion on the set of AFs characterizable through kernels presented in this book collapses to identity if we restrict ourselves to self-loop-free AFs. Consequently, known and the new results in this paper lead to the following observation:

Proposition 5.47. *For any self-loop-free AFs $\mathcal{F}, \mathcal{G}$, any $\Phi \in \{E, N, S\}$, and any semantics $\sigma \in \{stg, stb, ss, eg, ad, pr, id, gr, co\}$:*

$$\mathcal{F} = \mathcal{G} \text{ iff } \mathcal{F} \equiv_\Phi^\sigma \mathcal{G}.$$

Moreover, for $\sigma \in \{stg, ss, eg, ad, pr, id\}$

$$\mathcal{F} = \mathcal{G} \text{ iff } \mathcal{F} \equiv_L^\sigma \mathcal{G}.$$

This means, self-loop-free AFs are redundancy-free or in other words, all attacks may play a crucial role with respect to further evaluations provided that the expansions are normal or strong. In the introductory part of this book we noted that such kinds of expansions naturally occur if Dung-style AFs are (re-)instantiated by a deductive argumentation system where a new piece of information was added to the underlying knowledge base (compare Figure 1.1). We want to mention that there are some formalisms like classical logic-based frameworks where self-attacking arguments do not occur (cf. Theorem 4.13 in [Besnard and Hunter, 2001]). Other argumentation systems like ASPIC [Prakken, 2009] or a very simple formalism presented in [Caminada, 2005] "allow" self-defeating arguments. We refer the reader to [Prakken, 2009, Section 7] and [Caminada, 2005, Section 3] for examples of self-defeating arguments or [Gabbay, 2013, Chapter 18] for a detailed discussion of loops in argumentation.

As an aside, the result stated in Proposition 5.47 cannot be conveyed to local and weak expansion equivalence with respect to any considered semantics. This means, there are syntactically different and self-loop-free AFs which are local expansion equivalent (cf. Example 16 in [Oikarinen and Woltran, 2011]) or weak expansion equivalent (cf. Example 5.39).

Another particular property of all kernels obtained so far is that the same arguments have to be present and self-looping in either both or none of the compared frameworks.

Proposition 5.48. *For any AFs $\mathcal{F}$ and $\mathcal{G}$, any relation $\Phi \in \{E, N, S\}$, and any semantics $\sigma \in \{stg, stb, ss, eg, ad, pr, id, gr, co, na\}$:*

if $\mathcal{F} \equiv_\Phi^\sigma \mathcal{G}$, then $A(\mathcal{F}) = A(\mathcal{G})$, and moreover, for each $a \in A(\mathcal{F})$,

$$(a,a) \in R(\mathcal{F}) \text{ iff } (a,a) \in R(\mathcal{G}).$$

The same proposition can be given for $\sigma \in \{stg, ss, eg, ad, pr, id\}$ and local expansion equivalence.

Note that therefore the statements of Proposition 5.47 already hold if at least one of the two compared frameworks is self-loop-free.

Finally, in contrast to the statements of Proposition 5.47, we mention that for AFs with self-loops, many different AFs can have the same kernel, and thus are equivalent to each other. For instance, consider expansion equivalence w.r.t. stable semantics which is characterized by the *stb*-kernel (see Definition 5.1). Furthermore, let $\mathcal{F}$ be an AF possessing n arguments and m self-loops. Then there exist $2^{m(n-1)} - 1$ different AFs which are all expansion equivalent to $\mathcal{F}$ with respect to stable semantics. An AF possessing a huge number of equivalent AFs is given in Example 5.45.

5.6 Conclusions and Related Work

Studying equivalence notions between argumentation frameworks has gained increasing interest recently (see also [Wooldridge et al., 2006; Amgoud and Vesic, 2011]). In fact, due to the inherently nonmonotonic nature of argumentation frameworks, strong notions of equivalence give a handle to decide whether two frameworks represent the same knowledge even if this knowledge is not captured by the actual extensions, but can be made explicit by augmenting the framework under consideration. In contrast to other nonmonotonic formalisms where a huge number of equivalence notions inbetween standard and strong equivalence were studied, e.g. query equivalence [Shmueli, 1987] and uniform equivalence [Eiter and Fink, 2003] in case of logic programs, we are not aware of further studies apart from [Oikarinen and Woltran, 2011; Gaggl and Woltran, 2011] devoted to abstract argumentation.

In this chapter we mainly studied two new equivalence relations for AFs, namely normal and strong expansion equivalence which lie in-between standard equivalence and the recently proposed expansion equivalence [Oikarinen and Woltran, 2011]. The majority of the presented results was already published in [Baumann, 2012a]. Smaller pieces of the puzzle were published in [Baumann, 2011; Baumann and Woltran, 2014; Baumann and Brewka, 2013a]. We provided characterization theorems for ten prominent semantics, namely stage, stable, semi-stable, eager, admissible, preferred, ideal, grounded, complete and naive semantics. In particular, we showed that for any considered semantics, normal expansion equivalence coincides with expansion equivalence. Even more surprisingly we showed that stage, stable, semi-stable, eager and naive semantics do not "distinguish" between normal expansion and strong expansion equivalence. This was quite surprising as well as unexpected since the class of strong (normal) expansions is obviously a

Chapter 5. Notions of Equivalence and Replacement

proper subset of the class of normal (arbitrary) expansions. Except for naive semantics, the obtained characterization theorems are based on syntactical criteria. To determine whether two AFs are expansion, normal expansion or strong expansion equivalent with respect to a certain semantics σ it suffices to compare certain *kernels* of them (see Figure 5.18). A kernel of an AF $\mathcal{F}$ is itself an AF obtained from $\mathcal{F}$ by deleting certain attacks depending on the considered semantics σ. It has been shown that, if two AFs possess identical kernels under a certain semantics σ then they are inter-substitutable with respect to further evaluations in dynamic scenarios satisfying the concept of arbitrary, normal or strong expansions, respectively. Such replacement properties are essential for logical approaches in general, particularly for non-monotonic logics where this question becomes a non-trivial task.

Furthermore, we devote particular attention to the special role of self-loops with respect to characterizing equivalence notions. Roughly speaking, the presence of self-loops seems to be the crucial feature which separates syntactic equivalence from strong notions of equivalence. In other words, self-loop-free AFs are redundancy-free, i.e. all attacks may play a crucial role with respect to further evaluations provided that the expansions are normal or strong. Recent work by Lonc and Truszczyński [Lonc and Truszczyński, 2011] investigates equivalence relations on graphs in a very general setting. As part of future work we want to study to which extend the results of [Lonc and Truszczyński, 2011] can be applied to abstract argumentation.

In [Amgoud and Vesic, 2011] different notions of equivalence with respect to stable semantics of two logic-based argumentation systems is studied. More precisely, they studied the question when two systems, not necessarily built over the same knowledge base and/or not necessarily using the same attack definition, produce the same output with respect to stable semantics. Our results as well as the characterization theorems in [Oikarinen and Woltran, 2011; Gaggl and Woltran, 2011] are in some sense useless for their aim. The reason is that the authors concentrated on classical logic-based argumentation systems where self-attacking arguments provably do not occur (cf. Theorem 4.13 in [Besnard and Hunter, 2001]). Note that the results presented in this book are not restricted to a special instantiation like Tarskian logics but rather to **any** underlying logic. For instance, in case of ASPIC [Prakken, 2009] self-defeating arguments may occur and thus, identifying redundant attacks may simplify the evaluation of such systems.

Another mentionable work dealing with various notions of equivalence with regard to deductive argumentation is [Wooldridge et al., 2006]. Due to the use of a very basic definition of an argument the presented complexity results holds for a whole range of argumentation systems. They showed, for instance, that checking equivalence of argument sets is not computationally harder than checking equivalence of arguments. Both are co-NP-complete.

In [Baroni et al., 2012] so-called input/output argumentation frameworks are introduced, an approach to characterize the behavior of an argumentation framework as sort of a black box with a well-defined external interface. The paper defines the notion of semantics decomposability and analyzes complete, stable, grounded and preferred semantics in this regard. It turns out that, under grounded, complete, stable and credulous preferred semantics, input/output argumentation frameworks with the same behavior can be exchanged without affecting the results of the evaluation of other interacting arguments. Since replaceability is one of the main motivations for studying equivalence notions, we plan to explore connections between equivalence and decomposability in the near future.

For other future directions, we mention further semantics which have not investigated with respect to normal and strong expansion equivalence. One example is $cf2$ semantics [Baroni et al., 2005], and the recently introduced variant, $stage2$ semantics [Dvořák and Gaggl, 2012]. For both, expansion equivalence has been shown to coincide with syntactical equivalence [Dvořák and Gaggl, 2012; Gaggl and Woltran, 2012]) even if self-loops are permitted. It would be interesting to check whether this behaviour carries over to weaker notions of equivalence. A prominent semantics which has not been considered in terms of equivalence checking at all is the *resolution-based grounded* semantics [Baroni et al., 2011c]. Likewise, as we have done for stable and preferred semantics, characterizing weak expansion equivalence is on our agenda.

In [Baroni and Giacomin, 2007] several general criteria for comparing and evaluating semantics were introduced. This paper was an important step to classify semantics because until its publication comparisons between semantics were almost exclusively example driven. The results presented in this book motivate further criteria to compare argumentation semantics on an abstract level, for example, coincidence, containedness or incomparability (with respect to subset-relation) of equivalence relations. The study of such *equivalence-based criteria* as well as their relations to the criteria proposed in [Baroni and Giacomin, 2007] will be part of future work. Finally, the related notions of succinctness [Gaggl and Woltran, 2012] and regularity [Baumann, 2012b] deserve further attention.

Chapter 6

Cardinality Results

Stable extensions constitute one of the most important and well-researched semantics for abstract argumentation frameworks. Dung used the stable extension semantics in his original paper to relate AFs to Reiter's default logic, different forms of logic programming, and to solve the stable marriage problem, among others [Dung, 1995]. Alas, there are some fundamental questions to be asked about stable extension semantics which have yet remained unanswered.

Given an abstract argumentation framework for which the only thing we know is that it has n arguments and x attacks, how many stable extensions does it have at most? How many on average? For $x = 0$, without attacks, the case is quite clear – there will be exactly one stable extension, the set of all arguments. For $x = n^2$, the AF contains all possible attacks, in particular all self-attacks, and there will be no stable extension. But what happens in between, when $0 < x < n^2$? This chapter takes a step towards analytical and empirical answers to these questions.

In the considerable zoo of semantics for abstract argumentation, stable extension semantics is the only one for which extension existence is not guaranteed for finite AFs. While this is usually regarded as a weakness, there is an obvious benefit to it when AFs are used to model NP-complete problems, that do not necessarily possess a solution. In this setting, the fact that an NP problem instance encoded as an AF has no stable extension elegantly reflects the fact that the problem instance has no solution. Using other semantics, unsolvability would have to be represented by introducing new (meta-)language constructs.

NP problems typically have elements that are generating (that is, generate possible solution candidates) and elements that are constraining (that is, eliminate possible solution candidates). The classical example of an NP-complete problem is of course deciding the satisfiability of a given propositional formula in conjunctive normal form, the SAT problem. There, the propositional variables are the generating elements (since solution candidates are among all interpretations for the variables) while the disjunctive clauses are the constraining elements (they remove those interpretations not satisfying some clause). Can the same be said about arguments and attacks? Surely, arguments are generating, since extension candidates are sets of arguments. But are attacks always constraining?

Example 6.1. Consider the following scenario. The AFs $\mathcal{G}$ and $\mathcal{H}$ represent two specific ways to add an attack to the AF $\mathcal{F}$, namely adding an attack from a_2 to a_1 or adding an attack from a_1 to itself, respectively. Observe that $\mathcal{E}_{stb}(\mathcal{F}) = \{\{a_1\}\}$, $\mathcal{E}_{stb}(\mathcal{G}) = \{\{a_1\}, \{a_2\}\}$ and $\mathcal{E}_{stb}(\mathcal{H}) = \emptyset$. Thus, adding attacks to an AF may in general both increase or decrease the number of stable extensions.

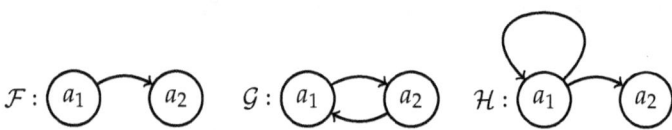

Figure 6.1: Generating vs. Constraining

The reason for this behavior is of course to be found in the definition of stable extensions: Roughly, to be a stable extension, a set has to satisfy two properties. It has to be conflict-free, and has to attack all arguments not in the set. Intuitively, the number of attacks in an AF correlates negatively with the number of conflict-free sets – the more attacks (that is, conflicts) there are, the less conflict-free sets are found. At the same time, the number of attacks correlates positively with the number of sets which attack all outsiders. So how will these two interleaved and counteracting forces come to terms in general?

In this chapter we present an analytical and empirical study of the maximal and average numbers of stable extensions in abstract argumentation frameworks. As one of the analytical main results, we prove a tight upper bound on the maximal number of stable extensions that depends only on the number of arguments in the framework. More interestingly, our empirical results indicate that the distribution of stable extensions as a function of the number of attacks in the framework seems to follow a universal pattern that is independent of the number of arguments.

The obtained results can be used to provide lower bounds for the minimal realizability of certain sets of extensions. Furthermore, counting techniques may yield upper bounds for algorithms computing extensions. Finally, the average number gives some guidance on how many extensions a given AF with n arguments and x attacks will have.

6.1 Background

Throughout the chapter we assume some familiarity with standard analysis, combinatorics and statistics. We start with the formal foundation, i.e. we introduce some well-known graph theoretical concepts and technical tools which are needed to prove the cardinality results.

Chapter 6. Cardinality Results

For a set X, a *(binary) relation* over X is any set $R \subseteq X \times X$. Special among these relations is the *identity* $id_X = \{(x,x) \mid x \in X\}$. A relation R over X is *irreflexive* iff $R \cap id_X = \emptyset$, that is, for each $x \in X$ we have $(x,x) \notin R$. It is *symmetric* iff for each $(x,y) \in R$ we have $(y,x) \in R$. The *inverse* of a relation R is given by $R^{-1} = \{(y,x) \mid (x,y) \in R\}$.

A *directed graph* is a pair (V,E) where V is a finite set and E a binary relation over V. The elements of V are called *nodes* and those of E are called *edges*. A directed graph is *symmetric* iff its edge relation E is symmetric. For a directed graph $G = (V,E)$, we denote by $sym(G) = (V, E \cup E^{-1})$ its symmetric version. Similarly, the irreflexive version of a graph $G = (V,E)$ is defined as $irr(G) = (V, E \setminus id_V)$.

An *undirected graph* is a pair (V,F) where V is as above and $F \subseteq \binom{V}{2} \cup \binom{V}{1}$ is a set of 2- and 1-element subsets of V, which represent the undirected edges. For a directed graph $G = (V,E)$, we denote by $und(G) = (V, \{\{u,v\} \mid (u,v) \in E\})$ its associated undirected graph. An undirected graph (V,F) is *simple* iff $F \subseteq \binom{V}{2}$. We denote by $\mathcal{G}_n$ the set of all simple graphs with n nodes.

For a simple graph $G = (V,F)$, a set $M \subseteq V$ is *independent* iff for all $u,v \in M$ we have $\{u,v\} \notin F$. A set $M \subseteq V$ is *maximal independent* iff it is independent and there is no proper superset of M which is independent. The set of all maximal independent sets of a simple graph G is denoted by $MIS(G)$.

Example 6.2. The following graphs illustrate the concepts of irreflexive, symmetric and undirected graphs. Observe that $MIS(und(sym(irr(\mathcal{F})))) = \{\{a_1,a_4\}, \{a_2,a_4\}, \{a_1,a_3,a_5\}, \{a_2,a_5\}\}$.

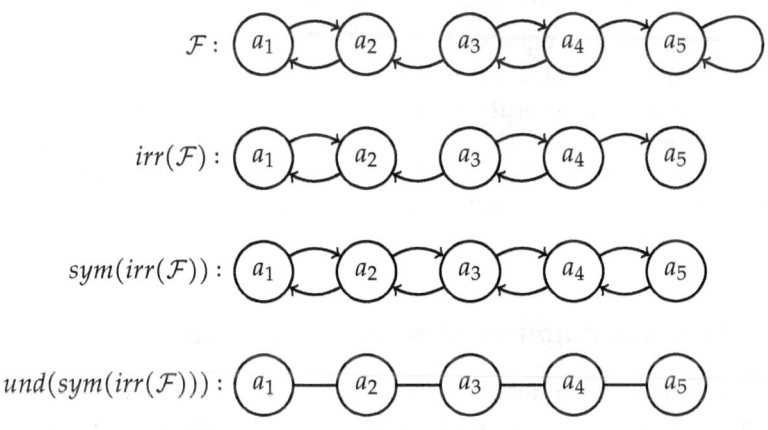

Figure 6.2: *Different Operators on Graphs*

For the purposes of this paper, we denote by $\mathcal{A}_n$ the set of all AFs with n arguments, and by $\mathcal{A}_{n,x}$ the set of all AFs with n arguments and x attacks.

There, not the precise arguments are of interest to us but only the *number* of arguments; we will implicitly assume that the n arguments can be numbered by $1,\ldots,n$. Once the arguments are fixed, however, we consider two AFs the same if and only if they have the same attack relation. So the AF with two arguments 1,2 where 1 attacks 2 is different from the AF with two arguments 1,2 where 2 attacks 1, although the two are isomorphic in a graph theoretic sense. This guarantees that all possible scenarios, that is, any arrangement of attacks for fixed numbers of arguments and attacks is considered.

A *full AF* is of the form $(A, A \times A)$ for some set A. Furthermore, we call an argumentation framework a y-AF iff it has exactly y stable extensions. Consider again Example 6.2. The AF $\mathcal{F}$ is a 2-AF since $\mathcal{F}$ has two stable extensions, namely $\mathcal{E}_{st}(\mathcal{F}) = \{\{a_1, a_4\}, \{a_2, a_4\}\}$. Moreover, we state that $\mathcal{F} \in \mathcal{A}_{5,7}$ and consequently, $\mathcal{F} \in \mathcal{A}_5$.

6.2 Analytical Results

In [Baroni et al., 2010] it was shown that counting the number of stable extensions of an argumentation framework is a computationally hard problem. The analysis of counting techniques may yield upper bounds for algorithms computing extensions. Furthermore, a fast counting algorithm gives a first advice on how controversial the information represented in an AF is. In this section, we contribute some analytical results to this direction of research.

For a fixed number n of arguments there are $|\mathcal{A}_n| = 2^{n^2}$ different AFs, since any attack relation whatsoever is possible and significant. Furthermore, if we additionally know that the AF in question possesses x attacks, then the total number of possibilities equals $|\mathcal{A}_{n,x}| = \binom{n^2}{x}$, the number of x-element subsets of an n^2-element set. This means that in principle, one may obtain numerically precise results by brute force for classes of AFs possessing a certain number of arguments and attacks. For example, specific classes of AFs could be enumerated and each element analyzed separately. But obviously, such an approach cannot provide a solution which is parametric in the numbers of arguments and attacks.

6.2.1 Maximal Number of Stable Extensions

What is the maximal number of stable extensions given an AF $\mathcal{F} = (A, R)$ with $|A| = n$ arguments? Since argumentation semantics choose their extensions from the set of subsets of A, we have $\mathcal{E}_{st}(\mathcal{F}) \subseteq 2^A$. This yields an immediate upper bound on the number of extensions for any semantics, namely $|\mathcal{E}_{st}(\mathcal{F})| \leq |2^A| = 2^n$. Can this quite naive bound be improved? In case of semantics satisfying I-maximality the answer is "yes." For short, I-maximality is fulfilled if no extension can be a proper subset of another (compare Definition 2.11). In other words, the cardinality of one of the largest $\subseteq$-antichains S being a subset of an n-element set gives a further upper bound on the num-

Chapter 6. Cardinality Results

ber of extensions.[1] The maximal cardinality of such antichains is given by Sperner's theorem [Sperner, 1928], namely $|S| = \binom{n}{\lfloor \frac{n}{2} \rfloor}$. By a straightforward calculation one may show that $\binom{n}{\lfloor \frac{n}{2} \rfloor} \leq \frac{2^n}{\sqrt{n}}$. Without any further knowledge about the considered semantics it is impossible to find better bounds.

Let us turn to stable semantics. In any case, we can achieve a high number of stable extensions by grouping. For instance, the maximal number of stable extensions for an AF possessing an even number $n = 2m$ of arguments is at least $2^m = 2^{\frac{n}{2}}$. Such a framework is given by grouping the arguments in pairs that mutually attack each other:

$$\mathcal{F} = (\{a_i, b_i \mid 1 \leq i \leq m\}, \{(a_i, b_i), (b_i, a_i) \mid 1 \leq i \leq m\})$$

Is grouping in pairs the best we can do?

Assume we group not in pairs but in groups of arbitrary size k such that all members of a single group attack each other. Then for n arguments the number of stable extensions is given by the following function:

$$f : \mathbb{N} \to \mathbb{N} \text{ where } f(k) = k^{\lfloor \frac{n}{k} \rfloor}$$

To approximate the maximum of $f(k)$ we calculate the extrema of the associated real-valued function

$$g : \mathbb{R} \to \mathbb{R} \text{ where } g(k) = k^{\frac{n}{k}} = e^{\frac{n}{k} \cdot \ln(k)}$$

For that, we have to solve the following equation:

$$k^{\frac{n}{k}} \left(-\frac{n}{k^2} \cdot \ln(k) + \frac{n}{k^2} \right) = k^{\frac{n}{k}} \cdot \frac{n}{k^2} \cdot (1 - \ln(k)) = 0$$

The only solution for this equation is that k equals Euler's number e. Of course, it is very difficult to arrange in groups of e when dealing with arguments. Nevertheless, the obtained result provides an upper bound for the initial problem – namely the value $g(e) = e^{\frac{n}{e}}$ – assuming that grouping is the best. We will see that the exact value is not far away.

On the path to the main theorem we start with two simple observations which hardly need a proof. Being aware of this fact, we still present them in the form of a proposition to be able to refer to them later on. For one, whenever a set E is a stable extension of $\mathcal{F}$, then E is also a stable extension in the symmetric and self-loop-free version of $\mathcal{F}$. Observe that the converse is not true in general.

Proposition 6.3. *For any AF $\mathcal{F} = (A, R)$ and any $E \in \mathcal{E}_{stb}(\mathcal{F})$ we have $E \in \mathcal{E}_{stb}(sym(irr(\mathcal{F})))$.*

For another, the second proposition establishes a simple relationship between stable extensions in symmetric AFs without self-loops and maximal independent sets in undirected graphs.

---
[1] A ⊆-antichain is a set of sets of which any two are mutually ⊆-incomparable.

Proposition 6.4. *For any symmetric and irreflexive AF $\mathcal{F} = (A, R)$ we have:*

$$E \in \mathcal{E}_{stb}(\mathcal{F}) \text{ iff } E \in MIS(und(\mathcal{F})).$$

Now we turn to the main theorem which is mainly based on a graph theoretical result by J.W. Moon and L. Moser from 1965 [Moon and Moser, 1965].[2] The theorem establishes a tight upper bound for the number of stable extensions of an AF with n arguments. The upper bound is obtained as a function σ_{max} of n.

Theorem 6.5. *For any natural number n, it holds that*

$$\max_{\mathcal{F} \in \mathcal{A}_n} |\mathcal{E}_{st}(\mathcal{F})| = \sigma_{max}(n)$$

where the function $\sigma_{max} : \mathbb{N} \to \mathbb{N}$ is defined by

$$\sigma_{max}(n) = \begin{cases} 1, & \text{if } n = 0 \text{ or } n = 1, \\ 3^s, & \text{if } n \geq 2 \text{ and } n = 3s, \\ 4 \cdot 3^{s-1}, & \text{if } n \geq 2 \text{ and } n = 3s + 1, \\ 2 \cdot 3^s, & \text{if } n \geq 2 \text{ and } n = 3s + 2. \end{cases}$$

Proof. The cases $n = 0$ and $n = 1$ are obvious; let $n \geq 2$.

"$\leq$": We already observed that for any AF $\mathcal{F}$, $\mathcal{E}_{stb}(\mathcal{F}) \subseteq \mathcal{E}_{stb}(sym(irr(\mathcal{F})))$ (Proposition 6.3). Consequently, $|\mathcal{E}_{stb}(\mathcal{F})| \leq |\mathcal{E}_{stb}(sym(irr(\mathcal{F})))|$ follows and

$$\max_{\mathcal{G} \in \mathcal{A}_n} |\mathcal{E}_{st}(\mathcal{G})| \leq \max_{\mathcal{G} \in \mathcal{A}_n} |\mathcal{E}_{stb}(sym(irr(\mathcal{G})))|$$

In the light of Proposition 6.4 we get

$$\max_{\mathcal{G} \in \mathcal{A}_n} |\mathcal{E}_{stb}(sym(irr(\mathcal{G})))| = \max_{\mathcal{G} \in \mathcal{A}_n} |MIS(und(sym(irr(\mathcal{G}))))|$$

Observe that the functions $irr(\cdot)$, $sym(\cdot)$ and $und(\cdot)$ do not change the number of nodes (respectively arguments). Consequently, we may estimate thus:

$$\max_{\mathcal{G} \in \mathcal{A}_n} |MIS(und(sym(irr(\mathcal{G}))))| \leq \max_{\mathcal{U} \in \mathcal{G}_n} |MIS(\mathcal{U})|.$$

This means, the value $\sigma_{max}(n)$ does not exceed the maximal number of maximal independent sets of simple undirected graphs of order n. Due to Theorem 1 in [Moon and Moser, 1965] these values are exactly given by the last three lines of the claimed value range of $\sigma_{max}(n)$.

"$\geq$": We define the following AFs.

- $A_2(i) = \{a_i, b_i\}$ and $A_3(i) = \{c_i, d_i, e_i\}$,

---

[2]Note that the original work deals with maximal cliques. The result can be equivalently formalized in terms of maximal independent sets as done in [Wood, 2011].

- $\mathcal{F}_2(i) = irr(A_2(i), A_2(i) \times A_2(i))$ and
  $\mathcal{F}_3(i) = irr(A_3(i), A_3(i) \times A_3(i))$.
- For $n = 3s$ consider $\mathcal{F}_{3s} = \bigcup_{i=1}^{s} \mathcal{F}_3(i)$.
- For $n = 3s + 1$ consider $\mathcal{F}_{3s+1} = (\bigcup_{i=1}^{2} \mathcal{F}_2(i)) \cup (\bigcup_{i=1}^{s-1} \mathcal{F}_3(i))$.
- Finally, in case of $n = 3s + 2$ consider $\mathcal{F}_{3s+2} = \mathcal{F}_2(1) \cup (\bigcup_{i=1}^{s} \mathcal{F}_3(i))$.

It is straightforward to verify that $|\mathcal{E}_{stb}(\mathcal{F}_{3s})| = 3^s$, $|\mathcal{E}_{stb}(\mathcal{F}_{3s+1})| = 4 \cdot 3^{s-1}$ and
$|\mathcal{E}_{stb}(\mathcal{F}_{3s+2})| = 2 \cdot 3^s$. □

For illustration we present here an instantiation of the presented prototypes, namely $\mathcal{F}_{10} = \mathcal{F}_{3\cdot 3+1} = (\bigcup_{i=1}^{2} \mathcal{F}_2(i)) \cup (\bigcup_{i=1}^{2} \mathcal{F}_3(i))$ which is graphically represented by the following figure:

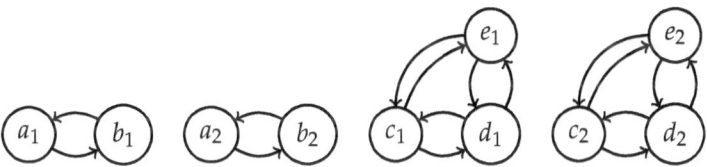

Figure 6.3: The AF $\mathcal{F}_{3\cdot 3+1}$

Observe that $|\mathcal{E}_{stb}(\mathcal{F}_{10})| = |\mathcal{E}_{stb}(\mathcal{F}_{3\cdot 3+1})| = 4 \cdot 3^2$. In general, the function σ_{max} looks more complicated than it is, because the numbers are slightly different depending on the remainder of n on division by 3. Here is a much simpler version.

Corollary 6.6 (Upper bound short cut). *For any natural number n, we find:*

$$\sigma_{max}(n) \leq 3^{\frac{n}{3}} \leq 1{,}4423^n.$$

As a final note we want to mention that it does not make much sense to ask for the minimal number of stable extensions, since for any $n > 0$ and $0 < x \leq n^2$ there are always AFs without stable extensions.

6.2.2 Average Number of Stable Extensions

What is the average number of stable extensions of argumentation frameworks with n arguments and x attacks? As in the case of the maximal number of stable extensions, the precise value is computable in principle. This is immediate from its formal definition:

Definition 6.7. The function $\bar{\sigma}(n, x)$ returns the average number of stable extensions of all AFs with n arguments and x attacks, and is defined thus:

$$\bar{\sigma} : \mathbb{N} \times \mathbb{N} \to \mathbb{R} \text{ where } \bar{\sigma}(n, x) = \frac{\sum_{\mathcal{F} \in \mathcal{A}_{n,x}} |\mathcal{E}_{st}(\mathcal{F})|}{\binom{n^2}{x}}$$

6.2. Analytical Results

While this definition makes it precise what we mean by "average number of stable extensions", computing this number for a given AF still remains as hard as computing all stable extensions of that AF.

But we are looking for a way to compute the number $\bar{\sigma}(n,x)$ *without actually inspecting the AF* except for determining the parameters n and x. This would be useful since the number n of arguments and the number x of attacks can be determined in linear time, and knowing $\bar{\sigma}(n,x)$ gives some guidance on how many extensions a given AF $\mathcal{F} \in \mathcal{A}_{n,x}$ will have.

The best-case scenario would be the specification of a closed-form function that returns the exact values of $\bar{\sigma}(n,x)$. Unfortunately, the combinatorial blowup even in case of small numbers of attacks turns this endeavor into a challenging task. Nevertheless, we were able to specify certain values. The following proposition presents some exact values of $\bar{\sigma}(n,x)$ given that the number of attacks x is close to 0 or close to n^2.

Proposition 6.8. *For any $n \in \mathbb{N}$, we have*

$$\bar{\sigma}(n,0) = 1$$

$$\bar{\sigma}(n,n^2 - 3) = \begin{cases} \frac{3 \cdot (n^2-n-1)}{(n+1) \cdot (n^2-2)}, & \text{if } n \geq 3, \\ 1 - \frac{1}{n}, & \text{if } n = 2 \\ 0, & \text{otherwise} \end{cases}$$

$$\bar{\sigma}(n,1) = \begin{cases} 1 - \frac{1}{n}, & \text{if } n \geq 1, \\ 0, & \text{otherwise} \end{cases}$$

$$\bar{\sigma}(n,n^2 - 2) = \begin{cases} \frac{2}{n+1}, & \text{if } n \geq 2, \\ 0, & \text{otherwise} \end{cases}$$

$$\bar{\sigma}(n,2) = \begin{cases} 1 - \frac{2n-2}{n^2+n}, & \text{if } n \geq 2, \\ 0, & \text{otherwise} \end{cases}$$

$$\bar{\sigma}(n,n^2 - 1) = \begin{cases} \frac{1}{n}, & \text{if } n \geq 1, \\ 0, & \text{otherwise} \end{cases}$$

$$\bar{\sigma}(n,n^2) = \begin{cases} 1, & \text{if } n = 0, \\ 0, & \text{otherwise} \end{cases}$$

Proof. The values of $\bar{\sigma}(n,0)$ and $\bar{\sigma}(n,n^2)$ are obvious. Consider $\bar{\sigma}(n,1) = 1 - \frac{1}{n}$. This can be seen as follows: If the associated attack is a self-loop, then we have no extensions. If it is not, then we have exactly one extension which is the union of all unattacked arguments. Obviously, we have $|\mathcal{A}_{n,1}| = \binom{n^2}{1} = n^2$ and furthermore, there are n different AFs in $\mathcal{A}_{n,1}$ possessing exactly one loop. Thus $\bar{\sigma}(n,1) = \frac{n^2-n}{n^2} = 1 - \frac{1}{n}$. Analogously one may prove $\bar{\sigma}(n,n^2 - 1) = \frac{1}{n}$.

We want to emphasize that the other values are non-trivial. To get an idea of the complexity of the remaining proofs we consider the value $\bar{\sigma}(n,n^2 - 3)$. Without loss of generality we may assume $n \geq 2$ since the number of attacks has to be non-negative. Furthermore we may even assume that $n \geq 3$ because if $n = 2$, then $\bar{\sigma}(n,n^2 - 3) = \bar{\sigma}(n,1)$ which is already solved. An AF $\mathcal{F} \in \mathcal{A}_{n,n^2-3}$ can be seen as the result of the following process: One starts with a full AF with n arguments. We then stepwise delete 3 attacks which are either loops or non-loops. We list now the probabilities to end up in an AF where k loops are deleted.

Chapter 6. Cardinality Results

$$P(k=3) = 1 \cdot \frac{n}{n^2} \cdot \frac{n-1}{n^2-1} \cdot \frac{n-2}{n^2-2}$$

$$P(k=2) = 3 \cdot \frac{n}{n^2} \cdot \frac{n-1}{n^2-1} \cdot \frac{n^2-n}{n^2-2}$$

$$P(k=1) = 3 \cdot \frac{n}{n^2} \cdot \frac{n^2-n}{n^2-1} \cdot \frac{n^2-n-1}{n^2-2}$$

We omit the consideration of $P(k=0)$ since such kind of frameworks do not possess an extension and thus does not contribute anything to $\bar{\sigma}(n, n^2-3)$. We list now the average number of extensions of AFs in $\mathcal{A}_{n,n^2-3}$ where k loops are deleted.

$$av(k=3) = 3$$

$$av(k=2) = 1 \cdot \frac{2(n-1)}{n^2-n} + 2 \cdot \frac{(n^2-n)-2(n-1)}{n^2-n}$$

$$= 2 \cdot \left(1 - \frac{1}{n}\right)$$

$$av(k=1) = 1 - \left(\frac{n-1}{n^2-n} + \frac{(n^2-n)-(n-1)}{n^2-n} \cdot \frac{n-1}{n^2-n-1}\right)$$

$$= \frac{n^2-3n+2}{n^2-n-1}$$

The average numbers can be seen as follows. If we delete exactly three loops we end up in an AF with 3 stable extensions, namely the singletons of the non-looping arguments. Consequently, $av(k=3) = 3$. If we delete 2 loops and 1 non-loop we either end up with 1 extension, namely if the deleted non-loop starts by a self-loop-free argument or 2 extensions otherwise. The probability of the former is $\frac{2(n-1)}{n^2-n}$. Since both cases are mutual exclusive and exhaustive we derive a probability of $\frac{(n^2-n)-2(n-1)}{n^2-n}$ for the latter case proving the claimed value of $av(k=2)$.

Consider now $av(k=1)$. Observe that the maximal number of extensions equals 1 because only 1 self-loop is deleted. In the following we call this argument arg. We specify now the probability that we end up in AF with zero stable extension. This is the case if at least one deleted non-loop starts by arg. The probability for the "first" non-loop is $\frac{n-1}{n^2-n}$. Furthermore, the probability for the "second" deleted non-loop to start by arg providing that the first one does not started by arg is given by $\frac{(n^2-n)-(n-1)}{n^2-n} \cdot \frac{n-1}{n^2-n-1}$. Thus, the claimed value for $av(k=1)$ follows. Finally, we have to sum up, that is,

$$\bar{\sigma}(n, n^2-3) = \sum_{i=1}^{3} P(k=i) \cdot av(k=i) = 3 \cdot \frac{n^2-n-1}{(n+1)(n^2-2)}$$

We omit the consideration of $\bar{\sigma}(n,2)$ and $\bar{\sigma}(n, n^2-2)$ since their treatment is similar in style to the above proof. □

It can be seen that the values of $\bar{\sigma}(n,1)$ and $\bar{\sigma}(n,2)$ do not give any indication on how $\bar{\sigma}(n,3)$ could look like, not even qualitatively. The same holds for $\bar{\sigma}(n,n^2-2)$ and $\bar{\sigma}(n,n^2-3)$, and potential informed guesses about $\bar{\sigma}(n,n^2-4)$. But having these exact values at hand we may consider the limit values for AFs with an increasing number of arguments. We have

$$\lim_{n\to\infty} \bar{\sigma}(n,0) = \lim_{n\to\infty} \bar{\sigma}(n,1) = \lim_{n\to\infty} \bar{\sigma}(n,2) = 1$$

On the other hand, we obtain

$$\lim_{n\to\infty} \bar{\sigma}(n,n^2) = \lim_{n\to\infty} \bar{\sigma}(n,n^2-1) = \lim_{n\to\infty} \bar{\sigma}(n,n^2-2) = \lim_{n\to\infty} \bar{\sigma}(n,n^2-3) = 0$$

This means that for increasing numbers of arguments, the average number of stable extensions in the case of very small numbers of attacks approaches from below to 1. In the case of very large numbers of attacks we have a convergence to 0 from above. So far, so good; but it is still unclear how many extensions there usually are in between. With an increasing number of attacks, does the average number of stable extensions just decrease in a monotone fashion? It turns out that this is a really hard problem.

Of course, we can look at simple special cases. For example, for $n = 2$, Proposition 6.8 yields the precise values for all possible numbers of attacks $0 \le x \le n^2 = 4$: an AF with 2 arguments and $0,1,2,3,4$ attacks will have an average number of $1, \frac{1}{2}, \frac{2}{3}, \frac{1}{2}, 0$ stable extensions, respectively. So while the number of attacks linearly increases, the average number of extensions first decreases, then increases and then decreases again. Qualitatively speaking, this means that for a fixed number of arguments, there are certain numbers of attacks where the average number of extensions is locally maximal or minimal, respectively.

We have seen in the proofs of the results above that already the closed-form solutions for values of $\bar{\sigma}(n,2)$ and $\bar{\sigma}(n,n^2-3)$ are quite hard to obtain. To nevertheless get an inkling of the characteristic distribution of stable extensions, we have set out to study the problem in an empirical way.

6.3 Empirical Results

As we have seen, combinatorial explosion stood in our way of mathematically analyzing the average number of stable extensions. While the same combinatorial explosions prevent us from an exhaustive empirical analysis of the average number of stable extensions, we can still use methods from descriptive statistics to draw some meaningful conclusions. The basic idea is simple: instead of computing the average number of stable extensions for *all* AFs in some class such as $\mathcal{A}_{n,x}$, we only analyze a uniformly drawn random sample $S \subseteq \mathcal{A}_{n,x}$ of a fixed size $|S|$. We thereby obtain a point estimation of the actual (hidden) parameter $\bar{\sigma}(n,x)$.

6.3.1 Experimental Setup

We wrote a program that randomly samples AFs with specific parameters and determines how many stable extensions they have. To create a random AF, we first set $A = \{1, \ldots, n\}$. To create attacks we then randomly select x elements from the set $A \times A$ with equal probability for each pair. Thus we obtain an AF $\mathcal{F} = (A, R) \in \mathcal{A}_{n,x}$. For a given n, this process is repeated for all $0 \leq x \leq n^2$. Now for each AF thus created, we determine the number of stable extensions as follows: We use the translation of Dung [Dung, 1995, Section 5] to transform the AF into a logic program. By [Dung, 1995, Theorem 62], the stable models of this logic program and the stable extensions of the AF are in one-to-one-correspondence. Using the answer set solver clingo [Gebser et al., 2011], we determine the number of stable models of the program and thus the number of stable extensions of the AF. So for a given n, we can empirically estimate the average number of stable extensions in each sample set of AFs with n arguments and x attacks for all $0 \leq x \leq n^2$.

6.3.2 Average Number of Stable Extensions

To check the experimental setup, we first ran the experiment with $n = 2$ and observed that the empirical results agreed with the predictions of Section 6.2.2. The results for $n = 20$ are depicted in a scatter plot, in Figure 6.4 on page 138; the results for $n = 50$ are plotted likewise in Figure 6.5, page 139.

The empirical data clearly vindicate our analytical predictions for very small and very large numbers of attacks. In between, the data furthermore confirm our predictions about the emergence of local minima and maxima. In addition to the experiments that are graphically depicted, we present the positions of these empirically obtained minima and maxima for several additional small n in Table 6.1.

For the local minimum and for small n, an approximation of the position $x_{\min}$ of the local minima from below is given by $n^2 - n \cdot \sqrt{n}$. More precisely – and astonishingly –, the position of the local maximum *always* coincides with $n^2 - n$. On an intuitive level, this suggests that removing n attacks from a full AF with n arguments quite probably leads to AFs for which *both adding and removing* attacks leads to a *decrease* in the number of stable extensions. To investigate this issue somewhat deeper, we next analyzed how the average number of stable extensions came about.

6.3.3 Number of AFs with at most one Stable Extension

The point estimator *sample mean* we used for approximating $\bar{\sigma}(n, x)$ does not per se tell us anything about the distribution of 0-AFs, 1-AFs, ..., y-AFs among the AFs sampled. Recall that a y-AF is an AF with exactly y stable extensions. In principle, an average number of 0.5 stable extensions could be obtained by a 50/50-ratio of 0-AFs to 1-AFs, or likewise by a 75/25-ratio of 0-AFs to 2-AFs. To find out what is the case, we extracted the absolute frequency of 0-AFs and 1-AFs from our results for $n = 50$ and plotted them in the stacked histogram (Figure 6.6) on page 141.

6.3. Empirical Results

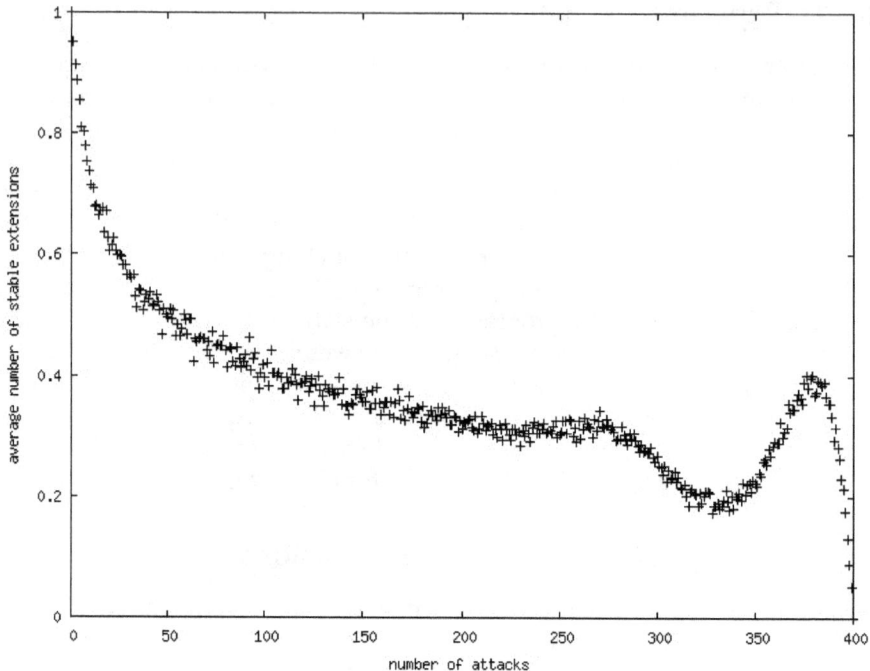

Figure 6.4: Average number of stable extensions of AFs with n = 20 arguments. The values have been obtained from a random sample of size 2500 for each possible number $0 \leq x \leq 400$ of attacks. (So the total sample size is 1 002 500.) We can see that there is a significant local minimum at $x_{\min} \approx 330$ and a local maximum at $x_{\max} \approx 380$.

The stacked histogram for $n = 20$ looks alike, indeed as much as the scatterplots in Figures 6.4 and 6.5 do. This suggests that there are certain recurring features in this distribution that are independent of the number n of arguments. It cannot be seen in the histogram, but we also observed that for any set of sampled AFs from $\mathcal{A}_{50,x}$ with $0 \leq x \leq 50^2$, there are typically more 1-AFs than 2-AFs, more 2-AFs than 3-AFs, and so on. This gives some hints about the sizes of the subclasses of 1-AFs, 2-AFs, ... in a given class $\mathcal{A}_{n,x}$.

We close the empirical section by presenting two conjectures supported by the obtained results. The first one is concerned with the cardinality of y-AFs for a fixed number n of arguments.

Conjecture 6.9. *For any natural numbers n, k and l with $0 < k < l \leq n$ we have:*

$$|\{\mathcal{F} \mid \mathcal{F} \in \mathcal{A}_n,\ \mathcal{F} \text{ is a } k\text{-AF}\}| \geq |\{\mathcal{G} \mid \mathcal{G} \in \mathcal{A}_n,\ \mathcal{G} \text{ is an } l\text{-AF}\}|.$$

The second conjecture claims that the average number of stable extensions of AFs is always located in between 0 and 1. Here is the precise formulation.

Conjecture 6.10. *For any natural numbers n and x with $0 < x < n^2$ we have:*

$$0 < \bar{\sigma}(n, x) < 1.$$

Chapter 6. Cardinality Results

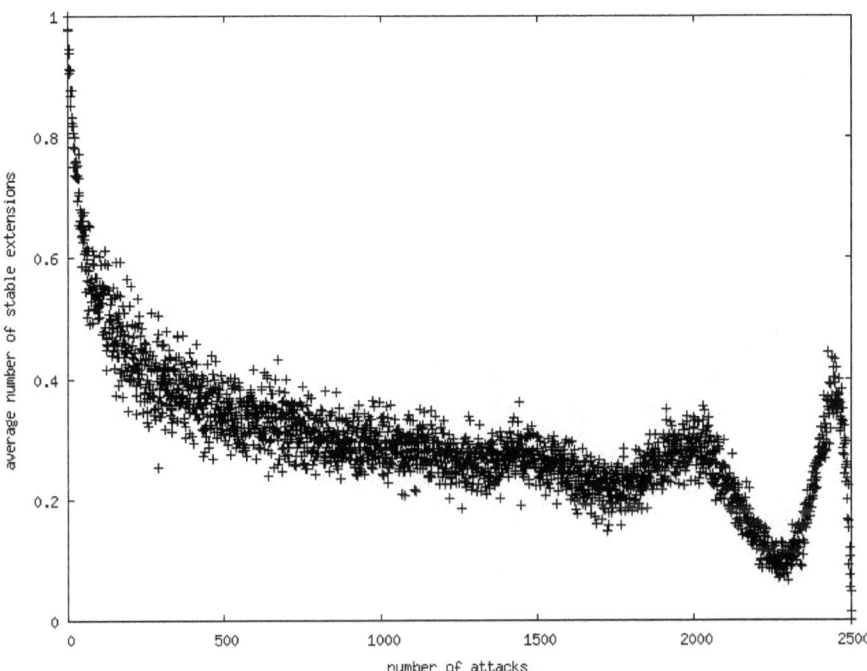

Figure 6.5: Average number of stable extensions of AFs with $n = 50$ arguments and sample size 400 for each $0 \leq x \leq 2500$. Again, there are significant extrema: a local minimum at $x_{\min} \approx 2250$ and a local maximum at $x_{\max} \approx 2450$. It even seems that there is another local maximum at $x'_{\max} \approx 2000$ and another local minimum before that, but the data are unreliable. $\left(\text{Recall that for } x = 2000 \text{ the number of AFs to sample from is } |A_{50,2000}| = \binom{2500}{2000} \geq \left(\frac{2500}{2000}\right)^{2000} \approx 6.6 \cdot 10^{193}.\right)$

6.4 Conclusions and Related Work

We have conducted a detailed analytical and empirical study on the maximal and average numbers of stable extensions in abstract argumentation frameworks. The presented results are already published in [Baumann and Strass, 2013]. First of all, we have proven a tight upper bound on the maximal number of stable extensions. For specific numbers of attacks, we have also given the precise average number of stable extensions in terms of closed-form expressions. As the calculation of these analytical values tends to be quite complex, we turned to studying the problem empirically. There, we obtained data about the distribution of stable extensions in samples of AFs which were randomly drawn with a uniform probability. Our empirical results offer new insights into the average number and also the distribution of stable extensions for AFs, given only the parameters n (number of arguments) and x (number of attacks).

6.4. Conclusions and Related Work

n	2	3	4	5	6	7	8	9	10
x_{min}	1	4	9	15	23	32	45	57	73
$n^2 - n \cdot \sqrt{n}$	1.17	3.80	8	13.82	21.30	30.48	41.37	54	68.38
e_{abs}	0.17	0.2	1	1.18	1.7	1.52	3.63	3	4.62
e_{rel}	0.17	0.04	0.11	0.08	0.07	0.05	0.08	0.05	0.06
x_{max}	2	6	12	20	30	42	56	72	90
$n^2 - n$	2	6	12	20	30	42	56	72	90
$e_{abs} = e_{rel}$	0	0	0	0	0	0	0	0	0

Table 6.1: Positions (at a specific number x of attacks) of empirically observed local minima and maxima (denoted by x_{min} or x_{max}, respectively) of the average number of stable extensions of AFs with n arguments. We additionally present the values of our analytical estimations. To approximate the position of the minima, we devised the function $n^2 - n \cdot \sqrt{n}$; for the maxima we obtained $n^2 - n$. The rows labelled by e_{abs} and e_{rel} show the absolute and relative error of these estimates.

We could not provide exhaustive theoretical explanations for the many empirical observations we have made, and consider this as one of the major future directions of this research. First and foremost we consider it important to work on proving or disproving the conjectures we explicitly formulated at the end of the previous section. Also the conjectured local maximum of the average number of stable extensions at $n^2 - n$ attacks deserves some attention. A possible way to tackle these conjectures may be to look at subclasses of AFs with special structural properties, such as having no self-loops, or more generally no cycles, those being symmetric, or the ones with a specific average connectivity. Finally, it is clear that many of the questions we asked about stable extension semantics can be asked about the other standard semantics.

Note that our results are not only of interest to the argumentation community: We have seen in the proof of Theorem 6.5 that there is a close relationship between stable extensions of AFs and maximal independent sets of undirected graphs. Indeed, maximal independent sets are sometimes called "stable sets" in the graph theory literature. In a sense, stable extensions represent a directed generalization of maximal independent sets, where the ⊆-maximality condition has been replaced by the condition that all nodes not in the set must be reached by a directed edge from the set. So there is also a

Chapter 6. Cardinality Results

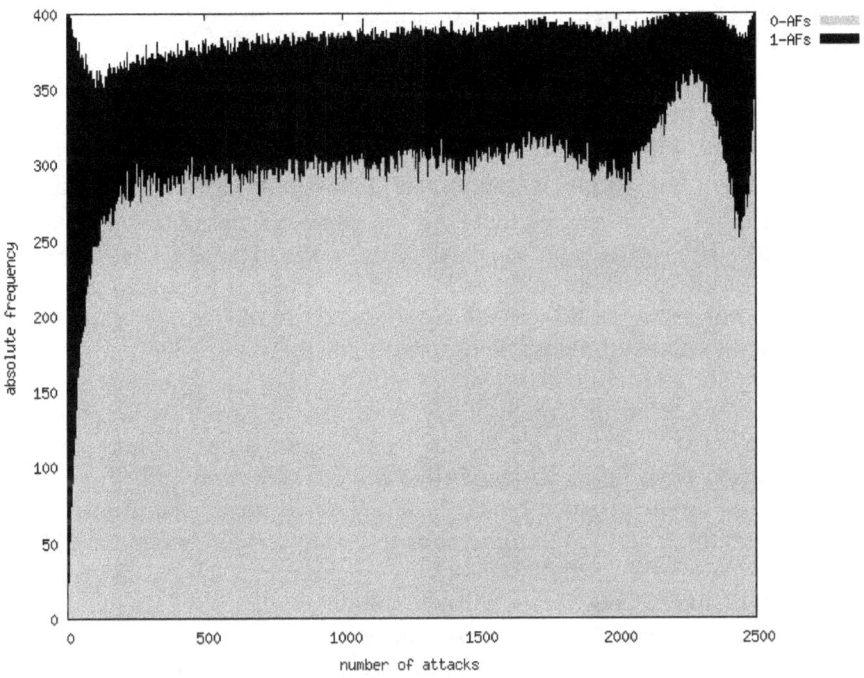

Figure 6.6: Absolute frequencies of 0-AFs (gray) and 1-AFs (black) among all AFs with $n = 50$ arguments and x attacks for $0 \leq x \leq n^2 = 2500$ with a total sample size of $1\,000\,400$. It is obvious from the histogram that the majority (at least two thirds) of all sampled AFs have no stable extension. Additionally, almost all AFs have at most one stable extension. The white area at the top consequently depicts the y-AFs for $y \geq 2$. For $x \approx 100 = 2n$, there is a meaningful number of such y-AFs, which however decreases with increasing x. (Note that the extremal graphs defined in Theorem 6.5 have n arguments and $2n$ attacks.) At $x \approx 2250$, where the average number of stable extensions has a local minimum, the absolute frequency of 0-AFs has a local maximum; furthermore at this position there are almost no y-AFs for $y \geq 2$. Conversely, at $x \approx 2450$ where the average number of stable extensions has a local maximum, the absolute frequency of 0-AFs has a local minimum; furthermore there are yet again y-AFs for $y \geq 2$.

graph theoretical significance to our results.

For abstract argumentation, our results show that – in the context of stable semantics – attacks cannot simply be thought of as constraining: adding an attack may sometimes increase and sometimes decrease the number of stable extensions. Although this might be obvious in general to argumentation researchers (AFs are, after all, a nonmonotonic formalism), for the first time we were able to present some precise numerical figures around this phenomenon.

The present work is also related to recent work on realizability in abstract argumentation [Dunne et al., 2013]. Realizability addresses the following question: given a set X of sets of arguments, is there an argumentation framework whose set of extensions exactly coincides with X? From the results, we immediately know that the answer is "no" if X involves n distinct arguments and the cardinality of X is greater than $3^{\frac{n}{3}}$. I-maximality and Sperner's theorem do not tell us that much: with $n = 6$ arguments, for example, I-maximality only guarantees that at least $2^{\frac{6}{2}} = 8$ extensions can be realized, while our construction shows that $3^{\frac{n}{3}} = 9$ is perfectly possible *and more than that is impossible*. Conversely, the cardinality of the extension-set X gives an indication of the minimal number of arguments needed to realize the extensions in X. For example, if there are 10 extensions to realize, we immediately know that we will need at least 7 arguments for that.

Our current results on the average number of stable extensions regard all possible AFs to occur equally likely. In future research, we want to look at AFs that occur "in practice," that is, from instantiations of more concrete argumentation languages. In [Modgil et al., 2013, Section 1.5], the authors acknowledge the need for a benchmark library in abstract argumentation. In particular, they mention that the library should contain benchmarks "that arise from real-world instantiations of argumentation." We consider the development of such a benchmark collection an important prerequisite for analyzing empirical properties of their instances.

A further related work is [Baroni et al., 2010]. Here the problem of counting without explicitly enumerating extensions was studied from a computational point of view. It was shown that in general counting the number of stable extensions of an argumentation framework is a computationally hard problem.

Chapter 7

Enforcing and Minimal Change

More recently several problems regarding *dynamic* aspects of abstract argumentation have been addressed in the literature [Boella et al., 2009; Cayrol et al., 2010; Bisquert et al., 2011; Liao et al., 2011]. One much cited problem among these concerns the acceptability of certain arguments and is called *enforcing problem* [Baumann and Brewka, 2010]. This is, in brief, the question whether it is possible, given a specific set of allowed operations, to modify a given AF such that a desired set of arguments becomes an extension or a subset of an extension of the modified AF. Several sufficient conditions under which enforcements are (im)possible were identified.

Consider the following snapshot of a dialogue among agents A and B depicted in Figure 7.1. Assume it is A's turn and his desired set of arguments is $E = \{a_1, a_2, a_3\}$. Furthermore, A and B are discussing under preferred semantics, which selects maximal conflict-free and self-defending sets of arguments.

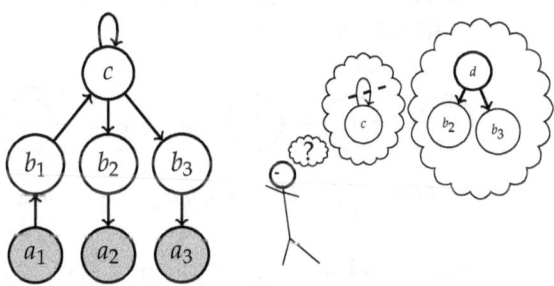

Figure 7.1: *Snapshot of a Dialogue*

In order to enforce E agent A may come up with new arguments which interact with the old ones (for example through introducing an argument which attacks b_2 and b_3) and/or question old arguments or attacks between them, respectively (for example through questioning the self-attack of c). Please note that firstly, in this scenario enforcing is possible and secondly, there are at least two different possibilities to achieve that. This observation leads us to the more general problem of *minimal change* [Baumann, 2012b]. To be more precise, we are not only interested whether enforcements are *possible*, but also in the *effort needed* to enforce a set of arguments. The numerical measure we will use for this effort corresponds to the number of modifications needed to transform the given AF into an AF in which E is enforced. The minimal number of additions or removals of attacks to reach such an enforcement is the so-called *characteristic* of E. Quite surprisingly, it was shown that, in case of certain semantics and modification types, there are local criteria to determine the minimal number, although infinitely many possibilities to modify a given AF exist.

The term *minimal change* traditionally concerns belief change, in particular belief revision (cf. the AGM approach [Alchourrón et al., 1985]). Our investigation shares some common ground with this area, in particular the central role of theories or knowledge bases which are most similar to the initial one. Nevertheless, there are important differences. First of all, completely different formalisms are used, e.g. classical logic in AGM and AFs here. Secondly, while belief revision aims at incorporating new beliefs, we are interested in enforcing already given but not yet accepted sets of arguments through adding new information. For an excellent elaboration of the relationships between argumentation and belief revision we refer the reader to [Falappa et al., 2009].

A further interesting and important question is the mutual replaceability or similarity between AFs in the light of achieving goals (i.e. enforcing desired sets). Without doubt being able to identify such similarities among argumentation scenarios is a big advantage for an agent. For instance, he may safely accept more counterattacks to his arguments knowing that they do not make it more difficult at all to reach his goals. How to decide whether two AFs are mutually replaceable with respect to minimal changes? Consider the following AFs which might stem from modeling the same scenario but where the underlying notions of attack are different or, putting it less abstract, the attack (a_3, a_1) is in question.

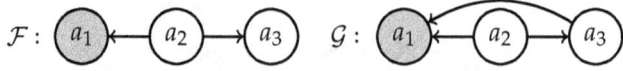

Figure 7.2: Minimal Change Equivalence

The frameworks $\mathcal{F}$ and $\mathcal{G}$ are on a par in a static, non-dynamical sense since both possess the unique preferred extension $\{a_2\}$. If we assume that the agents desired set of arguments is $E = \{a_1\}$ the situation becomes different because although, $\mathcal{F}$ and $\mathcal{G}$ are equivalent in the standard sense, the effort needed to getting a_1 accepted is different. To be more precise, using the results presented in this chapter it can be verified that the characteristic of E with respect to arbitrary expansions equals 1 in $\mathcal{F}$ and 2 in $\mathcal{G}$. This leads us to-

Chapter 7. Enforcing and Minimal Change

wards novel dynamic notions of equivalence which guarantee equal minimal efforts needed to enforce certain subsets, namely *minimal-D-equivalence* and the more general *minimal change equivalence*. We present characterization theorems for several Dung semantics and furthermore, we show the relations to standard and expansion equivalence in general as well as a complete picture for stable and preferred semantics.

Further motivations or theoretical and practical applications of our work are the following:

- *dialogue strategies*: Given an abstract argumentation scenario, assume that one agent reaches his goal if he enforces at least one of several desired sets of arguments. With the help of characteristics he may figure out which set is the closest one to being accepted in the current scenario. Furthermore, he may develop strategies which decrease or at least do not increase the characteristic.

- *monotonic vs. non-monotonic behaviour*: Argumentation is non-monotonic. Since every extension possesses a characteristic of zero, one may analyze and characterize dynamic scenarios where former extensions or at least accepted arguments survive, i.e. the characteristic remains zero. The characterization of monotonic parts contributes to a better understanding of the inherent non-monotonic formalism.

- *implicit vs. explicit information*: As demonstrated by the AFs depicted in Figure 7.2, knowledge about the extensions alone does not make explicit the implicit information with respect to minimal efforts needed to enforce certain subsets. In this sense minimal change equivalent AFs are insensitive to dynamics; in other words, they share the same implicit information.

A further, at first glance more theoretical problem we study in this chapter is the so-called *spectrum problem* [Baumann and Brewka, 2013b]. The name was chosen because of its similarity with the famous *Spektralproblem*[1] in model theory [Scholz, 1952]. Given a certain semantics σ and a modification type Φ, we study whether there is, for a given natural number n, an AF $\mathcal{F}$ and a set of arguments E such that n is the (σ, Φ)-characteristic of E with respect to $\mathcal{F}$. In other words, we want to determine the set of all natural numbers which may occur as (σ, Φ)-characteristics, the so-called (σ, Φ)-*spectrum*. This yields interesting insights into particular semantics. To mention one result, we will show that in case of semi-stable semantics and the addition of weak arguments (arguments which do not attack previous arguments) not each natural number may arise as the minimal effort needed to enforce a certain set E. In particular, the characteristic cannot be 1.

What makes our study even more interesting, as we believe, is the fact that it provides useful and at times surprising new insights into the interrelationships among the studied semantics. To this end, we perform our analysis

---

[1]Roughly speaking, Scholz investigated the possible sizes finite models of a first-order sentence may have.

in parallel for a whole group of semantics which we consider as some of the most important semantics for Dung frameworks. Rather than sets of values, spectra thus become sets of tuples of values.

7.1 Enforcing Problem

7.1.1 Conservative and Liberal Enforcements

As stated at the very beginning of this chapter enforcing an extension E means modifying an argumentation framework in such a way that E becomes one of its extensions or at least a subset of an extension. The modifications we are interested in here are normal expansions and changes in the semantics. The former correspond to additional arguments that are brought into play together with the attack relations among them and older arguments. The latter can be viewed as a switch in the applied proof standard, for instance a switch from stable semantics to the more cautious grounded semantics.

Definition 7.1. Let $\mathcal{F}, \mathcal{G}$ be AFs and σ, τ semantics. A pair $(\mathcal{G}, \tau)$ is called an $(\mathcal{F}, \sigma)$-*enforcement* of E if (1) $\mathcal{F} \leq_N \mathcal{G}$, and (2) there is a E', such that $E \subseteq E'$ and $E' \in \mathcal{E}_\tau(\mathcal{G})$. $(\mathcal{G}, \tau)$ is called

1. *strict* if $E \in \mathcal{E}_\tau(\mathcal{G})$,

2. *conservative* if $\sigma = \tau$,

3. *liberal* if $\sigma \neq \tau$,

4. *strong* (resp. *weak*) if $\mathcal{F} \leq_S \mathcal{G}$ (resp. $\mathcal{F} \leq_W \mathcal{G}$).

Whenever $\mathcal{F}$ and σ are clear from context we simply speak of enforcements of E. A *strict* enforcement explicitly demands that the desired set of arguments becomes an extension. Furthermore, if the considered semantics remains constant we call an enforcement *conservative*. Almost all existing papers dealing with belief revision consider a fixed semantics. *Liberal* enforcements change the semantics. As mentioned earlier, this may be interpreted as a change of proof standard or paradigm shift. Imagine a judicial proceeding. It is vitally important whether you are accused on the base of criminal or civil law. The required evidence is different and hence the acceptable sets of arguments differ.

Note that more general modifications like leaving out previous attack relations or adding further attacks between previous arguments are excluded by our definition. If these types of manipulation are allowed the problem becomes trivial because one may add or delete arguments and attack relations at will.

To familiarize the reader with enforcements we give two examples.

Example 7.2. Consider the following AF $\mathcal{F}$:

Chapter 7. Enforcing and Minimal Change

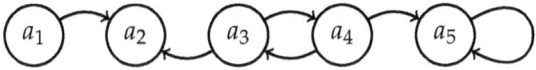

Figure 7.3: Liberal Enforcement

Let σ be stable semantics and $E = \{a_1, a_3\}$ the desired set of arguments. Obviously, $\mathcal{E}_{stb}(\mathcal{F}) = \{\{a_1, a_4\}\}$. How to enforce E? Define an enforcement $(\mathcal{G}, \tau)$ of E with $\mathcal{F} = \mathcal{G}$ and $\tau = pr$. Indeed, E becomes preferred in $\mathcal{G}$ and thus, E is enforced. The considered enforcement is strict and liberal.

Example 7.3. Consider $\mathcal{F} = (A, R) = (\{a_1, a_2, a_3\}, \{(a_1, a_2), (a_2, a_1), (a_2, a_3)\})$ depicted below:

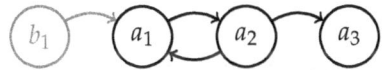

Figure 7.4: Conservative Enforcement

Let σ be grounded semantics and $E = \{a_2\}$ the desired set of arguments. The set E is not accepted because $\mathcal{E}_{gr}(\mathcal{F}) = \{\emptyset\}$. How to enforce E? Define an enforcement with $\mathcal{G} = (A \cup \{b_1\}, R \cup \{(b_1, a_1)\})$. Now, $\{\{b_1, a_2\}\} = \mathcal{E}_{gr}(\mathcal{F})$ and therefore E is enforced. The considered enforcement is non-strict, conservative and strong.

7.1.2 Impossibility Results

The two Examples 7.2 and 7.3 illustrate the possibility to enforce a desired set of arguments. In this subsection we study scenarios where strict enforcing is impossible. Instead of considering particular semantics we will use abstract principles like admissibility and reinstatement (compare Definition 2.11 and Figure 2.4 for an overview). Consequently, our results are general enough to cover even semantics which may be defined in the future.

We now show some useful interrelations between subsets of an AF, its normal expansions and abstract principles of semantics. The following properties are pretty obvious and hardly need any proof. Being aware of this fact, we still present them in the form of a proposition to be able to refer to them later on.

Proposition 7.4. *Given an AF $\mathcal{F} = (A, R)$ and a semantics σ.*

1. *If σ satisfies admissibility and $E \subseteq A$ does not defend all its elements in $\mathcal{F}$, then there is no strict conservative enforcement of E.*

2. *If σ satisfies reinstatement and $E \subseteq A$ does not contain all defended elements in $\mathcal{F}$, then there is no strict conservative weak enforcement of E.*

3. *If σ satisfies conflict-freeness and $E \subseteq A$ is conflicting, i.e. $(E, E) \in R$, then there is no conservative enforcement of E.*

Proof. 1. For any AF $\mathcal{G}$, such that $\mathcal{F} \leq_N \mathcal{G}$ we observe that E does not defend all its elements in $\mathcal{G}$ since normal expansions remain the former attack relation unchanged. Thus, $E \notin \mathcal{E}_\sigma(\mathcal{G})$ since σ satisfies admissibility.
2. Assume that E does not contain all defended arguments in $\mathcal{F}$. Without loss of generality let a be such an argument. Furthermore, let $\mathcal{G}$ be a weak expansion of $\mathcal{F}$. Thus, we deduce that all possible attackers of a are already elements of A. Consequently, E does not contain all defended arguments in $\mathcal{G}$ either. Thus, $E \notin \mathcal{E}_\sigma(\mathcal{G})$ since σ satisfies reinstatement.
3. Obvious. Any superset of a conflicting set is conflicting. □

The proposition above presents obvious necessary conditions for the enforcing problem. The following impossibility theorems are more sophisticated than the mentioned proposition. The first one shows limitations for exchanging believed with unattacking arguments.

Theorem 7.5. *(exchanging arguments) Given an AF $\mathcal{F} = (A, R)$ and*

1. *a semantics σ satisfying reinstatement,*

2. *a semantics τ, satisfying admissibility and conflict-freeness,*

3. *a set E such that $E \in \mathcal{E}_\sigma(\mathcal{F})$ and*

4. *two sets E', C such that $E' \subseteq E$, $C \subseteq A \setminus E$, $C \neq \emptyset$, $(C, A \setminus \{E' \cup C\}) \notin R$ and $E^* := E' \cup C \notin \mathcal{E}_\sigma(\mathcal{F})$,*

then there is no pair $(\mathcal{G}, \tau)$, such that $(\mathcal{G}, \tau)$ is a strict $(\mathcal{F}, \sigma)$-enforcement of E^*.

Proof. Suppose, towards a contradiction, that $\mathcal{F} \leq_N \mathcal{G}$ and $E^* \in \mathcal{E}_\tau(\mathcal{G})$. Since C is non-empty subset of $A \setminus E$ we deduce the existence of an argument c, such that $c \in C \wedge c \notin E$. Using that σ satisfies reinstatement and $E \in \mathcal{E}_\sigma(\mathcal{F})$ we deduce that c is not defended by E in $\mathcal{F}$. This means, there is an argument b, such that $b \in A$, $(b, c) \in R$ and $(E, b) \notin R$ (*). Since τ satisfies admissibility and $E^* \in \mathcal{E}_\tau(\mathcal{G})$ is assumed we conclude $(E^*, b) \in R \cup R(\mathcal{G})$. Note that $(E^*, b) \in R(\mathcal{G})$ is impossible since new attacks have to involve at least one new argument. Thus, $(E^*, b) \in R$ has to hold. Consequently, $(E', b) \in R$ or $(C, b) \in R$. The former disjunct is impossible because of (*) and the assumption $E' \subseteq E$. This means, $(C, b) \in R$ (+) has to hold. Obviously $b \notin E^*$ because τ satisfies conflict-freeness and $E^* \in \mathcal{E}_\tau(\mathcal{G})$ is assumed. This means, $b \in A \setminus \{E' \cup C\}$. Consequently, in contrast to (+) we deduce $(C, b) \notin R$ since $(C, A \setminus \{E' \cup C\}) \notin R$ is assumed. □

Intuitively, the theorem says the following: if the involved semantics satisfy the specified properties, then it is impossible to find a normal expansion which possesses an extension E^* composed of a subset of an old extension E and some formerly unaccepted arguments C, given no element of C attacks some element which is not in the new extension.

The second impossibility theorem demonstrates limitations for eliminating arguments of existing extensions.

Theorem 7.6. *(eliminating arguments) Given an AF $\mathcal{F} = (A, R)$ and*

Chapter 7. Enforcing and Minimal Change

1. a semantics σ satisfying admissibility and conflict-freeness,
2. a semantics τ satisfying reinstatement,
3. a set E such that $E \in \mathcal{E}_\sigma(\mathcal{F})$ and
4. a set C such that $C \subset E$, $(C, A \setminus E) \notin R$ and $E^* := E \setminus C \notin \mathcal{E}_\sigma(\mathcal{F})$,

then there is no pair $(\mathcal{G}, \tau)$, such that $(\mathcal{G}, \tau)$ is a strict weak $(\mathcal{F}, \sigma)$-enforcement of E^*.

Proof. To obtain a contradiction, we suppose that $\mathcal{F} \leq_W \mathcal{G}$ and $E^* \in \mathcal{E}_\tau(\mathcal{G})$. Observe that C has to be non-empty since $E \in \mathcal{E}_\sigma(\mathcal{F})$ and $E^* \notin \mathcal{E}_\sigma(\mathcal{F})$ are assumed. Thus, we deduce the existence of an argument c, such that $c \in C \wedge c \notin E^*$. Using that τ satisfies reinstatement and $E^* \in \mathcal{E}_\tau(\mathcal{G})$ we state that c is not defended by E^* in $\mathcal{G}$. This means, there is an argument b, such that $b \in A \cup A(\mathcal{G})$, such that $(b,c) \in R \cup R(\mathcal{G})$ and $(E^*, b) \notin R \cup R(\mathcal{G})$. Therefore, $(E^*, b) \notin R$ (*). Furthermore, $b \in A$ and $(b,c) \in R$ since $\mathcal{G}$ is assumed to be a weak expansion of $\mathcal{F}$. On the other hand, we deduce $(E, b) \in R$ because $c \in E$, $E \in \mathcal{E}_\sigma(\mathcal{F})$ and σ satisfies admissibility. Thus, $b \in A \setminus E$ (+) since σ also fulfills conflict-freeness. Note that $(E, b) \in R$ can be equivalently formulated as $(E^*, b) \in R \vee (C, b) \in R$. The former disjunct contradicts (*) and furthermore, in the light of the assumption $(C, A \setminus E) \notin R$ the latter disjunct is inconsistent with (+). □

Intuitively, the theorem says the following: if the involved semantics satisfy the specified properties, then it is impossible to find a weak expansion possessing an extension which is a proper subset of an old extension and not already an extension of the original AF, unless one of the arguments left out in the new extension attacks an element which was not in the old extension.

We want to point out that both Theorems 7.5 and 7.6 are not necessarily restricted to liberal enforcements although both refer to two semantics. Remember that stable, semi-stable, preferred, complete, grounded and ideal semantics fulfill reinstatement, admissibility and conflict-freeness simultaneously (compare Figure 2.4). Thus, for these semantics the impossibility of exchanging arguments as well as eliminating arguments apply to conservative enforcements.

7.1.3 Possibility Result

Consider the following simple AF $\mathcal{F}$:

Figure 7.5: Strict- vs. Non-strict Enforcements

Statement 1 of Proposition 7.4 as well as Theorem 7.6 imply that there is no strict conservative enforcement of $E = \{a_2\}$ given that the considered

7.1. Enforcing Problem

semantics satisfies admissibility. This means, if we do not change the considered semantics, then it is impossible that E becomes an extension of an AF $\mathcal{G}$ being a normal expansion of $\mathcal{F}$. What about the following weaker claim: Is there a non-strict enforcement of E, i.e. is there a normal expansion $\mathcal{G}$ of $\mathcal{F}$ such that E is a subset of an extension of $\mathcal{G}$? In the following theorem we will prove the positive answer for all semantics considered in this book.

Theorem 7.7. *Given $\sigma \in \{stg, stb, ss, eg, ad, pr, id, gr, co, na\}$. Let $\mathcal{F} = (A, R)$ be an AF and $C \subseteq A$. If C is conflict-free in $\mathcal{F}$, then there is a pair $(\mathcal{G}, \sigma)$ being a strong conservative $(\mathcal{F}, \sigma)$-enforcement of C. Moreover, for $\sigma \in \{stg, stb, ss, eg, pr, id, gr, co\}$ there exists a pair $(\mathcal{G}, \sigma)$ such that $|\mathcal{E}_\sigma(\mathcal{G})| = |\{E\}| = 1$ and $|E \smallsetminus C| = 1$.*

Proof. Let $A = \{a_1, ..., a_n\}$. Furthermore, $C \subseteq A$ and $C \in cf(\mathcal{F})$. The only nontrivial case is $C \notin \mathcal{E}_\sigma(\mathcal{F})$. In this case we conclude that $C = \{a_1, ..., a_i\}$ for some integer $i < n$. We define the following strong expansion $\mathcal{G}$ of $\mathcal{F}$:

$$\mathcal{G} = (A \cup \{b\}, R \cup \{(b, a_{i+1}), ..., (b, a_n)\})$$

We show that $E = C \cup \{b\} \in \mathcal{E}_\sigma(\mathcal{G})$ for any $\sigma \in \{stg, stb, ss, eg, ad, pr, id, gr, co, na\}$. Note that $E \in cf(\mathcal{G})$ is given by construction. Observe that $E \in \mathcal{E}_{stb}(\mathcal{G})$ since any argument not contained in E is attacked by E. By Proposition 2.7 we conclude $E \in \mathcal{E}_\sigma(\mathcal{G})$ for $\sigma \in \{stg, stb, ss, ad, pr, co, na\}$. We now prove that $|\mathcal{E}_{co}(\mathcal{G})| = |\mathcal{E}_{stg}(\mathcal{G})| = 1$. First, $|\mathcal{E}_{co}(\mathcal{G})| = 1$. Assume not, thus there is an $E' \in \mathcal{E}_{co}(\mathcal{G})$, such that $E \neq E'$. Observe that $b \in E'$ since b is unattacked in $\mathcal{G}$. Furthermore, all elements of C are defended by $\{b\}$ in $\mathcal{G}$. Consequently, $E \subseteq E'$. Moreover, $E \subset E'$ is impossible since any superset of E is conflicting in $\mathcal{G}$. Thus, $E = E'$ in contrast to the assumption. Second, $|\mathcal{E}_{stg}(\mathcal{G})| = 1$. Suppose, to derive a contradiction, that there is an $E' \in \mathcal{E}_{stg}(\mathcal{G})$, such that $E \neq E'$. Since $R^+_\mathcal{G}(E) = A(\mathcal{G})$ we conclude the same range for $E' \in \mathcal{G}$, i.e. $R^+_\mathcal{G}(E') = A \cup \{b\}$. Furthermore, $b \in E'$ since b is unattacked in $\mathcal{G}$. Since any stage extension is naive, i.e. maximal conflict-free (Proposition 2.7) we conclude $E \subseteq E'$. Again, $E \subset E'$ is impossible since any superset of E is conflicting in $\mathcal{G}$ by construction. Thus, $E = E'$ contradicting the assumption. Now, combining Propositions 2.7 and 2.8 we obtain $E \in \mathcal{E}_\sigma(\mathcal{G})$ for any $\sigma \in \{stg, stb, ss, eg, ad, pr, id, gr, co, na\}$ and $|\mathcal{E}_\tau(\mathcal{G})| = 1$ for any $\tau \in \{stg, stb, ss, eg, pr, id, gr, co\}$ concluding the proof. □

The theorem shows that whenever a set C is conflict-free we may enforce it by adding one additional argument b and certain attacks so that the union of C and b is the extension of the constructed AF. Moreover, $C \cup \{b\}$ is the unique extension in case of stage, stable, semi-stable, eager, preferred, ideal, grounded and complete semantics. We want to mention that in [Booth et al., 2013, Theorem 1] a similar result is proven. The authours consider the enforcing problem via formulas expressible in a certain logical labeling language.

It is important to emphasize that in special cases the enforcement of the desired set may be reached with less additional attack relations. Our construction shows the potential possibility only. To exemplify the standard construction consider the following example.

Example 7.8. Consider the black subframework $\mathcal{F} = (A, R)$ and the desired set of arguments $C = \{a_2, a_4\}$. Obviously, C is conflict-free in $\mathcal{F}$ and thus, Theorem 7.7 is applicable. Consider therefore the standard construction, namely the AF $\mathcal{G} = (A \cup \{b\}, R \cup \{(b, a_1), (b, a_3), (b, a_5)\})$ being a strong expansion of $\mathcal{F}$.

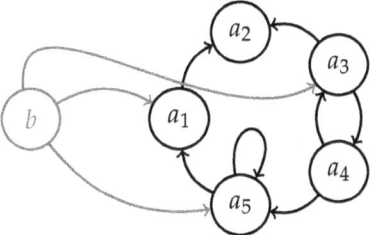

Figure 7.6: Standard Construction

We observe $C \cup \{b\} = \{a_2, a_4, b\} \in \mathcal{E}_\sigma(\mathcal{G})$ for any semantics $\sigma \in \{stg, stb, ss, eg, ad, pr, id, gr, co, na\}$ and furthermore, $|\mathcal{E}_\sigma(\mathcal{G})| = 1$ in case of $\sigma \in \{stg, stb, ss, eg, pr, id, gr, co\}$ in accordance with Theorem 7.7.

7.2 Minimal Change Problem

In the last section we studied the so-called *enforcing problem* in argumentation, i.e. the question whether it is possible to modify a given AF in such a way that a *desired* set of arguments becomes an extension or a subset of an extension. In particular, Theorem 7.7 gives a positive answer for all semantics considered in this book provided that the considered set of arguments is conflict-free. In this section we go an important step further. We are not only interested whether enforcements are *possible*, but also in the *effort needed* to enforce a set of arguments. In a nutshell, the *minimal change problem* [Baumann, 2012b] corresponds to the mathematical problem of determining the minimal number of modifications (additions or removals of attacks) needed to enforce a certain set.

Example 7.9 (Example 7.8 continued). Consider again the black subframework $\mathcal{F} = (A, R)$ and $C = \{a_2, a_4\}$. In Example 7.8 we have shown that enforcing is impossible via $\mathcal{G} = (A \cup \{b\}, R \cup \{(b, a_1), (b, a_3), (b, a_5)\})$.

7.2. Minimal Change Problem

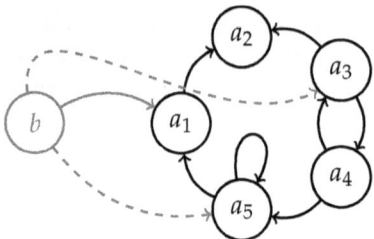

Figure 7.7: Alternative Construction

The attentive reader will have noticed that the dashed attacks (b, a_3) and (b, a_5) are not necessarily needed to enforce E. In particular, $C \cup \{b\} = \{a_2, a_4, b\} \in \mathcal{E}_\sigma(\mathcal{G}')$ for any semantics $\sigma \in \{stg, stb, ss, eg, ad, pr, id, gr, co, na\}$ where $\mathcal{G}' = (A \cup \{b\}, R \cup \{(b, a_1)\})$.

The example above shows that in special cases the enforcement of a desired set C may be reached with less additional attack relations in comparison to the standard construction. We want to emphasize that there is an infinite number of possibilities to modify a given argumentation scenario. Consequently, there is no brute force method solving the minimal change problem.

Our main contributions can be summarized as follows:

- Formalizing and characterizing the minimal change problem for weak, strong, normal, arbitrary expansions as well as arbitrary modifications for the stable, preferred, complete and admissible semantics.

- The most remarkable result is that the *characteristic* (representing the minimal change problem formally) does not change if we switch simultaneously between strong, normal or arbitrary expansions and preferred, complete or admissible semantics.

- For semi-stable semantics we show that its characteristic lies between the stable and preferred characteristic for all mentioned modification types.

7.2.1 Characteristics - A Formal Representation

To formalize the minimal change problem we have to introduce a numerical measure which indicates how far apart two argumentation scenarios are. We decided to count only added and removed attacks for a simple reason: the very nature of argumentation is the treatment of conflicting arguments. Adding or removing an isolated argument does not contribute at all to solving or increasing a given conflict, i.e. the conflicting information remains the same. This means, the decrease or increase of a conflict is directly linked to upcoming or disappearing attacks. The following definition takes this idea into account and can be formalized by the well-known symmetric difference $A \Delta B =_{def} (A \smallsetminus B) \cup (B \smallsetminus A)$.

Chapter 7. Enforcing and Minimal Change

Definition 7.10. The *distance* between two AFs $\mathcal{F}$ and $\mathcal{G}$ is a natural number defined by the following function

$$d : \mathscr{A} \times \mathscr{A} \to \mathbb{N} \quad (\mathcal{F}, \mathcal{G}) \mapsto |R(\mathcal{F}) \triangle R(\mathcal{G})|.$$

The following proposition states that the class of all AFs $\mathscr{A}$ together with the above defined distance d constitute a pseudometric space. Remember that in pseudometric spaces two distinct elements may have distance zero.

Proposition 7.11. $(\mathscr{A}, d)$ *is a pseudometric space.*

Proof. We have to prove the following three conditions: Let $\mathcal{F}, \mathcal{G}, \mathcal{H} \in \mathscr{A}$, then

1. $d(\mathcal{F}, \mathcal{F}) = 0$ (vanishing self-distance)
2. $d(\mathcal{F}, \mathcal{G}) = d(\mathcal{G}, \mathcal{F})$ (symmetry)
3. $d(\mathcal{F}, \mathcal{H}) \leq d(\mathcal{F}, \mathcal{G}) + d(\mathcal{G}, \mathcal{H})$ (triangle inequality)

We only prove the triangle inequality. Given $\mathcal{F}, \mathcal{G}, \mathcal{H} \in \mathscr{A}$. Observe that

$$\begin{aligned} R(\mathcal{F}) \triangle R(\mathcal{H}) &= (R(\mathcal{F}) \smallsetminus R(\mathcal{H})) \cup (R(\mathcal{H}) \smallsetminus R(\mathcal{F})) \\ &\subseteq (R(\mathcal{F}) \smallsetminus R(\mathcal{G})) \cup (R(\mathcal{G}) \smallsetminus R(\mathcal{H})) \\ &\quad \cup (R(\mathcal{H}) \smallsetminus R(\mathcal{G})) \cup (R(\mathcal{G}) \smallsetminus R(\mathcal{F})) \\ &= (R(\mathcal{F}) \triangle R(\mathcal{G})) \cup (R(\mathcal{G}) \triangle R(\mathcal{H})). \end{aligned}$$

Consequently,

$$\begin{aligned} d(\mathcal{F}, \mathcal{H}) &= |R(\mathcal{F}) \triangle R(\mathcal{H})| \\ &\leq |(R(\mathcal{F}) \triangle R(\mathcal{G})) \cup (R(\mathcal{G}) \triangle R(\mathcal{H}))| \\ &\leq |(R(\mathcal{F}) \triangle R(\mathcal{G}))| + |(R(\mathcal{G}) \triangle R(\mathcal{H}))| \\ &= d(\mathcal{F}, \mathcal{G}) + d(\mathcal{G}, \mathcal{H}). \end{aligned}$$

□

Example 7.12. The distances between the following three AFs are: $d(\mathcal{F}, \mathcal{G}) = 2$, $d(\mathcal{F}, \mathcal{H}) = 3$ and $d(\mathcal{G}, \mathcal{H}) = 5$.

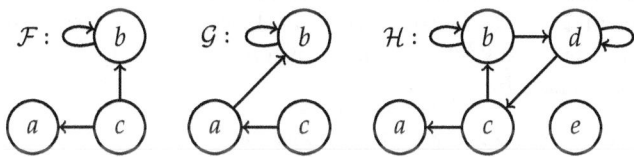

Figure 7.8: Distances

Now we are prepared to describe the minimal change problem formally. We define therefore a function, the so-called (σ, Φ)-*characteristic*[2]. The σ indicates the considered semantics and Φ describes the possible modifications.

---

[2] If the semantics σ and the modification type Φ are clear from the context (or unimportant) we only refer to characteristic. The same holds for the later defined value functions.

7.2. Minimal Change Problem

Definition 7.13. Given a semantics σ, a binary relation $\Phi \subseteq \mathcal{A} \times \mathcal{A}$ and an AF $\mathcal{F}$. The (σ, Φ)-*characteristic* of a set $C \subseteq A(\mathcal{F})$ is a natural number or infinity defined by the following function

$$N^{\mathcal{F}}_{\sigma,\Phi} : \wp(A(\mathcal{F})) \to \mathbb{N}_\infty \quad C \mapsto \begin{cases} 0, & \exists C' : C \subseteq C' \text{ and } C' \in \mathcal{E}_\sigma(\mathcal{F}) \\ k, & k = \min\{d(\mathcal{F}, \mathcal{G}) \mid (\mathcal{F}, \mathcal{G}) \in \Phi, N^{\mathcal{G}}_{\sigma,\Phi}(C) = 0\} \\ \infty, & \text{otherwise.} \end{cases}$$

The following proposition constitutes a first relation between different characteristics if certain subset properties between the considered semantics and/or the kinds of modifications are fulfilled.

Proposition 7.14. *Let σ, τ be semantics and Φ, Ψ binary relations over $\mathcal{A}$, such that $\sigma \subseteq \tau$ and $\Phi \subseteq \Psi$. For any AF $\mathcal{F}$, $N^{\mathcal{F}}_{\sigma,\Phi} \geq N^{\mathcal{F}}_{\tau,\Psi}$.*

Proof. We have to show that for every $C \in \wp(A(\mathcal{F}))$, $N^{\mathcal{F}}_{\sigma,\Phi}(C) \geq N^{\mathcal{F}}_{\tau,\Psi}(C)$. Let $N^{\mathcal{F}}_{\sigma,\Phi}(C) = k$ with $k \in \mathbb{N}_\infty$. If $k = \infty$ nothing needs to be shown. Let $0 < k < \infty$. Thus, there is an AF $\mathcal{G}$, such that $(\mathcal{F}, \mathcal{G}) \in \Phi$, $N^{\mathcal{G}}_{\sigma,\Phi}(C) = 0$ and $d(\mathcal{F}, \mathcal{G}) = k$. Since $\Phi \subseteq \Psi$ was assumed we observe $(\mathcal{F}, \mathcal{G}) \in \Psi$. $N^{\mathcal{G}}_{\sigma,\Phi}(C) = 0$ implies the existence of a set C', such that $C \subseteq C'$ and $C' \in \mathcal{E}_\sigma(\mathcal{G})$. Hence, $C' \in \mathcal{E}_\tau(\mathcal{G})$ because $\sigma \subseteq \tau$ was assumed. We thus have found an AF $\mathcal{G}$, such that $d(\mathcal{F}, \mathcal{G}) = k$, $(\mathcal{F}, \mathcal{G}) \in \Psi$ and $N^{\mathcal{G}}_{\tau,\Psi}(C) = 0$. Consequently, $\min\{d(\mathcal{F}, \mathcal{G}) \mid (\mathcal{F}, \mathcal{G}) \in \Psi \text{ and } N^{\mathcal{G}}_{\tau,\Psi}(C) = 0\} \leq k$ which proves $k \geq N^{\mathcal{F}}_{\tau,\Psi}(C)$. In case of $k = 0$, $N^{\mathcal{F}}_{\tau,\Psi}(C) = 0$ follows immediately since $\sigma \subseteq \tau$ was assumed. □

It is well-known that the considered semantics satisfy the following subset relation, $stb \subseteq ss \subseteq pr \subseteq co \subseteq ad$ (compare Proposition 2.7). Hence, we obtain the following corollary.

Corollary 7.15. *For any AF $\mathcal{F}$ and any binary relation Φ over $\mathcal{A}$,*

$$N^{\mathcal{F}}_{stb,\Phi} \geq N^{\mathcal{F}}_{ss,\Phi} \geq N^{\mathcal{F}}_{pr,\Phi} \geq N^{\mathcal{F}}_{co,\Phi} \geq N^{\mathcal{F}}_{ad,\Phi}.$$

A similar result may be obtained for different kinds of modifications (compare Definition 2.13). Note that strong and weak expansions are incomparable, i.e. there is no subset relation between them. U denotes the universal relation over $\mathcal{A}$, i.e. $U = \mathcal{A} \times \mathcal{A}$.

Corollary 7.16. *For any AF $\mathcal{F}$ and any semantics σ,*

$$N^{\mathcal{F}}_{\sigma,W}, N^{\mathcal{F}}_{\sigma,S} \geq N^{\mathcal{F}}_{\sigma,N} \geq N^{\mathcal{F}}_{\sigma,E} \geq N^{\mathcal{F}}_{\sigma,U}.$$

The following proposition strengthens the result of Corollary 7.15. It shows that the preferred, complete and admissible characteristics coincide.

Proposition 7.17. *For any $\mathcal{F}$ and any binary relation Φ over $\mathcal{A}$,*

$$N^{\mathcal{F}}_{stb,\Phi} \geq N^{\mathcal{F}}_{ss,\Phi} \geq N^{\mathcal{F}}_{pr,\Phi} = N^{\mathcal{F}}_{co,\Phi} = N^{\mathcal{F}}_{ad,\Phi}.$$

Chapter 7. Enforcing and Minimal Change

Proof. We have to show that for every $C \subseteq A(\mathcal{F})$, $N^{\mathcal{F}}_{stb,\Phi}(C) \geq N^{\mathcal{F}}_{ss,\Phi}(C) \geq N^{\mathcal{F}}_{pr,\Phi}(C) = N^{\mathcal{F}}_{co,\Phi}(C) = N^{\mathcal{F}}_{ad,\Phi}(C)$. In consideration of Corollary 7.15 it suffices to show that $N^{\mathcal{F}}_{pr,\Phi}(C) \leq N^{\mathcal{F}}_{ad,\Phi}(C)$. First note that the assertion follows immediately in case of $N^{\mathcal{F}}_{ad,\Phi}(C) = \infty$. Assume now $N^{\mathcal{F}}_{ad,\Phi}(C) = k < \infty$. If $k = 0$, then there is a set C', such that $C \subseteq C'$ and $C' \in \mathcal{E}_{ad}(\mathcal{F})$. Either C' is a subset maximal admissible extension, i.e. $C' \in \mathcal{E}_{pr}(\mathcal{F})$ or not, hence there is a C'', such that $C \subseteq C' \subset C''$ and $C'' \in \mathcal{E}_{pr}(\mathcal{F})$. In both cases, $N^{\mathcal{F}}_{pr,\Phi}(C) = 0$. If $k > 0$ we conclude the existence of an AF $\mathcal{G}$, such that $d(\mathcal{F},\mathcal{G}) = k$, $(\mathcal{F},\mathcal{G}) \in \Phi$ and $N^{\mathcal{G}}_{ad,\Phi}(C) = 0$. Consequently, $N^{\mathcal{G}}_{pr,\Phi}(C) = 0$ (compare subcase $k = 0$). Hence, $\min\{d(\mathcal{F},\mathcal{G}) \mid (\mathcal{F},\mathcal{G}) \in \Phi \text{ and } N^{\mathcal{G}}_{pr,\Phi}(C) = 0\} \leq k$ concluding the proof. $\square$

7.2.2 Value Functions - Local Criteria

The characteristic is a precisely defined numerical value, but as yet we have not provided any means to actually compute this value. This is the question we address in the following three subsections. The significant challenge here is to define a function whose values are equal to the considered characteristics and, furthermore, whose values can be established in a **finite** number of steps based on rather simple properties of the underlying AF.

7.2.2.1 The (σ, W)-value

The first characteristics we are interested in are characteristics with respect to weak expansions. This means, what is the minimal number with respect to the distance d if only the addition of weaker arguments, i.e. arguments which do not attack previous arguments, is allowed. It turns out that there are only two possibilities, namely either a desired set C is already contained in an extension, i.e. the characteristic equals zero, or C is unenforceable, i.e. the characteristic equals infinity.

Definition 7.18. Given an AF $\mathcal{F}$. The (σ, W)-value ($\sigma \in \{stb, ad\}$) of a set $C \subseteq A(\mathcal{F})$ is zero or infinity defined by the following function

$$V^{\mathcal{F}}_{\sigma,W} : \wp(A(\mathcal{F})) \to \{0, \infty\} \qquad C \mapsto \begin{cases} 0, & \exists C' : C \subseteq C' \text{ and } C' \in \mathcal{E}_\sigma(\mathcal{F}) \\ \infty, & \text{otherwise.} \end{cases}$$

Proposition 7.19. *For any AF $\mathcal{F}$ and any semantics $\sigma \in \{stb, ad\}$, $V^{\mathcal{F}}_{\sigma,W} \geq N^{\mathcal{F}}_{\sigma,W}$.*

Proof. Let $C \in \wp(A(\mathcal{F}))$. In case of $V^{\mathcal{F}}_{\sigma,W}(C) = \infty$ the inequality is obviously fulfilled. If $V^{\mathcal{F}}_{\sigma,W}(C) = 0$, then there is a set C' such that $C \subseteq C'$ and $C' \in \mathcal{E}_\sigma(\mathcal{F})$. In consideration of Definition 7.13 we conclude $N^{\mathcal{F}}_{\sigma,\leq^N_W}(C) = 0$ concluding the proof. $\square$

Proposition 7.20. *For any AF $\mathcal{F}$ and any semantics $\sigma \in \{stb, ad\}$, $V^{\mathcal{F}}_{\sigma,W} \leq N^{\mathcal{F}}_{\sigma,W}$.*

7.2. Minimal Change Problem

Proof. Let $C \in \mathcal{P}(A(\mathcal{F}))$ and $V^{\mathcal{F}}_{\sigma,W}(C) = k$. For $k = 0$ we have nothing to show since $0 \leq N^{\mathcal{F}}_{\sigma,W}(C)$ holds. Let $k = \infty$. Hence, there is no superset C' of C with the property $C' \in \mathcal{E}_\sigma(\mathcal{F})$ (*). Assume now the existence of an AF $\mathcal{G}$ with $\mathcal{F} \leq_W \mathcal{G}$ and a set C'', such that $C \subseteq C''$ and $C'' \in \mathcal{E}_\sigma(\mathcal{G})$. Applying statement 2 of splitting theorem 4.10 we have $C'' \cap A(\mathcal{F}) \in \mathcal{E}_\sigma(\mathcal{F})$ contradicting (*). Consequently, $N^{\mathcal{F}}_{\sigma,W}(C) = \infty$. □

Now we are prepared to show the main characterization theorem for stable, preferred, complete and admissible semantics for the class of weak expansions.

Theorem 7.21. *For any AF $\mathcal{F}$ and any semantics $\sigma \in \{pr, co, ad\}$,*

$$N^{\mathcal{F}}_{stb,W} = V^{\mathcal{F}}_{stb,W} \text{ and } N^{\mathcal{F}}_{ad,W} = V^{\mathcal{F}}_{\sigma,W}.$$

Proof. Combine Propositions 7.17, 7.19 and 7.20. □

Interestingly, semi-stable semantics does possess values between zero and infinity and is therefore not adequately characterized by the (stb, W)- or (ad, W)-values. Consider therefore the following example.

Example 7.22. Consider the AF $\mathcal{F}$ and the desired set of arguments $\{a_1\}$. We observe that the (stb, W)-value of $\{a_1\}$ equals infinity since there are no supersets which are stable extensions in $\mathcal{F}$. In case of admissible semantics we deduce $V^{\mathcal{F}}_{ad,W}(\{a_1\}) = 0$ since $\{a_1\}$ is admissible in $\mathcal{F}$. Furthermore, $\{a_1\}$ and all its proper supersets are not semi-stable in $\mathcal{F}$. This means, $N^{\mathcal{F}}_{ss,W}(\{a_1\}) \neq 0$.

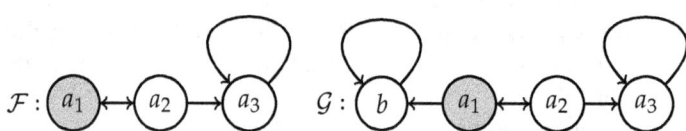

Figure 7.9: Inbetween Zero and Infinity

Consider now the weak expansion $\mathcal{G}$ of $\mathcal{F}$. It shows that $N^{\mathcal{F}}_{ss,W}(\{a_1\}) \leq 2$ since $\{a_1\}$ is semi-stable in $\mathcal{G}$ and $d(\mathcal{F}, \mathcal{G}) = 2$.

7.2.2.2 The (σ, S)-value

How does the situation change if we consider strong expansions? Since the class of strong expansions provides the possibility to attack or defend former arguments there should be a greater range of the characteristics than in case of weak expansions. We will see that this is indeed the case. Furthermore, we will show that the strong expansion characteristics are a lower bound for their corresponding weak expansion terms. Quite surprisingly, we were able to show that the characteristics of arbitrary expansions as well as normal expansions coincide with the strong expansion characteristic. This means, the additional feature of arbitrary extensions in contrast to strong expansions,

Chapter 7. Enforcing and Minimal Change

namely bringing into play new attacks between existing arguments or new attacks from old argument to new arguments, is *useless* in the sense of achieving a lower value for the minimal change problem.

We now start with a first approximation for the characteristics with respect to strong expansions. As a by-product of the standard construction (compare Example 7.8) in the proof of the possibility theorem 7.7 we state the following upper bound. We want to mention that the assertion does not necessarily hold if considering infinite AFs.

Corollary 7.23. *Given an AF $\mathcal{F}$ and a set $C \in cf(\mathcal{F})$. For any semantics $\sigma \in \{stb, ss, pr, co, ad\}$, $\infty > |A(\mathcal{F}) \smallsetminus C| \geq N_{\sigma,S}^{\mathcal{F}}(C)$.*

This means, whenever a set C is conflict-free we may enforce C in finitely many steps. As shown in Example 7.9 we may stay below the upper bound $|A(\mathcal{F}) \smallsetminus C|$. In this example the upper bound according to the above corollary equals $3 = |A(\mathcal{F}) \smallsetminus C|$. On the other hand the alternative construction depicted in Figure 7.7 proved a minimal effort of at most 1.

Inspired by this observation we give the following strong expansion values. We use $R_{\mathcal{F}}^+(C)$ for $C \cup \{b \mid (C,b) \in R(\mathcal{F})\}$ and analogously, $R_{\mathcal{F}}^-(C)$ for $C \cup \{b \mid (b,C) \in R(\mathcal{F})\}$.

Definition 7.24. *Given an AF $\mathcal{F}$. The (σ, S)-value ($\sigma \in \{stb, ad\}$) of a set $C \subseteq A(\mathcal{F})$ is a natural number or infinity defined by the following function*

$$V_{\sigma,S}^{\mathcal{F}} : \mathcal{P}(A(\mathcal{F})) \to \mathbb{N}_\infty \quad C \mapsto \min\left(\{|\sigma(\mathcal{F},C')| \mid C \subseteq C' \text{ and } C' \in cf(\mathcal{F})\} \cup \{\infty\}\right)$$

where $ad(\mathcal{F}, C') = R_{\mathcal{F}}^-(C') \smallsetminus R_{\mathcal{F}}^+(C')$ and $stb(\mathcal{F}, C') = A(\mathcal{F}) \smallsetminus R_{\mathcal{F}}^+(C')$.

As a direct consequence of the definition above we may state that the weak expansion value is greater than or equal to the strong expansion value.

Corollary 7.25. *For any AF $\mathcal{F}$ and any semantics $\sigma \in \{stb, ad\}$, $V_{\sigma,W}^{\mathcal{F}} \geq V_{\sigma,S}^{\mathcal{F}}$.*

Now we will show that the (σ, S) - value does not exceed the (σ, E) - characteristic. We therefore prove some technical properties first.

Proposition 7.26. *Given two AFs $\mathcal{F}, \mathcal{G}$, such that $\mathcal{F} \leq_E \mathcal{G}$ and $d(\mathcal{F}, \mathcal{G}) = n$ and a semantics $\sigma \in \{stb, ad\}$. For any $C \subseteq C' \subseteq A(\mathcal{F})$, $C \subseteq D \subseteq A(\mathcal{G})$ and any $G \subseteq A(\mathcal{G}) \smallsetminus A(\mathcal{F})$,*

1. $V_{\sigma,S}^{\mathcal{F}}(C) \leq V_{\sigma,S}^{\mathcal{F}}(C')$, (monotonicity[1])

2. $\sigma(\mathcal{F}, C) \subseteq \sigma(\mathcal{F}^{A(\mathcal{G})}, C \cup G)$, (subset property)

3. $V_{\sigma,S}^{\mathcal{F}}(C) \leq V_{\sigma,S}^{\mathcal{F}^{A(\mathcal{G})}}(C \cup G)$, (monotonicity[2])

4. $V_{\sigma,S}^{\mathcal{F}}(C) \leq V_{\sigma,S}^{\mathcal{F}^{A(\mathcal{G})}}(D)$, (monotonicity[3])

5. *if $|\sigma(\mathcal{G}, D)| = k$, then $|\sigma(\mathcal{F}^{A(\mathcal{G})}, D)| \leq k + n$ and* (cardinality relations)

6. *if $V_{\sigma,S}^{\mathcal{F}}(C) = k$, then $V_{\sigma,S}^{\mathcal{G}}(C) \geq \max\{k - n, 0\}$.* (greatest lower bound)

7.2. Minimal Change Problem

where $\mathcal{F}^{A(\mathcal{G})} = (A(\mathcal{G}), R(\mathcal{F}))$.

Proof. ad 1.) Assume $V_{\sigma,S}^{\mathcal{F}}(C') = k$. If $k = \infty$ we have nothing to show. Let $k < \infty$. Thus, there is a set C'', such that $C' \subseteq C''$, $C'' \in cf(\mathcal{F})$ and $|\sigma(\mathcal{F}, C'')| = k$. Since $C \subseteq C'$ ($\subseteq C''$) is assumed, we deduce $V_{\sigma,S}^{\mathcal{F}}(C) \leq k$.

ad 2.) Let $a \in stb(\mathcal{F}, C)$, i.e. $a \in A(\mathcal{F}) \setminus R_{\mathcal{F}}^{+}(C)$. Since $R(\mathcal{F}) = R(\mathcal{F}^{A(\mathcal{G})})$ we deduce $R_{\mathcal{F}^{A(\mathcal{G})}}^{+}(C \cup G) = R_{\mathcal{F}}^{+}(C) \cup G$. Hence, $a \notin R_{\mathcal{F}^{A(\mathcal{G})}}^{+}(C \cup G)$ and consequently, $a \in A(\mathcal{G}) \setminus R_{\mathcal{F}^{A(\mathcal{G})}}^{+}(C \cup G) = stb(\mathcal{F}^{A(\mathcal{G})}, C \cup G)$ follows. Note that $R_{\mathcal{F}^{A(\mathcal{G})}}^{-}(C \cup G) = R_{\mathcal{F}}^{-}(C) \cup G$ holds. Since G and $R_{\mathcal{F}}^{-}(C), R_{\mathcal{F}}^{+}(C)$ are disjoint we get $R_{\mathcal{F}^{A(\mathcal{G})}}^{-}(C \cup G) \setminus R_{\mathcal{F}^{A(\mathcal{G})}}^{+}(C \cup G) = R_{\mathcal{F}}^{-}(C) \setminus R_{\mathcal{F}}^{+}(C)$ and this means, in case of admissible semantics we have even shown equality (instead of the subset property).

ad 3.) Since $R(\mathcal{F}) = R(\mathcal{F}^{A(\mathcal{G})})$ we deduce $C \in cf(\mathcal{F})$ iff $C \cup G \in cf(\mathcal{F}^{A(\mathcal{G})})$. Hence, $V_{\sigma,S}^{\mathcal{F}}(C) = \infty$ iff $V_{\sigma,S}^{\mathcal{F}^{A(\mathcal{G})}}(C \cup G) = \infty$. Let $V_{\sigma,S}^{\mathcal{F}}(C) = k < \infty$ and assume $V_{\sigma,S}^{\mathcal{F}^{A(\mathcal{G})}}(C \cup G) = k^* < k$ (proof by contradiction). Thus, there is a set $C' \cup G' \in cf(\mathcal{F}^{A(\mathcal{G})})$, such that $C \subseteq C' \subseteq A(\mathcal{F})$, $G \subseteq G' \subseteq A(\mathcal{G}) \setminus A(\mathcal{F})$ and $|\sigma(\mathcal{F}^{A(\mathcal{G})}, C' \cup G')| = k^*$. Applying statement 2 (subset property) we deduce $|\sigma(\mathcal{F}, C')| \leq k^*$ and consequently, $V_{\sigma,S}^{\mathcal{F}}(C) \leq k^* < k$ contradicting the assumption.

ad 4.) Without loss of generality we may assume that $D = C \cup F \cup G$, such that $F \subseteq A(\mathcal{F}) \setminus C$ and $G \subseteq A(\mathcal{G}) \setminus A(\mathcal{F})$. Let us assume that the claim does not hold, i.e. $V_{\sigma,S}^{\mathcal{F}}(C) > V_{\sigma,S}^{\mathcal{F}^{A(\mathcal{G})}}(D)$ (A1). Applying the already proven statement 3 we get $V_{\sigma,S}^{\mathcal{F}^{A(\mathcal{G})}}(D) \geq V_{\sigma,S}^{\mathcal{F}}(C \cup F)$. Furthermore, we derive $V_{\sigma,S}^{\mathcal{F}}(C \cup F) \geq V_{\sigma,S}^{\mathcal{F}}(C)$ by statement 1. Thus,

$$V_{\sigma,S}^{\mathcal{F}}(C) \overset{A1}{>} V_{\sigma,S}^{\mathcal{F}^{A(\mathcal{G})}}(C \cup F \cup G) \overset{11.3.}{\geq} V_{\sigma,S}^{\mathcal{F}}(C \cup F) \overset{11.1.}{\geq} V_{\sigma,S}^{\mathcal{F}}(C).$$

Since we have derived a contradiction $V_{\sigma,S}^{\mathcal{F}}(C) > V_{\sigma,S}^{\mathcal{F}}(C)$ the original claim is proven.

ad 5.) At first note that $d(\mathcal{F}, \mathcal{F}^{A(\mathcal{G})}) = 0$ holds by definition. Hence, $d(\mathcal{F}, \mathcal{G}) = n$ implies $d(\mathcal{F}^{A(\mathcal{G})}, \mathcal{G}) = n$ (Proposition 7.11, triangle inequality). Since $\mathcal{F} \leq_E \mathcal{G}$ is assumed we deduce that $R(\mathcal{G})$ contains $R(\mathcal{F}) = R(\mathcal{F}^{A(\mathcal{G})})$ plus n further attacks. Obviously, removing one attack may increase the number of unattacked attackers of D by one. The same holds for the number of elements which are not attacked by D. Consequently, $|\sigma(\mathcal{F}^{A(\mathcal{G})}, D)| \leq k + n$ follows.

ad 6.) Let $V_{\sigma,S}^{\mathcal{F}}(C) = \infty$, i.e. $C \notin cf(\mathcal{F})$. Since $\mathcal{F} \leq_E \mathcal{G}$ is assumed we deduce $D \notin cf(\mathcal{G})$ for any $D \supseteq C$. Consequently, $V_{\sigma,S}^{\mathcal{G}}(C) = \infty$ which proves $V_{\sigma,S}^{\mathcal{G}}(C) \geq \max\{k-n, 0\}$. Consider now $V_{\sigma,S}^{\mathcal{F}}(C) = k < \infty$ (A1). Note that in case of $d(\mathcal{F}, \mathcal{G}) = n \geq k$ we have nothing to show because $\max\{k-n, 0\} = 0$. Hence, without loss of generality we may assume that $n < k$. Let us further assume that $V_{\sigma,S}^{\mathcal{G}}(C) \geq \max\{k-n, 0\}$ does not hold (proof by contradiction),

Chapter 7. Enforcing and Minimal Change

i.e. $V_{\sigma,S}^{\mathcal{G}}(C) = k^* < k - n$ (A2). This implies the existence of a set C', such that $C \subseteq C'$, $C' \in cf(\mathcal{G})$ and $|\sigma(\mathcal{G}, C')| = k^*$. Applying the already proven statement 5 we deduce $|\sigma(\mathcal{F}^{A(\mathcal{G})}, C')| \leq k^* + n$. Consequently, $V_{\sigma,S}^{\mathcal{F}^{A(\mathcal{G})}}(C') \leq k^* + n$ (R1). Thus,

$$k \stackrel{A1}{=} V_{\sigma,S}^{\mathcal{F}}(C) \stackrel{11.4.}{\leq} V_{\sigma,S}^{\mathcal{F}^{A(\mathcal{G})}}(C') \stackrel{R1}{\leq} k^* + n \stackrel{A2}{<} k - n + n = k.$$

Since we have derived a contradiction $k < k$ the original claim follows. □

Now we are prepared to show that the strong expansion values do not exceed the (arbitrary) expansion characteristics.

Proposition 7.27. *For any AF $\mathcal{F}$ and any semantics $\sigma \in \{stb, ad\}$, $V_{\sigma,S}^{\mathcal{F}} \leq N_{\sigma,E}^{\mathcal{F}}$.*

Proof. It suffices to show that for any $C \in \wp(A(\mathcal{F}))$, $V_{\sigma,S}^{\mathcal{F}}(C) = k$ implies $N_{\sigma,E}^{\mathcal{F}}(C) \not< k$. If $k = \infty$, then no superset C' of C is conflict-free in $\mathcal{F}$ and even not conflict-free in any $\mathcal{G}$ with $\mathcal{F} \leq_E \mathcal{G}$, i.e. $C' \notin cf(\mathcal{G})$. Consequently, $C' \notin \mathcal{E}_\sigma(\mathcal{G})$ which proves $N_{\sigma,E}^{\mathcal{F}}(C) = \infty$.

Let us assume now $V_{\sigma,S}^{\mathcal{F}}(C) = k < \infty$ and $N_{\sigma,E}^{\mathcal{F}}(C) = k^* < k$ (proof by contradiction). Assume $k^* = 0$. Hence, $k \geq 1$ has to hold. $N_{\sigma,E}^{\mathcal{F}}(C) = 0$ implies the existence of a set C', such that $C \subseteq C'$ and $C' \in \mathcal{E}_\sigma(\mathcal{F})$ (in particular $C' \in cf(\mathcal{F})$). Thus, $\sigma(\mathcal{F}, C') = \emptyset$ and consequently, $V_{\sigma,S}^{\mathcal{F}}(C) = 0 = k$ contradicting $k \geq 1$. Consider now $k^* > 0$. Hence, there is an AF $\mathcal{G}$, such that $\mathcal{F} \leq_E \mathcal{G}$, $d(\mathcal{F}, \mathcal{G}) = k^*$ and $N_{\sigma,E}^{\mathcal{G}}(C) = 0$. This implies the existence of a set C', such that $C \subseteq C'$ and $C' \in \mathcal{E}_\sigma(\mathcal{G})$. Consequently, $V_{\sigma,S}^{\mathcal{G}}(C) = 0$ which contradicts $V_{\sigma,S}^{\mathcal{G}}(C) \geq k - k^* > 0$ (statement 6, Proposition 7.26). □

The next proposition proves that the (σ, S)-value is greater than or equal to the (σ, S)-characteristic.

Proposition 7.28. *For any AF $\mathcal{F}$ and $\sigma \in \{stb, ad\}$, $V_{\sigma,S}^{\mathcal{F}} \geq N_{\sigma,S}^{\mathcal{F}}$.*

Proof. It suffices to show that for any $C \in \wp(A(\mathcal{F}))$, $V_{\sigma,S}^{\mathcal{F}}(C) = k$ implies $N_{\sigma,S}^{\mathcal{F}}(C) \leq k$. If $k = \infty$ we have nothing to show. Let k be finite. $V_{\sigma,S}^{\mathcal{F}}(C) = k$ means that there is set C', such that $C \subseteq C'$, $C' \in cf(\mathcal{F})$ and $|\sigma(\mathcal{F}, C')| = k$. In case of $k = 0$ nothing needs to be shown because $C' \in \mathcal{E}_\sigma(\mathcal{F})$ follows and hence, $N_{\sigma,S}^{\mathcal{F}}(C) = 0$. Let $k > 0$. Thus, there are k unattacked attackers $a_1, \ldots, a_k$ of C' in $\mathcal{F}$ ($\sigma = ad$) or k unattacked elements $a_1, \ldots, a_k$ in $\mathcal{F}$ ($\sigma = st$), respectively. Let b be a fresh argument. We define the AF $\mathcal{G} = (A(\mathcal{F}) \cup \{b\}, R(\mathcal{F}) \cup \{(b, a_1), \ldots, (b, a_k)\})$. We observe $d(\mathcal{F}, \mathcal{G}) = k$, $\mathcal{F} \leq_S \mathcal{G}$ and $C' \cup \{b\} \in \mathcal{E}_\sigma(\mathcal{G})$, i.e. $N_{\sigma,S}^{\mathcal{G}}(C) = 0$. Hence, $N_{\sigma,S}^{\mathcal{F}}(C) \leq k$ concluding the proof. □

The following theorem shows that the stable value and admissible value adequately determine the strong expansion characteristics for stable or preferred, complete and admissible semantics, respectively. Furthermore, this characteristic does not vary if we shift from strong expansions to normal or arbitrary expansions and vice versa.

7.2. Minimal Change Problem

Theorem 7.29. *For any AF $\mathcal{F}$, any semantics $\sigma \in \{pr, co, ad\}$ and any $\Phi \in \{E, N, S\}$,*

$$N^{\mathcal{F}}_{stb,\Phi} = V^{\mathcal{F}}_{stb,S} \text{ and } N^{\mathcal{F}}_{\sigma,\Phi} = V^{\mathcal{F}}_{ad,S}.$$

Proof. Let $\sigma \in \{stb, ad\}$. Applying Corollary 7.16 and Proposition 7.19 and 7.20 we get $V^{\mathcal{F}}_{\sigma,S} \geq N^{\mathcal{F}}_{\sigma,S} \geq N^{\mathcal{F}}_{\sigma,N} \geq N^{\mathcal{F}}_{\sigma,E} \geq V^{\mathcal{F}}_{\sigma,S}$. Hence, $N^{\mathcal{F}}_{\sigma,\Phi} = V^{\mathcal{F}}_{\sigma,S}$ for each $\Phi \in \{E, N, S\}$. Together with Proposition 7.17 we conclude $V^{\mathcal{F}}_{ad,S} = N^{\mathcal{F}}_{\tau,\Phi}$ for $\tau \in \{pr, co, ad\}$ completing the proof. $\square$

A valuable side-effect of the theorem above is that we have now clarified the relation between the classes of weak and strong expansions in case of stable, preferred, complete and admissible semantics. Remember that there is no subset relation between them since they are completely different concepts. Nevertheless, the following proposition states that the weak expansion characteristic is greater than or equal to the strong expansion characteristic for the considered semantics. Here is the whole picture.

Proposition 7.30. *For any AF $\mathcal{F}$ and any semantics $\sigma \in \{stb, pr, co, ad\}$,*

$$N^{\mathcal{F}}_{\sigma,W} \geq N^{\mathcal{F}}_{\sigma,S} \geq N^{\mathcal{F}}_{\sigma,N} \geq N^{\mathcal{F}}_{\sigma,E} \geq N^{\mathcal{F}}_{\sigma,U}.$$

Proof. Combine Corollaries 7.16, 7.25 and Theorem 7.29. $\square$

Consider the following example showing that neither the stable value nor the admissible value adequately determine the semi-stable characteristic with respect to strong expansions.

Example 7.31. Consider the AF $\mathcal{F}$ and the desired set of arguments $\{a_1\}$. The (stb, S)-value of $C = \{a_1\}$ in $\mathcal{F}$ equals 2 since firstly, b, a_3 and a_4 are unattacked by C and secondly, there is no conflict-free superset of C. In case of admissible semantics we have $R^+_{\mathcal{F}}(C) \setminus R^-_{\mathcal{F}}(C) = \emptyset$ and hence $V^{\mathcal{F}}_{ad,S}(\{a_1\}) = 0$. Since C is not semi-stable and does not have proper and conflict-free supersets in $\mathcal{F}$ we conclude $N^{\mathcal{F}}_{ss,W}(\{a_1\}) \neq 0$.

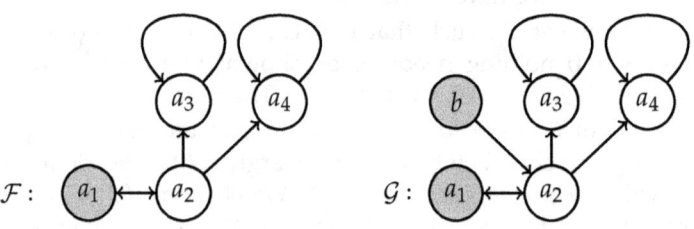

Figure 7.10: Inadequateness of Stable and Admissible Value

Consider now the strong expansion $\mathcal{G}$ of $\mathcal{F}$. One may easily check that $C \cup \{b\}$ is semi-stable in $\mathcal{G}$. Hence, $N^{\mathcal{F}}_{ss,S}(C) = 1$. Thus, neither the stable value nor the admissible value can be used to determine the semi-stable characteristic with respect to strong expansions.

Chapter 7. Enforcing and Minimal Change

7.2.2.3 The $(\sigma, \mathcal{U})$-value

We now turn to arbitrary modifications, i.e. in contrast to arbitrary expansion we even allow deleting attacks between previous arguments. What consequences does this have for the minimal change problem? As a first and important difference we will show that any desired set of arguments may be enforced by a finite manipulation, i.e. the $(\sigma, \mathcal{U})$-characteristic has to be finite for all considered semantics. The following proposition gives us an upper bound.

Proposition 7.32. *Given an AF $\mathcal{F}$ and a set $C \subseteq A(\mathcal{F})$. For any semantics $\sigma \in \{stb, ss, pr, co, ad\}$, $\infty > |R(\mathcal{F}) \cap (C \times C)| + |A(\mathcal{F}) \setminus C| \geq N^{\mathcal{F}}_{\sigma, \mathcal{U}}(C)$.*

Proof. First we delete all inner conflicts of C ($= |R(\mathcal{F}) \cap (C \times C)|$). Hence, C is conflict-free now. Second, we attack all outer arguments of C ($= |A(\mathcal{F}) \setminus C|$). Obviously, C stable in the resulting AF and therefore semi-stable, preferred, complete and admissible (compare Proposition 2.7). □

Inspired by the proposition above we present the following definitions which adequately determine the characteristics with respect to arbitrary modifications.

Definition 7.33. *Given an AF $\mathcal{F}$. The $(\sigma, \mathcal{U})$-value ($\sigma \in \{stb, ad\}$) of a set $C \subseteq A(\mathcal{F})$ is a natural number defined by the following function*

$$V^{\mathcal{F}}_{\sigma, \mathcal{U}} : \mathcal{P}(A(\mathcal{F})) \to \mathbb{N} \qquad C \mapsto \min\left(\{|R(\mathcal{F})_{\downarrow C'}| + |\sigma(\mathcal{F}, C')| \mid C \subseteq C' \subseteq A(\mathcal{F})\}\right),$$

where $R(\mathcal{F})_{\downarrow C'} = R(\mathcal{F}) \cap (C' \times C')$.

To prove that the defined values do not exceed the characteristics with respect to arbitrary modifications we have to show some technical properties first.

Proposition 7.34. *Given two AFs $\mathcal{F}, \mathcal{G}$ and a semantics $\sigma \in \{stb, ad\}$. For any $C \subseteq D \subseteq A(\mathcal{F}) \cap A(\mathcal{G})$, $E \subseteq A(\mathcal{G}) \setminus A(\mathcal{F})$ where $D \cup E \in \mathcal{E}_\sigma(\mathcal{G})$ and $S \subseteq A(\mathcal{F}) \setminus A(\mathcal{G})$ a set of isolated arguments in $\mathcal{F}$,*

1. $R(\mathcal{G}) \Delta R(\mathcal{F}^*) = R(\mathcal{G}) \Delta R(\mathcal{G}^*) \mathbin{\dot\cup} R(\mathcal{G}^*) \Delta R(\mathcal{F}^*)$, *(partition[1])*
2. $R(\mathcal{G}) \Delta R(\mathcal{F}) = R(\mathcal{G}) \Delta R(\mathcal{G}^*) \mathbin{\dot\cup} R(\mathcal{G}^*) \Delta R(\mathcal{F}^*) \mathbin{\dot\cup} R(\mathcal{F}^*) \Delta R(\mathcal{F})$, *(partition[2])*
3. $|R(\mathcal{G}^*)_{\downarrow D}| + |\sigma(\mathcal{G}^*, D)| \leq d(\mathcal{G}, \mathcal{G}^*)$, *(inequality[1])*
4. $|R(\mathcal{F}^*)_{\downarrow D}| + |\sigma(\mathcal{F}^*, D)| \leq d(\mathcal{G}, \mathcal{F}^*)$, *(inequality[2])*
5. $|R(\mathcal{F})_{\downarrow D}| + |ad(\mathcal{F}, D)| \leq d(\mathcal{G}, \mathcal{F})$, *(inequality[3])*
6. $|R(\mathcal{F})_{\downarrow D}| + |stb(\mathcal{F}, D)| \leq d(\mathcal{G}, \mathcal{F}) + |I|$, *(inequality[4])*
7. $|st(\mathcal{F}, D \cup S)| = |stb(\mathcal{F}, D)| - |S|$ *(isolated arguments)*

where

7.2. Minimal Change Problem

- $\mathcal{F}^* = (A(\mathcal{F}) \cap A(\mathcal{G}), R(\mathcal{F})_{\downarrow(A(\mathcal{F}) \cap A(\mathcal{G}))})$,
- $\mathcal{G}^* = (A(\mathcal{F}) \cap A(\mathcal{G}), R(\mathcal{G})_{\downarrow(A(\mathcal{F}) \cap A(\mathcal{G}))})$ and
- $I = \{i \mid i \in A(\mathcal{F}) \setminus A(\mathcal{G}), i \text{ isolated in } \mathcal{F}\}$.

Proof. ad 1.-2.) These assertions can be checked in a straightforward manner. Take into account that the following equations hold: $R(\mathcal{G}) \Delta R(\mathcal{G}^*) = R(\mathcal{G})_{\downarrow(A(\mathcal{G}) \setminus A(\mathcal{F}))}$, $R(\mathcal{F}) \Delta R(\mathcal{F}^*) = R(\mathcal{F})_{\downarrow(A(\mathcal{F}) \setminus A(\mathcal{G}))}$ and $R(\mathcal{G}^*) \Delta R(\mathcal{F}^*) = (R(\mathcal{G}) \Delta R(\mathcal{F}))_{\downarrow(A(\mathcal{F}) \cap A(\mathcal{G}))}$.

ad 3.) Since $D \cup E \in \mathcal{E}_\sigma(\mathcal{G})$ is assumed we deduce $|R(\mathcal{G})_{\downarrow(D \cup E)}| + |\sigma(\mathcal{G}, D \cup E)| = 0$. Hence, $|R(\mathcal{G}^*)_{\downarrow D}|$ has to be zero too (conflict-freeness). Assume $n = d(\mathcal{G}, \mathcal{G}^*)$ and let us consider the AF $\mathcal{G}' = (A(G), R(\mathcal{G}^*))$. We observe $n = d(\mathcal{G}, \mathcal{G}')$ and furthermore, $D \cup E$ possess at most n unattacked attackers in $\mathcal{G}'$ ($\sigma = ad$) or unattacked arguments in $\mathcal{G}'$ ($\sigma = stb$), respectively since n attacks were deleted. This means, $|\sigma(\mathcal{G}', D \cup E)| \leq n$. The attack-relation $R(\mathcal{G}^*)$ does not contain attacks which involve arguments in $A(\mathcal{G}) \setminus A(\mathcal{G}^*)$. Consequently, $ad(\mathcal{G}^*, D) = ad(\mathcal{G}', D \cup E)$ and $stb(\mathcal{G}^*, D) \subseteq stb(\mathcal{G}', D \cup E)$. In both cases, $|R(\mathcal{G}^*)_{\downarrow D}| + |\sigma(\mathcal{G}^*, D)| \leq n = d(\mathcal{G}, \mathcal{G}^*)$ follows.

ad 4.) Let us assume $d(\mathcal{G}, \mathcal{F}^*) = n$. In consideration of statement 1 there are some $k_1, k_2 \in \mathbb{N}$, such that $k_1 + k_2 = n$, $k_1 = d(\mathcal{G}, \mathcal{G}^*)$ and $k_2 = d(\mathcal{G}^*, \mathcal{F}^*)$. Applying statement 3 it follows $|R(\mathcal{G}^*)_{\downarrow D}| + |\sigma(\mathcal{G}^*, D)| \leq k_1$. Furthermore, $k_2 = l_1 + l_2$ for some $l_1, l_2 \in \mathbb{N}$, such that l_1 equals the number of deleted attacks during the transformation from $\mathcal{G}^*$ to $\mathcal{F}^*$, namely $|R(\mathcal{G}^*) \setminus R(\mathcal{F}^*)|$ and l_2 equals the number of added attacks during the transformation from $\mathcal{G}^*$ to $\mathcal{F}^*$, namely $|R(\mathcal{F}^*) \setminus R(\mathcal{G}^*)|$. Consider therefore $\mathcal{G}' = (A(\mathcal{G}^*), R(\mathcal{G}^*) \cap R(\mathcal{F}^*))$. Obviously, $d(\mathcal{G}^*, \mathcal{G}') = l_1$ and $d(\mathcal{G}', \mathcal{F}^*) = l_2$. We obtain, $R(\mathcal{G}')_{\downarrow D} = R(\mathcal{G}^*)_{\downarrow D}$ and $|\sigma(\mathcal{G}', D)| \leq |\sigma(\mathcal{G}^*, D)| + l_1$ since deleted attacks do not increase the number of inner conflicts of D but may increase the number of unattacked arguments or unattacked attackers by at most l_1. Furthermore, $|R(\mathcal{F}^*)_{\downarrow D}| + |\sigma(\mathcal{F}^*, D)| \leq |R(\mathcal{G}')_{\downarrow D}| + |\sigma(\mathcal{G}', D)| + l_2$ since added attacks may either increase inner conflicts of D or the number of unattacked attackers by at most l_2. Altogether we have shown that $|R(\mathcal{F}^*)_{\downarrow D}| + |\sigma(\mathcal{F}^*, D)| \leq k_1 + l_1 + l_2 = k_1 + k_2 = n = d(\mathcal{G}, \mathcal{F}^*)$ concluding the proof.

ad 5.-6.) Let us assume $d(\mathcal{G}, \mathcal{F}) = n$. Applying statement 1 and 2 there are some $k_1, k_2 \in \mathbb{N}$, such that $k_1 + k_2 = n$, $k_1 = d(\mathcal{G}, \mathcal{F}^*)$ and $k_2 = d(\mathcal{F}^*, \mathcal{F})$. By statement 4 we get $|R(\mathcal{F}^*)_{\downarrow D}| + |\sigma(\mathcal{F}^*, D)| \leq k_1$. Note that k_2 equals the number of added attacks during the transformation from $\mathcal{F}^*$ to $\mathcal{F}$, namely $|R(\mathcal{F}) \setminus R(\mathcal{F}^*)|$. This means, $R(\mathcal{F})_{\downarrow D} = R(\mathcal{F}^*)_{\downarrow D}$ since the new attacks do contain attacks in $D \times D$. In case of admissible semantics we obtain $|ad(\mathcal{F}, D)| \leq |ad(\mathcal{F}^*, D)| + k_2$ because at most k_2 new unattacked attackers may arise. In case of stable semantics we have to take into account that new isolated arguments (which are distance-neutral) may arise. This means, an upper bound for $A(\mathcal{F}) \setminus A(\mathcal{G})$ is given by $k_2 + |I|$ where $I = \{i \mid i \in A(\mathcal{F}) \setminus A(\mathcal{G}), i \text{ isolated in } \mathcal{F}\}$. Consequently, $|stb(\mathcal{F}, D)| \leq |stb(\mathcal{F}^*, D)| + k_2 + |I|$. Altogether, $|R(\mathcal{F})_{\downarrow D}| + |ad(\mathcal{F}, D)| \leq k_1 + k_2 = n = d(\mathcal{G}, \mathcal{F})$ and $|R(\mathcal{F})_{\downarrow D}| + |stb(\mathcal{F}, D)| \leq k_1 + k_2 + |I| = n + |I| = d(\mathcal{G}, \mathcal{F}) + |I|$ is shown.

Chapter 7. Enforcing and Minimal Change

ad 7.) Note that D and S are disjoint. Since S only contains isolated arguments we have $S \cup S' = stb(\mathcal{F}, D)$ for some S' disjoint from S. Consequently, $S' = stb(\mathcal{F}, D \cup S)$ and the assertion follows. □

Proposition 7.35. *For any AF $\mathcal{F}$ and any semantics $\sigma \in \{stb, ad\}$, $V_{\sigma,\mathcal{U}}^{\mathcal{F}} \leq N_{\sigma,\mathcal{U}}^{\mathcal{F}}$.*

Proof. Given a set $C \in \mathcal{P}(A(\mathcal{F}))$ and $V_{\sigma,\mathcal{U}}^{\mathcal{F}}(C) = k$. Let us assume $N_{\sigma,\mathcal{U}}^{\mathcal{F}}(C) = k^* < k$ (proof by contradiction). Hence, there is an AF $\mathcal{G}$, such that $d(\mathcal{F},\mathcal{G}) = k^*$ and $N_{\sigma,\mathcal{U}}^{\mathcal{G}}(C) = 0$. This means, there is a set C' with the property $C \subseteq C'$ and $C' \in \mathcal{E}_\sigma(\mathcal{G})$. Without loss of generality $C' = D \cup E$ where $C \subseteq D \subseteq A(\mathcal{F}) \cap A(\mathcal{G})$ and $E \subseteq A(\mathcal{G}) \setminus A(\mathcal{F})$. Hence, in case of admissible semantics we deduce $|R(\mathcal{F})_{\downarrow D}| + |ad(\mathcal{F}, D)| \leq d(\mathcal{F},\mathcal{G}) = k^*$ (statement 5, Proposition 7.34). Since $C \subseteq D$ holds we conclude $k = V_{ad,\mathcal{U}}^{\mathcal{F}}(C) \leq k^*$ contradicting the assumption $k^* < k$. In case of stable semantics we get $|R(\mathcal{F})_{\downarrow D}| + |stb(\mathcal{F}, D)| \leq d(\mathcal{F},\mathcal{G}) + |I|$ where $I = \{i \mid i \in A(\mathcal{F}) \setminus A(\mathcal{G}), i \text{ isolated in } \mathcal{F}\}$ (statement 6, Proposition 7.34). Obviously, $|R(\mathcal{F})_{\downarrow D \cup I}| = |R(\mathcal{F})_{\downarrow D}|$ since all i's in I are isolated. Furthermore, in consideration of statement 7, Proposition 7.34 it follows $|R(\mathcal{F})_{\downarrow D \cup I}| + |stb(\mathcal{F}, D \cup I)| \leq d(\mathcal{F},\mathcal{G}) + |I| - |I| = d(\mathcal{F},\mathcal{G}) = k^*$. Again, since $C \subseteq D \cup I$ holds we deduce $k = V_{stb,\mathcal{U}}^{\mathcal{F}}(C) \leq k^*$ contradicting the assumption $k^* < k$. □

Proposition 7.36. *For any AF $\mathcal{F}$ and $\sigma \in \{stb, ad\}$, $V_{\sigma,\mathcal{U}}^{\mathcal{F}} \geq N_{\sigma,\mathcal{U}}^{\mathcal{F}}$.*

Proof. Given a set $C \in \mathcal{P}(A(\mathcal{F}))$ and $V_{\sigma,\mathcal{U}}^{\mathcal{F}}(C) = k$. This means, there is a superset C' of C, such that $|R(\mathcal{F})_{\downarrow C'}| + |\sigma(\mathcal{F}, C')| = k$. In case of $k = 0$ we deduce $C' \in \mathcal{E}_\sigma(\mathcal{F})$ and hence, $N_{\sigma,\mathcal{U}}^{\mathcal{F}}(C) = 0$. Let $k > 0$. We may assume $|R(\mathcal{F})_{\downarrow C'}| = l$ and $|\sigma(\mathcal{F}, C')| = m$ for some $l, m \in \mathbb{N}$ with $k = l + m$. This means C' has l inner attacks and furthermore, there are m unattacked attackers $a_1, ..., a_m$ of C' in $\mathcal{F}$ ($\sigma = ad$) or m unattacked elements $a_1, ..., a_m$ in $\mathcal{F}$ ($\sigma = stb$) respectively. Since $k > 0$ is assumed we conclude $C' \neq \emptyset$ in case of $\sigma = ad$. Let $c \in C'$. We define the AF $\mathcal{G} = (A(\mathcal{F}), (R(\mathcal{F}) \setminus R(\mathcal{F})_{\downarrow C'}) \cup \{(c, a_1), ..., (c, a_m)\})$. We observe $d(\mathcal{F},\mathcal{G}) = l + m = k$ and $C' \in \mathcal{E}_{ad}(\mathcal{G})$, i.e. $N_{ad,\mathcal{U}}^{\mathcal{G}}(C) = 0$. Hence, $N_{ad,\mathcal{U}}^{\mathcal{F}}(C) \leq k$. The same construction holds for stable semantics if $C' \neq \emptyset$. Consider $C' = \emptyset$. Hence, $l = 0$ and $m = A(\mathcal{F}) > 0$. Let $c \in A(\mathcal{F})$. We define $\mathcal{G} = (A(\mathcal{F}), (R(\mathcal{F}) \setminus \{(c,c)\}) \cup \{(c, a_1), ..., (c, a_m)\})$. In any case $d(\mathcal{F},\mathcal{G}) \leq m = k$. Obviously, $\{c\} \in \mathcal{E}_{stb}(\mathcal{G})$, i.e. $N_{stb,\mathcal{U}}^{\mathcal{G}}(\{c\}) = 0$. Thus, $N_{stb,\mathcal{U}}^{\mathcal{F}}(\emptyset) \leq k$ concluding the proof. □

Theorem 7.37. *For any AF $\mathcal{F}$ and any semantics $\sigma \in \{pr, co, ad\}$,*

$$N_{stb,\mathcal{U}}^{\mathcal{F}} = V_{stb,\mathcal{U}}^{\mathcal{F}} \text{ and } N_{\sigma,\mathcal{U}}^{\mathcal{F}} = V_{ad,\mathcal{U}}^{\mathcal{F}}.$$

Proof. Combine 7.17, 7.35 and 7.36. □

The following AFs prove that the $(ss, \mathcal{U})$-characteristic is not adequately characterized by the $(stb, \mathcal{U})$- or $(ad, \mathcal{U})$-values.

7.2. Minimal Change Problem

Example 7.38 (Example 7.31 continued). Consider again the AF $\mathcal{F}$ and the desired set of arguments $\{a_1\}$ depicted below. The (stb, U)-value of $\{a_1\}$ equals 2 because: firstly, $|R(\mathcal{F})_{\downarrow\{a_1,a_2\}}| + |stb(\mathcal{F},\{a_1,a_2\})| = 2$ and secondly, there is no other superset C of $\{a_1\}$, such that $|R(\mathcal{F})_{\downarrow C}| + |stb(\mathcal{F}, C)| < 2$. In case of admissible semantics we deduce $V^{\mathcal{F}}_{ad,U}(\{a_1\}) = 0$ since $\{a_1\}$ is admissible in $\mathcal{F}$. On the other hand $N^{\mathcal{F}}_{ss,U}(\{a_1\}) \neq 0$ since $\{a_1\}$ and all its proper supersets are not semi-stable.

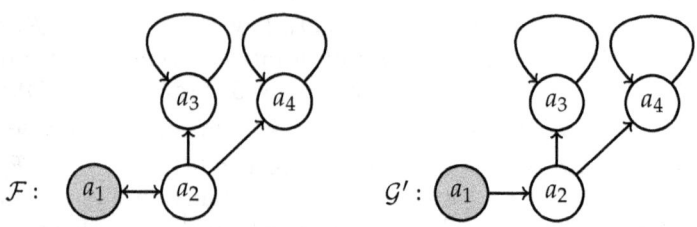

Figure 7.11: Inadequateness of Stable and Admissible Value

Consider now the AF $\mathcal{G}'$ resulting from $\mathcal{F}$ by deleting the attack (a_2, a_1). Now, $N^{\mathcal{F}}_{ss,U}(\{a_1\}) = 1$ since $\{a_1\}$ is semi-stable in $\mathcal{G}'$ and $d(\mathcal{F}, \mathcal{G}') = 1$.

7.2.3 Summary of Results

The following table gives a comprehensive overview over results presented in this section. In particular, we summarize the main characterization theorems with respect to weak (Theorem 7.21), strong, normal and arbitrary (Theorem 7.29) expansions as well as arbitrary modifications (Theorem 7.37). Note that the values may shrink from above to below and from left to right.

$N^{\mathcal{F}}_{\sigma,\Phi}$	W	S	N	E	U
stb	$V^{\mathcal{F}}_{st,W}$	$V^{\mathcal{F}}_{st,S}$	$V^{\mathcal{F}}_{st,S}$	$V^{\mathcal{F}}_{st,S}$	$V^{\mathcal{F}}_{st,U}$
pr	$V^{\mathcal{F}}_{ad,W}$	$V^{\mathcal{F}}_{ad,S}$	$V^{\mathcal{F}}_{ad,S}$	$V^{\mathcal{F}}_{ad,S}$	$V^{\mathcal{F}}_{ad,U}$
co	$V^{\mathcal{F}}_{ad,W}$	$V^{\mathcal{F}}_{ad,S}$	$V^{\mathcal{F}}_{ad,S}$	$V^{\mathcal{F}}_{ad,S}$	$V^{\mathcal{F}}_{ad,U}$
ad	$V^{\mathcal{F}}_{ad,W}$	$V^{\mathcal{F}}_{ad,S}$	$V^{\mathcal{F}}_{ad,S}$	$V^{\mathcal{F}}_{ad,S}$	$V^{\mathcal{F}}_{ad,U}$

Figure 7.12: State of the Art: Characterizing Characteristics

On a final note, our results and/or introduced techniques can contribute to solve open problems addressed in several works dealing with dynamics.

Chapter 7. Enforcing and Minimal Change

In [Baroni and Giacomin, 2007] for instance the authors say: "Moreover, ..., we are interested in questions about the change needed to change an argument from being accepted to rejected, or vice versa." or slightly different versions like in [Cayrol et al., 2008], "How to make the minimal change to a given argumentation framework so that it has a unique non-empty extension?".

7.3 Spectrum Problem

In the last section we studied the so-called *minimal change problem* [Baumann, 2012b] which can be formulated as follows: what is the minimal number of modifications (additions or removals of attacks) needed to reach an enforcement? This value, called *characteristic*, depends on the underlying semantics σ and type of allowed modifications Φ. As we have shown, there are local criteria, so-called *value functions* to determine the minimal number (see Figure 7.12 for an overview).

In this section we proceed with a formal analysis of characteristics. In particular, given a certain semantics σ and a modification type Φ, we study whether there is, for a given natural number n, an AF $\mathcal{F}$ and a set of arguments E such that n is the (σ, Φ)-characteristic of E with respect to $\mathcal{F}$. In other words, we want to determine the set of all natural numbers which may occur as (σ, Φ)-characteristics, the so-called (σ, Φ)-*spectrum*. This is what we call the *spectrum problem* [Baumann and Brewka, 2013b]. The characterization of spectra yields interesting insights into particular semantics. To mention one result, we will show that in case of semi-stable semantics and the addition of weak arguments (arguments which do not attack previous arguments) not each natural number may arise as the minimal effort needed to enforce a certain set E. In particular, the characteristic cannot be 1.

What makes our study even more interesting, as we believe, is the fact that it provides useful and at times surprising new insights into the interrelationships among the studied semantics. To this end, we perform our analysis in parallel for a whole group of semantics which we consider as some of the most important semantics for Dung frameworks. Rather than sets of values, spectra thus become sets of tuples of values. Appropriate properties of the spectra like *m.d.s.-completeness* and *coherence* will help us to identify such relationships.

7.3.1 Formal Properties of Spectra

In the following we consider n-tuples of semantics and modification types and ask whether some $\mathcal{F}$ and C possess a given n-tuple of characteristics simultaneously. A tuple of characteristics satisfying this condition is called a *fibre*. A fibre is said to be *finite* if all entries are natural numbers. The set of all fibres provides important insights on how close or far apart the characteristics of a set C may be. That is why this set is called the *spectrum*. Here is the formal definition.

7.3. Spectrum Problem

Definition 7.39. Given n semantics $\sigma_1,...,\sigma_n$ and n binary relations $\Phi_1,...,\Phi_n \subseteq \mathscr{A} \times \mathscr{A}$. The $(\sigma_1, \Phi_1, ..., \sigma_n, \Phi_n)$-spectrum is a set of n-tuples (so-called *fibres*) defined as follows:

$$S_{(\sigma_i, \Phi_i)_{i=1}^n} = \{(k_1,...,k_n) \mid \exists \mathcal{F} \in \mathscr{A}\ \exists C \subseteq A(\mathcal{F}) : N_{\sigma_i, \Phi_i}^{\mathcal{F}}(C) = k_i \text{ for all } i \in \{1,...,n\}\}.$$

For convenience, if $\Phi_1 = ... = \Phi_n$ we simply write $(\sigma_1,...,\sigma_n, \Phi)$-spectrum or $S_{(\sigma_1,...,\sigma_n, \Phi)}$. These are exactly the types of spectra which we will consider in this chapter. In particular, we consider (stb, ss, pr, Φ)-spectra where Φ equals normal, strong, weak or arbitrary expansions. Remember that we have already shown that for any modification type Φ the characteristics with respect to preferred, complete and admissible semantics coincide (Proposition 7.17). Thus, an analysis of $S_{(stb,ss,co,\Phi)}$ and $S_{(stb,ss,ad,\Phi)}$ is implicitly given.

We first introduce some basic properties spectra may possess.

Definition 7.40. A spectrum $S_{(\sigma_i, \Phi_i)_{i=1}^n}$ is

1. *m.d.s.* iff any finite fibre $(k_1,...,k_n) \in S_{(\sigma_i, \Phi_i)_{i=1}^n}$ is a monotonic decreasing sequence,

2. *m.d.s.-complete* iff $S_{(\sigma_i, \Phi_i)_{i=1}^n}$ is m.d.s. and $\{(k_1,...,k_n) \in \mathbb{N}^n \mid k_1 \geq ... \geq k_n\} \subseteq S_{(\sigma_i, \Phi_i)_{i=1}^n}$,

3. *coherent* iff there is no fibre $(k_1,...,k_n) \in S_{(\sigma_i, \Phi_i)_{i=1}^n}$, such that $k_i = \infty$ and $k_j \neq \infty$ for some indices $1 \leq i, j \leq n$ and

4. *positive* iff any fibre $(k_1,...,k_n) \in S_{(\sigma_i, \Phi_i)_{i=1}^n}$ is finite.

These properties are interesting for the following reasons: if a spectrum for semantics $\sigma_1,...,\sigma_n$ is *m.d.s.*, then we know that whenever enforcing is possible for all of them it is at least as difficult using σ_i as it is using σ_j given that $i < j$. If it is *m.d.s.-complete* we know in addition that it can in fact be arbitrarily more difficult. *Coherence* means that whether some C is enforceable or not does not depend on the choice of the considered semantics. *Positive* means each set C can actually be enforced.

A few relationships among these properties are clear by definition. First, an m.d.s.-complete spectrum is m.d.s. and second, a positive spectrum is coherent. Further interpretations of the introduced properties are given in the following subsections.

7.3.2 The (stb, ss, pr, Φ)-Spectrum ($\Phi \in \{E, N, S\}$)

In this subsection we will characterize the (stb, ss, pr)-spectra with respect to strong, normal and arbitrary expansions. In Corollary 7.15 it was shown that the stable (semi-stable) characteristic exceeds the semi-stable (preferred) characteristic w.r.t. any binary relation over the set of all finite AFs. Consequently, the considered spectra are m.d.s.

Chapter 7. Enforcing and Minimal Change

Quite surprisingly, the following proposition shows that the mentioned spectra are even m.d.s.-complete, i.e. the stable (semi-stable) characteristic may take values which exceed the semi-stable (preferred) characteristic by **any** natural number. In a sense this result is negative as it tells us that information about the characteristic of one semantics does not help in determining the characteristic of the other semantics: even if we know that the characteristic with respect to preferred semantics for a certain set C is, say, 1 (i.e., only 1 additional attack is needed), there is no possibility to give an upper bound of the characteristic with respect to semi-stable or stable semantics. The result underlines the independence of the considered semantics with respect to the minimal change problem. It indicates that the choice of the considered semantics may influence the characteristic dramatically, even though the considered semantics possess many similarities.

Proposition 7.41. *For any $\Phi \in \{E, N, S\}$, $\mathcal{S}_{(stb,ss,pr,\Phi)}$ is m.d.s.-complete.*

Proof. Let $\Phi \in \{E, N, S\}$ and $k, l, m \in \mathbb{N}$, such that $k \geq l \geq m$. Hence, we may assume that $l = m + n$ and $k = m + n + o$ for some $n, o \in \mathbb{N}$. If we may construct AFs $\mathcal{F}$ and corresponding sets $C \subseteq A(\mathcal{F})$, such that $N^{\mathcal{F}}_{stb,\Phi}(C) = m + n + o$, $N^{\mathcal{F}}_{ss,\Phi}(C) = m + n$ and $N^{\mathcal{F}}_{pr,\Phi}(C) = m$, then $(k, l, m) \in \mathcal{S}_{(stb,ss,pr,\Phi)}$ follows. Thus, $\mathcal{S}_{(stb,ss,pr,\Phi)}$ is shown to be m.d.s.-complete. We define the AF $\mathcal{F}_{m,n,o} = (A_{m,n,o}, R_{m,n,o})$ where

$A_{m,n,o} = \{a\} \cup \{b_j \mid 1 \leq j \leq m\} \cup \{c_j, d_j, e_j \mid 1 \leq j \leq n\} \cup \{f_j \mid 1 \leq j \leq o\}$ and

$R_{m,n,o} = \{(b_j, a), (b_j, b_j) \mid 1 \leq j \leq m\} \cup \{(c_j, d_j), (d_j, c_j), (e_j, e_j) \mid 1 \leq j \leq n\} \cup$
$\{(d_j, e_i) \mid 1 \leq j, i \leq n\} \cup \{(d_j, b_i) \mid 1 \leq j \leq n, 1 \leq i \leq m\} \cup$
$\{(d_j, d_i) \mid j \neq i, 1 \leq j, i \leq n\} \cup \{(f_j, f_j) \mid 1 \leq j \leq o\} \cup$
$\{(d_j, f_i) \mid 1 \leq j \leq n, 1 \leq i \leq o\}$.

Note that if a subindex equals zero, then there are no corresponding arguments and attacks. For the sake of clarity we present here an instantiation of the presented scheme, namely $\mathcal{F}_{3,2,4}$ depicted in Figure 7.13.

The grey highlighted arguments belong to the set $C_2 = \{a, c_1, c_2\}$ which is an instantiation of the scheme $C_n = \{a\} \cup \{c_j \mid 1 \leq j \leq n\}$. We claim that $N^{\mathcal{F}_{m,n,o}}_{stb,\Phi}(C_n) = m + n + o$, $N^{\mathcal{F}_{m,n,o}}_{ss,\Phi}(C_n) = m + n$ and $N^{\mathcal{F}_{m,n,o}}_{pr,\Phi}(C_n) = m$. By construction C_n is conflict-free in $\mathcal{F}_{m,n,o}$. Furthermore, C_n does not have proper conflict-free supersets (*). Applying the characterization theorem 7.29 we obtain $N^{\mathcal{F}_{m,n,o}}_{pr,\Phi}(C_n) = V^{\mathcal{F}_{m,n,o}}_{ad,S}(C_n) = \left| R^{-}_{\mathcal{F}_{m,n,o}}(C_n) \setminus R^{+}_{\mathcal{F}_{m,n,o}}(C_n) \right| = |\{b_j \mid 1 \leq j \leq m\}| = m$ because these arguments are not counterattacked by C_n. In case of stable semantics $N^{\mathcal{F}_{m,n,o}}_{stb,\Phi}(C_n) = V^{\mathcal{F}_{m,n,o}}_{stb,S}(C_n) = \left| A(\mathcal{F}_{m,n,o}) \setminus R^{+}_{\mathcal{F}_{m,n,o}}(C_n) \right| =$
$|\{b_j \mid 1 \leq j \leq m\} \cup \{e_j \mid 1 \leq j \leq n\} \cup \{f_j \mid 1 \leq j \leq o\}| = m + n + o$ since exactly these arguments are not attacked by C_n.

To see that $N^{\mathcal{F}_{m,n,o}}_{ss,\Phi}(C_n) = m + n$ is much more difficult. At first we will show that $N^{\mathcal{F}_{m,n,o}}_{ss,E}(C_n) \geq m + n$ and finally, $N^{\mathcal{F}_{m,n,o}}_{ss,S}(C_n) \leq m + n$. Consequently,

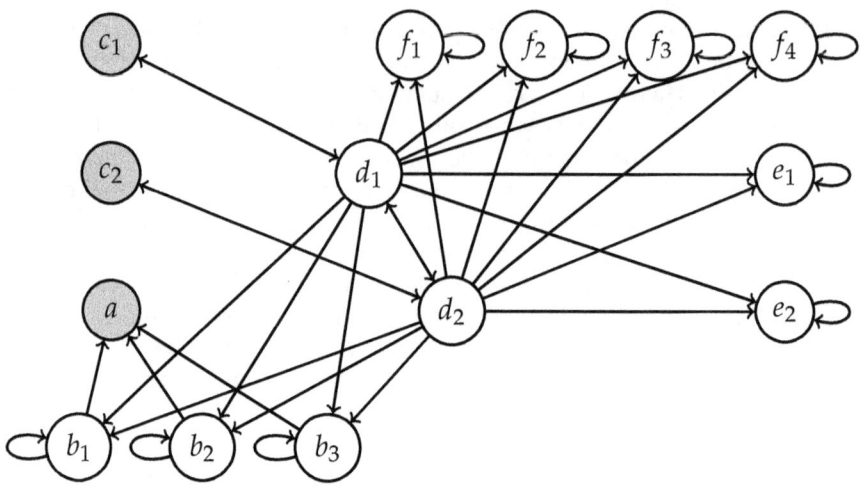

Figure 7.13: The AF $\mathcal{F}_{3,2,4}$

$N_{ss,\Phi}^{\mathcal{F}_{m,n,o}}(C_n) = m+n$ for any $\Phi \in \{E, N, S\}$ is proven (Corollary 7.16). Consider the n conflict-free sets $S_n^1, ..., S_n^n$ where $S_n^j = \{a\} \cup \{c_i \mid 1 \le i \le n\} \smallsetminus \{c_j\} \cup \{d_j\}$. We observe that $C_n \not\subseteq S_n^j$ for $n \ge 1$ and furthermore, $R_{\mathcal{F}_{m,n,o}}^+(C_n) \subset R_{\mathcal{F}_{m,n,o}}^+(S_n^j) = A_{m,n,o}$. Assume now $N_{ss,E}^{\mathcal{F}_{m,n,o}}(C_n) = l' < m+n$. Hence, there is an AF $\mathcal{G}$, such that $d(\mathcal{F}_{m,n,o}, \mathcal{G}) = l'$, $\mathcal{F}_{m,n,o} \le \mathcal{G}$ and furthermore, there is a conflict-free superset C_n' of C_n with the property $C_n' \in \mathcal{E}_{ss}(\mathcal{G})$. In consideration of (*) we deduce that $C_n' = C_n \cup G$ where G is a set of fresh arguments. Since any semi-stable extension is admissible we conclude that each b_j has to be attacked by C_n'. This means at least m additional attacks of $\mathcal{G}$ are required for this task.

Let us consider now the remaining $l'' < n = \left|\{S_n^j \mid 1 \le j \le n\}\right|$ new attacks. The set S_n^j that we look for satisfies the following conditions: 1. for any $g \in G$, $(d_j, g), (g, d_j) \notin R(\mathcal{G})$, 2. $(d_j, d_j) \notin R(\mathcal{G})$, 3. for any $i \ne j$, $(c_i, d_j), (d_j, c_i) \notin R(\mathcal{G})$ as well as $(a, d_j), (d_j, a) \notin R(\mathcal{G})$ and 4. for any $g \in A(\mathcal{G}) \smallsetminus \{A(\mathcal{F}_{m,n,o}) \cup G\}$, $(c_j, g), (g, d_j) \notin R(\mathcal{G})$. Since any new attack may eliminate at most one potential candidate we deduce that there is indeed such a S_n^j satisfying 1. - 4. We will show now that $S_n^j \cup G \in \mathcal{E}_{ad}(\mathcal{G})$ and $R_{\mathcal{G}}^+(C_n \cup G) \subset R_{\mathcal{G}}^+(S_n^j \cup G)$ contradicting $C_n' \in \mathcal{E}_{ss}(\mathcal{G})$. Let us consider the range $R_{\mathcal{G}}^+(C_n \cup G)$. Obviously, there is an index i, such that $e_i \notin R_{\mathcal{G}}^+(C_n \cup G)$ since $l'' < n$ was assumed. Note that $e_i \in R_{\mathcal{G}}^+(S_n^j \cup G)$ by construction of $\mathcal{F}_{m,n,o}$ and S_n^j. Furthermore, in consideration of the first part of condition 4. (c_j does not "reach" further arguments) we immediately conclude that $R_{\mathcal{G}}^+(C_n \cup G) \subseteq R_{\mathcal{G}}^+(S_n^j \cup G)$. Altogether, $R_{\mathcal{G}}^+(C_n \cup G) \subset R_{\mathcal{G}}^+(S_n^j \cup G)$ has to hold. Furthermore, $S_n^j \cup G$ is conflict-free in $\mathcal{G}$ for two reasons, first S_n^j satisfies conditions 1. - 3. and second, $C_n \cup G$ is assumed to be admissible and in particular, conflict-free in $\mathcal{G}$. Assume now that $S_n^j \cup G \notin \mathcal{E}_{ad}(\mathcal{G})$. This means, there is argument $g \in A(\mathcal{G})$ which attacks $S_n^j \cup G$

without being counterattacked. Since conflict-freeness is already shown and $A_{\mathcal{F}_{m,n,o}} \subseteq R_{\mathcal{G}}^+(S_n^j \cup G)$ obviously holds, we deduce $g \in A(\mathcal{G}) \setminus \{A_{m,n,o} \cup G\}$. In consideration of the second part of condition 4. $((g, d_j) \notin R(\mathcal{G}))$ it follows that g attacks some c_i with $i \neq j$ or an argument $g' \in G$. Since $C_n \cup G$ is assumed to be admissible in $\mathcal{G}$ there is an argument $c' \in C_n \cup G$, such that $(c', g) \in R(\mathcal{G})$. If $c' \in G$, then obviously $c' \in S_n^j \cup G$. If $c' \in C_n$, then $c' \in S_n^j \cup G$ because the second part of condition 4. $((c_j, g) \notin R(\mathcal{G}))$ guarantees $c_j \neq c'$. This means, under the assumption $N_{ss,E}^{\mathcal{F}_{m,n,o}}(C_n) = l' < m + n$ we derived a contradiction, namely $C_n' \in \mathcal{E}_{ss}(\mathcal{G}) \wedge C_n' \notin \mathcal{E}_{ss}(\mathcal{G})$. Hence, $N_{ss,E}^{\mathcal{F}_{m,n,o}}(C_n) \geq m + n$ is shown.

Let us prove now that $N_{ss,S}^{\mathcal{F}_{m,n,o}}(C_n) \leq m + n$. Consider therefore a fresh argument c and the AF $\mathcal{G}_{m,n} = (A_{m,n,o} \cup \{c\}, R_{m,n,o} \cup \{(c, b_j) \mid 1 \leq j \leq m\} \cup \{(c, d_j) \mid 1 \leq j \leq n\})$. One can easily verify that $C_n \cup \{c\} \in \mathcal{E}_{ss}(\mathcal{G}_{m,n})$ and furthermore, $\mathcal{F} \leq_S \mathcal{G}_{m,n}$. Since $d(\mathcal{F}_{m,n,o}, \mathcal{G}_{m,n}) = m + n$ we conclude $N_{ss,S}^{\mathcal{F}_{m,n,o}}(C_n) \leq m + n$. Finally, $N_{ss,\Phi}^{\mathcal{F}_{m,n,o}}(C_n) = m + n$ for any $\Phi \in \{E, N, S\}$ is proven. □

The following proposition shows that the spectrum $\mathcal{S}_{(stb,ss,pr,\Phi)}$ is coherent, i.e. any fibre either possesses finite values or all values equal infinity. This means, under the considered semantics it is impossible that a set C may be enforced with respect to a semantics σ and simultaneously, C is not enforceable with respect to another semantics τ. Furthermore, we show that the considered spectra are not positive, i.e. there are unenforceable sets.

Proposition 7.42. *For any $\Phi \in \{E, N, S\}$, $\mathcal{S}_{(stb,ss,pr,\Phi)}$ is coherent but not positive.*

Proof. Given $\Phi \in \{E, N, S\}$. First, we will prove the coherence of $\mathcal{S}_{(stb,ss,pr,\Phi)}$. Since $\mathcal{S}_{(stb,ss,pr,\Phi)}$ is already shown to be m.d.s.-complete it suffices to prove that for any fibre $(k, l, m) \in \mathcal{S}_{(stb,ss,pr,\Phi)}$, if $m < \infty$, then $l < \infty$ and if $l < \infty$, then $k < \infty$. Let $m < \infty$. Hence there is an AF $\mathcal{F}$ and a set $C \subseteq A(\mathcal{F})$, such that $N_{pr,\Phi}^{\mathcal{F}}(C) = m$. This means, C has to be conflict-free in $\mathcal{F}$. Applying Corollary 7.23 we deduce $l = N_{ss,S}^{\mathcal{F}}(C) \leq |A(\mathcal{F}) \setminus C| < \infty$. Since $N_{ss,S}^{\mathcal{F}}(C) \geq N_{ss,N}^{\mathcal{F}}(C) \geq N_{ss,E}^{\mathcal{F}}(C)$ (compare Corollary 7.16) holds we are done. In the same way one may show that $l < \infty$ implies $k < \infty$.

To prove that $\mathcal{S}_{(stb,ss,pr,\Phi)}$ is not positive it suffices to construct a non-finite fibre. Consider therefore $\mathcal{F} = (\{a\}, \{(a, a)\})$ and $C = \{a\}$. Since C does not possess conflict-free supersets we deduce $N_{ad,\Phi}^{\mathcal{F}}(C) = \infty$ (Theorem 7.29). Furthermore, by Proposition 7.17 we get $(\infty, \infty, \infty) \in \mathcal{S}_{(stb,ss,pr,\Phi)}$ concluding the proof. □

The following Theorem summarizes the earlier results. Note that the listed properties fully characterize the considered spectra. This means, it is decidable whether an arbitrary fibre belongs to the considered spectra.

Theorem 7.43. *For any $\Phi \in \{E, N, S\}$, $\mathcal{S}_{(stb,ss,pr,\Phi)}$ is coherent, m.d.s.-complete but not positive.*

7.3.3 Properties of the (stb, ss, pr, W)-Spectrum

We start with an example showing some first and notable differences between the coinciding spectra with respect to normal, strong and arbitrary expansions and the spectrum with respect to weak expansions considered in this section.

Example 7.44 (Example 7.22 continued). The AFs below exemplify that the (stb, ss, pr, W)-spectrum is not coherent since $N^{\mathcal{F}}_{stb,W}(\{a_1\}) = \infty$ (unenforceable) and $N^{\mathcal{F}}_{pr,W}(\{a_1\}) = 0$ (already accepted). Furthermore, $1 \le N^{\mathcal{F}}_{ss,W}(\{a_1\}) \le 2$ because $\{a_1\}$ and all its proper supersets are not semi-stable in $\mathcal{F}$ but $\{a_1\}$ is semi-stable in $\mathcal{G}$.

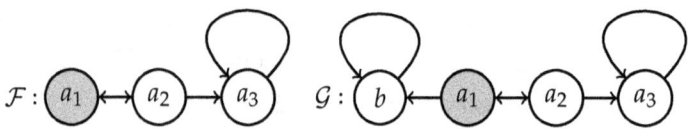

Figure 7.14: Non-coherence of the Weak Spectrum

Unfortunately, (up to now) there are no characterization theorems for semi-stable semantics. Nevertheless, using the following impossibility result it is shown that $N^{\mathcal{F}}_{ss,W}(\{a_1\}) = 2$ holds. This means, if a desired set of arguments D is not already contained in a semi-stable extension of the initial framework, then the minimal effort needed to enforce D is at least 2 in case of weak expansions.

Proposition 7.45. $(n, 1, m) \notin \mathcal{S}_{(stb, ss, pr, W)}$ for each $n, m \in \mathbb{N}_\infty$.

Proof. Since n, m are assumed to be arbitrary natural numbers or ∞ it suffices to prove that $(1) \notin \mathcal{S}_{(ss,W)}$. Assume $(1) \in \mathcal{S}_{(ss,W)}$, i.e. there is an AF $\mathcal{F}$ and a set C with the property $N^{\mathcal{F}}_{ss,W}(C) = 1$. This means there is an AF $\mathcal{G}$, such that $\mathcal{F} \le_W \mathcal{G}$, $d(\mathcal{F}, \mathcal{G}) = 1$ and a set $C' \supseteq C$ with $C' \in \mathcal{E}_{ss}(\mathcal{G})$. Without loss of generality $C' = D \cup E$ where $C \subseteq D \subseteq A(\mathcal{F})$ and $E \subseteq A(\mathcal{G}) \setminus A(\mathcal{F})$. Since every semi-stable extension is admissible we deduce $D \in \mathcal{E}_{ad}(\mathcal{F})$. Furthermore, $N^{\mathcal{F}}_{ss, \le_W}(C) = 1 \neq 0$ implies there is an admissible set D' in $\mathcal{F}$, such that $R^+_{\mathcal{F}}(D) \subset R^+_{\mathcal{F}}(D')$ (*). We will show now that $D \cup E \notin \mathcal{E}_{ss}(\mathcal{G})$ by proof by cases. Let (d, e) be the new attack. Note that $e \in A(\mathcal{G}) \setminus A(\mathcal{F})$ is implied since $\mathcal{F} \le_W \mathcal{G}$ is assumed. Furthermore, $d \in D$ and $e \in E$ is impossible since $D \cup E \in cf(\mathcal{G})$.

1st case: Let $d \in D \setminus D'$ and $e \notin E$. We observe $R^+_{\mathcal{G}}(D \cup E) = R^+_{\mathcal{F}}(D) \cup E \cup \{e\}$. Furthermore, E only contains isolated arguments in $\mathcal{G}$ and hence, $D' \cup E \in \mathcal{E}_{ad}(\mathcal{G})$. Because of (*) and $d \in D \setminus D'$ we conclude e is defended by D' in $\mathcal{G}$. Thus, $D' \cup E \cup \{e\} \in \mathcal{E}_{ad}(\mathcal{G})$ and obviously, $R^+_{\mathcal{G}}(D' \cup E \cup \{e\}) = R^+_{\mathcal{F}}(D') \cup E \cup \{e\}$. In consideration of (*) it follows that $R^+_{\mathcal{F}}(D) \cup E \cup \{e\} = R^+_{\mathcal{G}}(D \cup E) \subset R^+_{\mathcal{G}}(D' \cup E \cup \{e\})$ and hence, $D \cup E \notin \mathcal{E}_{ss}(\mathcal{G})$ is shown.

2nd case: Let $d \in D \cap D'$ and $e \notin E$. Consequently, $D' \cup E \in \mathcal{E}_{ad}(\mathcal{G})$ and furthermore, $R^+_{\mathcal{G}}(D \cup E) = R^+_{\mathcal{F}}(D) \cup E \cup \{e\} \subset^{(*)} R^+_{\mathcal{F}}(D') \cup E \cup \{e\} = R^+_{\mathcal{G}}(D' \cup E)$ contradicting $D \cup E \in \mathcal{E}_{ss}(\mathcal{G})$.

3rd case: Let $d \in D' \setminus D$ and $e \notin E$. Again, $D' \cup E \in \mathcal{E}_{ad}(\mathcal{G})$ holds and furthermore, $R_\mathcal{G}^+(D \cup E) = R_\mathcal{F}^+(D) \cup E \subset^{(*)} R_\mathcal{F}^+(D') \cup E \cup \{e\} = R_\mathcal{G}^+(D' \cup E)$ in contradiction to $D \cup E \in \mathcal{E}_{ss}(\mathcal{G})$.

4th case: Let $d \in D' \setminus D$ and $e \in E$. Hence, $D' \cup (E \setminus \{e\}) \in \mathcal{E}_{ad}(\mathcal{G})$. Furthermore, $R_\mathcal{G}^+(D \cup E) = R_\mathcal{F}^+(D) \cup E \subset^{(*)} R_\mathcal{F}^+(D') \cup (E \setminus \{e\}) \cup \{e\} = R_\mathcal{F}^+(D) \cup E = R_\mathcal{G}^+(D' \cup E)$. Consequently, $D \cup E \notin \mathcal{E}_{ss}(\mathcal{G})$ is shown.

5th case: Let $d \in A(\mathcal{F}) \setminus (D' \cup D)$ and $e \in E$. Thus, D has to counterattack d in $\mathcal{F}$ since $D \cup E$ is assumed to be admissible in $\mathcal{G}$. Let $d' \in D$ be the counterattacker of d. If $d' \in D'$ we conclude $D' \cup E \in \mathcal{E}_{ad}(\mathcal{G})$. If not, it follows the existence of an argument $d'' \in D'$, such that $(d'', d) \in R(\mathcal{F})$ since $(*)$ is assumed. Again, we get $D' \cup E \in \mathcal{E}_{ad}(\mathcal{G})$. In both cases, $R_\mathcal{G}^+(D \cup E) = R_\mathcal{F}^+(D) \cup E \subset^{(*)} R_\mathcal{F}^+(D') \cup E = R_\mathcal{G}^+(D' \cup E)$ contradicting $D \cup E \in \mathcal{E}_{ss}(\mathcal{G})$.

6th case: Let $d \in A(\mathcal{F}) \setminus (D' \cup D)$ and $e \notin E$. Consequently, $D' \cup E \in \mathcal{E}_{ad}(\mathcal{G})$ and thus, $R_\mathcal{G}^+(D \cup E) = R_\mathcal{F}^+(D) \cup E \subset^{(*)} R_\mathcal{F}^+(D') \cup E = R_\mathcal{G}^+(D' \cup E)$ in contradiction to $D \cup E \in \mathcal{E}_{ss}(\mathcal{G})$.

7th case: Let $d, e \in A(\mathcal{G}) \setminus A(\mathcal{F})$. Since $d(\mathcal{F}, \mathcal{G}) = 1$ it follows that $e \notin E$. Consequently, $D' \cup E \in \mathcal{E}_{ad}(\mathcal{G})$ and furthermore, $R_\mathcal{G}^+(D \cup E) = R_\mathcal{F}^+(D) \cup R_\mathcal{G}^+(E) \subset^{(*)} R_\mathcal{F}^+(D') \cup R_\mathcal{G}^+(E) = R_\mathcal{G}^+(D' \cup E)$ contradicting $D \cup E \in \mathcal{E}_{ss}(\mathcal{G})$. □

The proposition above and its usage for the illustrated problem, namely determining the characteristic in a certain argumentation scenario, underline that the investigation of spectra reveals important insights into the minimal change problem. The following impossibility result reveals a further surprising interrelation between the considered semantics, namely that for any $\mathcal{F}$ and any set of arguments C it is impossible that C is already contained in a preferred extension yet unenforceable using semi-stable semantics.

Proposition 7.46. $(\infty, \infty, 0) \notin \mathcal{S}_{(stb, ss, pr, W)}$.

Proof. We will show the stronger result, namely $(\infty, 0) \notin \mathcal{S}_{(ss, pr, W)}$. Assume $(\infty, 0) \in \mathcal{S}_{(ss, pr, W)}$, i.e. there is an AF $\mathcal{F}$ and a set C with the property $N^\mathcal{F}_{pr, W}(C) = 0$ and $N^\mathcal{F}_{ss, W}(C) = \infty$. This means, there exists a set $C' \supseteq C$ with $C' \in \mathcal{E}_{pr}(\mathcal{F})$. Since all considered AFs are assumed to be finite we deduce $C' = \{c'_1, ..., c'_n\}$ for some $n \in \mathbb{N}$. Let $D = \{d_1, ..., d_n\}$ be a set of fresh arguments and consider $\mathcal{G} = (A(\mathcal{F}) \cup D, R(\mathcal{F}) \cup \{(d_i, d_i), (c'_i, d_i) \mid 1 \leq i \leq n\})$. Obviously, $d(\mathcal{F}, \mathcal{G}) = 2n$ and $\mathcal{F} \leq_W \mathcal{G}$. Furthermore, the range of C' in $\mathcal{G}$ includes the set D and obviously, no proper subset of C' possess this property too. Consequently, there is no $C'' \in \mathcal{E}_{ad}(\mathcal{G})$, such that $R_\mathcal{G}^+(C') \subset R_\mathcal{G}^+(C'')$ because C' is also preferred in $\mathcal{G}$. Hence, $C' \in \mathcal{E}_{ss}(\mathcal{G})$ contradicting the assumption. □

In the light of Prop. 7.46 the corresponding question about the fibres $(\infty, \infty, \infty), (\infty, 0, 0)$ and $(0, 0, 0)$ arises. The following proposition gives the (positive) answer:

Proposition 7.47. $\{(\infty, \infty, \infty), (\infty, 0, 0), (0, 0, 0)\} \subseteq \mathcal{S}_{(stb, ss, pr, W)}$.

Proof. Consider the AFs $\mathcal{F}_1 = (\{a\}, \{(a, a)\})$, $\mathcal{F}_2 = (\{a, b\}, \{(b, b)\})$ and $\mathcal{F}_3 = (\{a\}, \emptyset)$. In consideration of Theorem 7.21 and Definition 7.18 one may easily verify that the set $\{a\}$ possesses the claimed fibres with respect to the AFs $\mathcal{F}_1$, $\mathcal{F}_2$ and $\mathcal{F}_3$. □

We have already shown that the minimal effort with respect to semi-stable semantics and weak expansions needed to enforce a desired set C cannot be 1. This raises the question about other natural numbers lying between 2 and ∞. The following proposition proves that there are infinitely many numbers n between 2 and ∞, such that $(\infty, n, 0)$ is a fibre of the (stb, ss, pr, W)-spectrum.

Proposition 7.48. *For any natural number $n \in \mathbb{N}$ there exists $k \in \mathbb{N}$, such that $n \leq k \leq 2n$ and $(\infty, k, 0) \in \mathcal{S}_{(stb,ss,pr,W)}$.*

Proof. We define the AF $\mathcal{F}_{\infty,n,0} = (A_{\infty,n,0}, R_{\infty,n,0})$ where

$$A_{\infty,n,0} = \{c_j, d_j, e_j \mid 1 \leq j \leq n\} \text{ and}$$

$$R_{\infty,n,0} = \{(c_j, d_j), (d_j, c_j), (d_j, e_j)(e_j, e_j) \mid 1 \leq j \leq n\} \cup \{(d_i, e_j) \mid 1 \leq i, j \leq n\}.$$

For the sake of clarity we present here an instantiation of the presented scheme, namely $\mathcal{F}_{\infty,3,0}$.

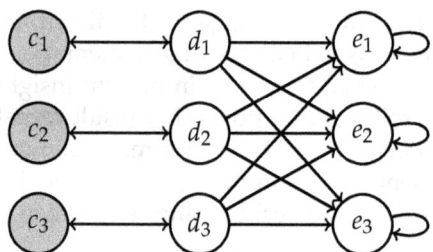

Figure 7.15: The AF $\mathcal{F}_{\infty,3,0}$

The grey highlighted arguments belong to the set $C_3 = \{c_1, c_2, c_3\}$ which is an instantiation of the scheme $C_n = \{c_j \mid 1 \leq j \leq n\}$ (D_n, E_n are defined analogously). We claim that $N_{stb,W}^{\mathcal{F}_{\infty,n,0}}(C_n) = \infty$ and $N_{pr,W}^{\mathcal{F}_{\infty,n,0}}(C_n) = 0$. We observe that no superset of C_n is stable in $\mathcal{F}_{\infty,n,0}$ and furthermore, C_n itself is preferred in $\mathcal{F}_{\infty,n,0}$. Consequently (Theorem 7.21, Definition 7.18), the characteristics of C_n in case of stable and preferred semantics hold as claimed.

Consider now the semi-stable semantics. At first we will show that $N_{ss,W}^{\mathcal{F}_{\infty,n,0}}(C_n) \geq n$. Suppose, to derive a contradiction, that $N_{ss,W}^{\mathcal{F}_{\infty,n,0}}(C_n) = n' < n$. This means, there is an AF $\mathcal{G}$, such that $d(\mathcal{F}_{\infty,n,0}, \mathcal{G}) = n'$, $\mathcal{F}_{\infty,n,0} \leq_W \mathcal{G}$ and furthermore, there is a superset C'_n of C_n, such that $C'_n \in \mathcal{E}_{ss}(\mathcal{G})$. We deduce that $C'_n = C_n \cup G$ where G is a set of fresh arguments since we consider weak expansions and furthermore, C_n does not possess proper supersets which are conflict-free in $\mathcal{F}_{\infty,n,0}$.

Since $n' < n$ is assumed it follows that there has to be an index j, such that $c_j \in C_n$ does not possess attacks to arguments in $A(\mathcal{G}) \setminus A_{\infty,n,0}$ (1) and $d_j \in D_n$ does not possess attacks to arguments in G (2). Consider $S_n^j = \{c_i \mid$

$1 \leq i \leq n\} \setminus \{c_j\} \cup \{d_j\}$. Obviously, $R^+_{\mathcal{F}_{\infty,n,0}}(C_n) \subset R^+_{\mathcal{F}_{\infty,n,0}}(S_n^j) = A_{\infty,n,0}$ (3). We will show now that $S_n^j \cup G$ is admissible in $\mathcal{G}$ and it possesses a strictly greater range than C_n' in $\mathcal{G}$. Since we assumed $C_n' \in \mathcal{E}_{ss}(\mathcal{G})$, (2) and we are considering weak expansions the conflict-freeness of $S_n^j \cup G$ in $\mathcal{G}$ is implied. Furthermore, admissibility of $S_n^j \cup G$ in $\mathcal{G}$ holds because S_n^j is admissible in $\mathcal{F}_{\infty,n,0}$ and all potential attackers of arguments in G are counterattacked by at least one argument in $S_n^j \cup G$ (d_j counterattacks any e_i, any d_i where $i \neq j$ is counterattacked by c_i, an attacker $g' \in A(\mathcal{G}) \setminus \{A_{\infty,n,0} \cup G\}$ is counterattacked by some $g \in G$ or some $c_i \in S_n^j$ because of the admissibility of C_n' and property (1)). Finally, $R^+_{\mathcal{G}}(C_n \cup G) \subset R^+_{\mathcal{G}}(S_n^j \cup G)$ has to hold because of properties (1) and (3). This contradicts the assumption that $C_n \cup G$ is semi-stable in $\mathcal{G}$.

Let us prove now that $N^{\mathcal{F}_{\infty,n,0}}_{ss,S}(C_n) \leq 2n$. Let $C_n' = \{c_1', ..., c_n'\}$ a set of fresh arguments and consider $\mathcal{G} = (A_{\infty,n,0} \cup C_n', R_{\infty,n,0} \cup \{(c_i', c_i'), (c_i, c_i') \mid 1 \leq i \leq n\})$. Obviously, $d(\mathcal{F}_{\infty,n,0}, \mathcal{G}) = 2n$ and $\mathcal{F}_{\infty,n,0} \leq_W \mathcal{G}$. One can easily verify that $C_n \in \mathcal{E}_{ss}(\mathcal{G})$. Finally, $N^{\mathcal{F}_{\infty,n,0}}_{ss,W}(C_n) \leq 2n$ is shown. □

It is an open question whether each number greater than 1 can appear as the characteristic of semi-stable semantics in a fibre, i.e. whether $\{(\infty, k, 0) \mid 2 \leq k < \infty\} \subseteq \mathcal{S}_{(stb,ss,pr,W)}$. We would like to recall that it is already shown that in case of stable and preferred semantics, either a desired set C is already contained in an extension or C is not enforceable (compare Definition 7.18 and Theorem 7.21). Consequently, an affirmative answer of the open question would imply a complete characterization of the (stb, ss, pr, W)-spectrum.

7.3.4 A Note on the $(stb, ss, pr, \mathcal{U})$-Spectrum

We use $\mathcal{U}$ to denote the universal relation among argumentation frameworks. In other words, we allow for arbitrary modifications including deletions of attacks and arguments. What consequences does this have for the corresponding spectrum? In contrast to the other considered spectra the $(stb, ss, pr, \mathcal{U})$-spectrum is the first one proven to be positive. This means there are no cases where the enforcing of a certain set D is impossible. Furthermore, the $(stb, ss, pr, \mathcal{U})$-spectrum is m.d.s. in analogy to the spectra with respect to arbitrary, normal and strong expansions.

Proposition 7.49. *The spectrum $\mathcal{S}_{(stb,ss,pr,\mathcal{U})}$ is positive and m.d.s.*

Proof. Both properties follow immediately by applying Proposition 7.32 (positive) and Corollary 7.15 (m.d.s.). □

A detailed analysis of the $(stb, ss, pr, \mathcal{U})$-spectrum is part of future work. Due to the multitude of possibilities to modify a certain argumentation scenario if arbitrary modifications are allowed it is a hard task to show further properties. We want to mention that we conjecture that the considered spectrum is m.d.s.-complete (but were unable to find a proof so far).

7.4 Minimal Change Equivalence

Expansion equivalence (see Chapter 5) between two AFs guarantees their mutual replaceability for *any* dynamic scenario without loss of information. In contrast to taking into account any dynamic scenario we are now interested in an equivalence notion which corresponds to the very nature of a dispute where (counter)arguments are put forward with the objective to convince the participants of a certain opinion D. Two AFs are called *minimal-D-equivalent* if the minimal effort needed to enforce D is the same for both. If minimal-D-equivalence holds for every subset D of the AFs in question we call them *minimal change equivalent* [Baumann, 2012b].

In this section we present characterization theorems for the newly introduced equivalence notions in case of stable, preferred, complete and admissible semantics. Furthermore, we study the relation between minimal change equivalence and expansion or standard equivalence in general, i.e. rather than considering specific semantics we provide our results for a whole range of semantics satisfying certain abstract principles. More specifically, we provide a complete picture about the relationships among all equivalence notions (so-called *equivalence zoo*) studied in this book for two of the most relevant semantics of Dung-style AFs, namely stable and preferred semantics [Baumann and Brewka, 2013a]. It turns out that minimal change equivalence naturally fits into this equivalence zoo, although its definition includes a graph-theoretical distance function and therefore an arithmetic aspect in contrast to standard or expansion equivalence as well as the considered intermediate variants.

7.4.1 Formal Definitions and General Results

We start with the formal definitions. Minimal-D-equivalence between two AFs guarantees that the minimal effort needed to convince the participants of a certain opinion D (a set of arguments) is identical. Minimal change equivalent AFs possess this property for every subset of the considered AFs.

Definition 7.50. Two AFs $\mathcal{F}$ and $\mathcal{G}$ are

1. *minimal-D-equivalent* (in symbols: $\mathcal{F} \equiv_\Phi^{\sigma,D} \mathcal{G}$) or

2. *minimal change equivalent* (in symbols: $\mathcal{F} \equiv_\Phi^{\sigma,MC} \mathcal{G}$),

with respect to a semantics σ and a binary relation $\Phi \subseteq \mathcal{A} \times \mathcal{A}$ iff

1. $D \subseteq A(\mathcal{F})$ and $N_{\sigma,\Phi}^{\mathcal{F}}(D) = N_{\sigma,\Phi}^{\mathcal{G}}(D)$ and

2. for any D, such that $D \subseteq A(\mathcal{F})$ or $D \subseteq A(\mathcal{G})$, $N_{\sigma,\Phi}^{\mathcal{F}}(D) = N_{\sigma,\Phi}^{\mathcal{G}}(D)$.

A few properties are clear by definition: First, minimal change equivalence guarantees sharing the same arguments (in contrast to minimal-D-equivalence) and second, minimal change equivalence implies minimal-D-equivalence for all subsets D.

Chapter 7. Enforcing and Minimal Change

In the following we are interested in characterization theorems for the introduced equivalence notions as well as their relation to standard and expansion equivalence. In Section 7.2.2 we already proved that the introduced value-functions adequately determine the corresponding characteristics and, therefore, the minimal change problem. Now we take advantage of this work. The value-functions provide us with a procedure for deciding whether two AFs are minimal change equivalent or not since they are based on local criteria of the underlying AF and thus can be established in a finite number of steps.

The following two theorems provide characterizations for stable, preferred, complete and admissible semantics with respect to weak, strong, normal and arbitrary expansions as well as arbitrary modifications.

Theorem 7.51. Let $\sigma \in \{stb, ad\}$ and $\Phi \in \{W, S, U\}$. For any AFs $\mathcal{F}$ and $\mathcal{G}$,

1. $\mathcal{F} \equiv_\Phi^{\sigma,D} \mathcal{G}$ iff $D \subseteq A(\mathcal{F})$ and $V_{\sigma,\Phi}^{\mathcal{F}}(D) = V_{\sigma,\Phi}^{\mathcal{G}}(D)$,

2. $\mathcal{F} \equiv_\Phi^{\sigma,MC} \mathcal{G}$ iff for any D, such that $D \subseteq A(\mathcal{F})$ or $D \subseteq A(\mathcal{G})$, $V_{\sigma,\Phi}^{\mathcal{F}}(D) = V_{\sigma,\Phi}^{\mathcal{G}}(D)$.

Proof. Consider Definition 7.50 and the characterization theorems 7.21, 7.29 and 7.37. □

Remember that some values are even strong enough to characterize the minimal change problem for more than one semantics and/or different classes of expansions (see Figure 7.12 for an overview).

Theorem 7.52. Let $\Phi \in \{N, E\}$. For any AFs $\mathcal{F}$ and $\mathcal{G}$,

1. $\mathcal{F} \equiv_\Phi^{stb,D} \mathcal{G}$ iff $D \subseteq A(\mathcal{F})$ and $V_{stb,S}^{\mathcal{F}}(D) = V_{stb,S}^{\mathcal{G}}(D)$,

2. $\mathcal{F} \equiv_\Phi^{stb,MC} \mathcal{G}$ iff for any D, such that $D \subseteq A(\mathcal{F})$ or $D \subseteq A(\mathcal{G})$, $V_{stb,S}^{\mathcal{F}}(D) = V_{stb,S}^{\mathcal{G}}(D)$.

Let $\sigma \in \{pr, co\}$. For any AFs $\mathcal{F}$ and $\mathcal{G}$,

3. $\mathcal{F} \equiv_W^{\sigma,D} \mathcal{G}$ iff $D \subseteq A(\mathcal{F})$ and $V_{ad,W}^{\mathcal{F}}(D) = V_{ad,W}^{\mathcal{G}}(D)$,

4. $\mathcal{F} \equiv_W^{\sigma,MC} \mathcal{G}$ iff for any D, such that $D \subseteq A(\mathcal{F})$ or $D \subseteq A(\mathcal{G})$, $V_{ad,W}^{\mathcal{F}}(D) = V_{ad,W}^{\mathcal{G}}(D)$.

Let $\sigma \in \{pr, co\}$ and $\Phi \in \{N, E\}$. For any AFs $\mathcal{F}$ and $\mathcal{G}$,

5. $\mathcal{F} \equiv_\Phi^{\sigma,D} \mathcal{G}$ iff $D \subseteq A(\mathcal{F})$ and $V_{ad,S}^{\mathcal{F}}(D) = V_{ad,S}^{\mathcal{G}}(D)$,

6. $\mathcal{F} \equiv_\Phi^{\sigma,MC} \mathcal{G}$ iff for any D, such that $D \subseteq A(\mathcal{F})$ or $D \subseteq A(\mathcal{G})$, $V_{ad,S}^{\mathcal{F}}(D) = V_{ad,S}^{\mathcal{G}}(D)$.

Let $\sigma \in \{pr, co\}$. For any AFs $\mathcal{F}$ and $\mathcal{G}$,

7.4. Minimal Change Equivalence

7. $\mathcal{F} \equiv_{\mathcal{U}}^{\sigma,D} \mathcal{G}$ iff $D \subseteq A(\mathcal{F})$ and $V_{ad,\mathcal{U}}^{\mathcal{F}}(D) = V_{ad,\mathcal{U}}^{\mathcal{G}}(D)$,

8. $\mathcal{F} \equiv_{\mathcal{U}}^{\sigma,MC} \mathcal{G}$ iff for any D, such that $D \subseteq A(\mathcal{F})$ or $D \subseteq A(\mathcal{G})$, $V_{ad,\mathcal{U}}^{\mathcal{F}}(D) = V_{ad,\mathcal{U}}^{\mathcal{G}}(D)$.

Proof. Combine Definition 7.50 and characterization theorems 7.21, 7.29 and 7.37. □

Example 7.53. The following AFs exemplify the novel equivalence notions.

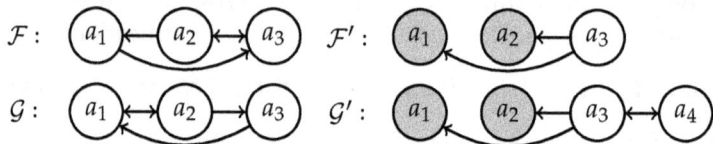

Figure 7.16: Minimal Change Equivalence and Minimal-$\{a_1, a_2\}$-Equivalence

The AFs $\mathcal{F}$ and $\mathcal{G}$ are minimal change equivalent with respect to preferred semantics and strong expansions, i.e. $\mathcal{F} \equiv_S^{pr,MC} \mathcal{G}$. This can be seen as follows: First, $V_{ad,S}^{\mathcal{F}}(\{a_1\}) = V_{ad,S}^{\mathcal{G}}(\{a_1\}) = V_{ad,S}^{\mathcal{F}}(\{a_3\}) = V_{ad,S}^{\mathcal{G}}(\{a_3\}) = 1$, second, $V_{ad,S}^{\mathcal{F}}(\{a_2\}) = V_{ad,S}^{\mathcal{G}}(\{a_2\}) = V_{ad,S}^{\mathcal{F}}(\emptyset) = V_{ad,S}^{\mathcal{G}}(\emptyset) = 0$ and finally, all other sets $E \subseteq \{a_1, a_2, a_3\}$ do not possess conflict-free supersets in $\mathcal{F}$ or $\mathcal{G}$ and thus take the value infinity. Similarly, one can show that $\mathcal{F}' \equiv_S^{pr,\{a_1,a_2\}} \mathcal{G}'$.

We turn now to the relation between minimal change equivalence and expansion or standard equivalence in general, i.e. rather than considering specific semantics we provide our results for a whole range of semantics satisfying certain abstract principles.

To get the first insight into the relations consider again the minimal change equivalent AFs $\mathcal{F}$ and $\mathcal{G}$ depicted in Figure 7.16. One main result in [Oikarinen and Woltran, 2011] is that expansion equivalence is immediately linked to the existence of self-loops, in particular it collapses to syntactic equivalence in case of self-loop-free AFs (compare Proposition 5.47). This means, $\mathcal{F}$ and $\mathcal{G}$ are not expansion equivalent which indicates that minimal change equivalence is possibly weaker than strong equivalence. The following theorem shows that the assertion holds for any semantics σ which satisfy that expansion equivalent AFs have to share the same arguments. Let us call this property *regularity*. Note that regularity is a very weak criterion. One can imagine that for reasonable defined semantics it is always possible to find an AF $\mathcal{H}$, such that $\mathcal{E}_\sigma(\mathcal{F} \cup \mathcal{H}) \neq \mathcal{E}_\sigma(\mathcal{G} \cup \mathcal{H})$ if $A(\mathcal{F}) \neq A(\mathcal{G})$. All semantics considered in this book satisfy regularity (see Proposition 5.48).

Theorem 7.54. *For any AFs $\mathcal{F}, \mathcal{G}$, any regular semantics σ and any binary $\Phi \in \{E, N, S, W\}$,*

$$\mathcal{F} \equiv_E^\sigma \mathcal{G} \Rightarrow \mathcal{F} \equiv_\Phi^{\sigma,MC} \mathcal{G}.$$

Proof. Towards a contradiction, assume $\mathcal{F} \equiv_E^\sigma \mathcal{G}$ and $\mathcal{F} \not\equiv_\Phi^{\sigma,MC} \mathcal{G}$. Since σ is assumed to be regular it follows $A(\mathcal{F}) = A(\mathcal{G})$. Hence, we deduce the existence of a subset E, such that $N_{\sigma,\Phi}^\mathcal{F}(E) \neq N_{\sigma,\Phi}^\mathcal{G}(E)$. Let $N_{\sigma,\Phi}^\mathcal{F}(E) = k_1$ and $N_{\sigma,\Phi}^\mathcal{G}(E) = k_2$ where $k_1, k_2 \in \mathbb{N}_\infty$ and without loss of generality $k_1 < k_2$. Consequently, there are an AF $\mathcal{H}$ and a set $E' \subseteq A(\mathcal{H})$, such that $(\mathcal{F}, \mathcal{H}) \in \Phi$, $d(\mathcal{F}, \mathcal{H}) = k_1$ and $E \subseteq E' \in \mathcal{E}_\sigma(\mathcal{H})$. Without loss of generality there exists an AF $\mathcal{H}'$, such that $|R(\mathcal{H}')| = k_1$, $R(\mathcal{F}) \cap R(\mathcal{H}') = \emptyset$ and $\mathcal{H} = \mathcal{F} \cup \mathcal{H}'$ (compare Definition 2.13). Since $\mathcal{F} \equiv_E^\sigma \mathcal{G}$ is assumed we conclude $E' \in \mathcal{E}_\sigma(\mathcal{G} \cup \mathcal{H}')$. It can be easily seen that $(\mathcal{G}, \mathcal{G} \cup \mathcal{H}') \in \Phi$ and $d(\mathcal{G}, \mathcal{G} \cup \mathcal{H}') \leq k_1$. Thus, $k_2 = N_{\sigma,\Phi}^\mathcal{G}(E) \leq k_1$ which contradicts the assumption $k_1 < k_2$. This means, $\mathcal{F} \equiv_E^\sigma \mathcal{G} \Rightarrow \mathcal{F} \equiv_\Phi^{\sigma,MC} \mathcal{G}$ is shown. □

Since extensions possess a characteristic of zero, one might expect minimal change equivalence to imply standard equivalence. However, it turns out that this implication does not hold in general but can be shown for semantics satisfying *I-maximality* [Baroni and Giacomin, 2007]. Remember I-maximality is fulfilled, if no extension can be a proper subset of another one (compare Definition 2.11).

Theorem 7.55. *For any AFs $\mathcal{F}$, $\mathcal{G}$, any semantics σ satisfying I-maximality and any binary $\Phi \in \{U, E, N, S, W\}$,*

$$\mathcal{F} \equiv_\Phi^{\sigma,MC} \mathcal{G} \Rightarrow \mathcal{F} \equiv^\sigma \mathcal{G}.$$

Proof. Suppose, to derive a contradiction, that $\mathcal{F} \equiv_\Phi^{\sigma,MC} \mathcal{G}$ and $\mathcal{F} \not\equiv^\sigma \mathcal{G}$. By Definition 7.50 minimal change equivalence implies $A(\mathcal{F}) = A(\mathcal{G})$. Without loss of generality we may assume the existence of a subset E, such that $E \in \mathcal{E}_\sigma(\mathcal{F})$ and $E \notin \mathcal{E}_\sigma(\mathcal{G})$. This means, $N_{\sigma,\Phi}^\mathcal{F}(E) = 0$ and thus, $N_{\sigma,\Phi}^\mathcal{G}(E) = 0$ since $\mathcal{F}$ and $\mathcal{G}$ are assumed to be minimal change equivalent. Consequently, there is a proper superset E' of E, such that $E' \in \mathcal{E}_\sigma(\mathcal{G})$. This means, $N_{\sigma,\Phi}^\mathcal{G}(E') = 0$ and therefore $N_{\sigma,\Phi}^\mathcal{F}(E') = 0$. Again, there has to be a superset E'' of E', such that $E'' \in \mathcal{E}_\sigma(\mathcal{F})$. Observe that $E \subset E''$ is implied which contradicts the I-maximality of σ. □

Remember that stable, semi-stable, stage, preferred, grounded, ideal, eager and naive semantics satisfy I-maximality whereas complete and admissible does not (see Figure 2.4). We want to conclude our study with two AFs showing that we cannot drop the I-maximality criterion in Theorem 7.55.

Example 7.56. We have $\mathcal{F} \not\equiv^{co} \mathcal{G}$ since $\{a_4\} \in \mathcal{E}_{co}(\mathcal{G})$ and $\{a_4\} \notin \mathcal{E}_{co}(\mathcal{F})$. On the other hand, $\mathcal{F} \equiv_E^{co,MC} \mathcal{G}$ since: First, for all $E \subseteq \{a_2, a_4\}$, $V_{ad,S}^\mathcal{F}(E) = V_{ad,S}^\mathcal{G}(E) = 0$ because $\{a_2, a_4\}$ is complete in $\mathcal{F}$ and $\mathcal{G}$. Furthermore, $V_{ad,S}^\mathcal{F}(\{a_3\}) = V_{ad,S}^\mathcal{G}(\{a_3\}) = 1$, and finally, all other sets $E \subseteq \{a_1, a_2, a_3, a_4\}$ (such that $E \not\subseteq \{a_2, a_4\}$ and $E \neq \{a_3\}$) do not possess conflict-free supersets in $\mathcal{F}$ or $\mathcal{G}$ and thus take the value infinity.

7.4. Minimal Change Equivalence

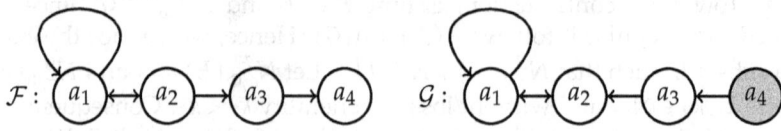

Figure 7.17: Necessity of I-maximality

7.4.2 Stable and Preferred Semantics: Analyzing the Equivalence Zoo

In Section 5.5.2 (Theorem 5.46) we studied relationships between expansion, normal expansion, strong expansion, weak expansion, local expansion and standard equivalence for all semantics considered in this book. In this section we provide a complete analysis including the family of minimal change equivalence relations for two of the most relevant semantics of Dung-style AFs, namely stable and preferred semantics. Besides the general case, i.e. considering arbitrary AFs, we also provide results for two special cases, namely the case where the AFs do not contain self-loops, i.e. attacks of the form (a,a) for some argument a, and the case where two AFs have the same arguments. In the interest of readability we present our results not only in terms of propositions, but also graphically.

7.4.2.1 Stable Semantics: The Full Picture

The following proposition characterizes stable semantics in the unrestrictive case, i.e. we consider arbitrary AFs.

Proposition 7.57. *For any AFs $\mathcal{F}$ and $\mathcal{G}$,*

- $\mathcal{F} \equiv_E^{stb} \mathcal{G} \Leftrightarrow \mathcal{F} \equiv_N^{stb} \mathcal{G} \Leftrightarrow \mathcal{F} \equiv_S^{stb} \mathcal{G} \Rightarrow \mathcal{F} \equiv_L^{stb} \mathcal{G} \Rightarrow \mathcal{F} \equiv_W^{stb} \mathcal{G} \Rightarrow \mathcal{F} \equiv^{stb} \mathcal{G}$,

- $\mathcal{F} \equiv_E^{stb,MC} \mathcal{G} \Leftrightarrow \mathcal{F} \equiv_N^{stb,MC} \mathcal{G} \Leftrightarrow \mathcal{F} \equiv_S^{stb,MC} \mathcal{G} \Rightarrow \mathcal{F} \equiv_W^{stb,MC} \mathcal{G} \Rightarrow \mathcal{F} \equiv_W^{stb} \mathcal{G}$,

- $\mathcal{F} \equiv_E^{stb} \mathcal{G} \Leftrightarrow \mathcal{F} \equiv_E^{stb,MC} \mathcal{G}$.

Figure 7.18 describes the results for stable semantics graphically. In case of stable semantics only local expansion equivalence and the family of minimal change equivalence relations are unrelated. For any other two equivalence relations we have at least one implication chain. In particular, the different forms of minimal change equivalence *are shown to be intermediate forms between strong expansion and weak expansion equivalence.*

Proof. In Theorem 5.46 it was already shown that $\mathcal{F} \equiv_E^{stb} \mathcal{G} \Leftrightarrow \mathcal{F} \equiv_N^{stb} \mathcal{G} \Leftrightarrow \mathcal{F} \equiv_S^{stb} \mathcal{G} \Rightarrow \mathcal{F} \equiv_L^{stb} \mathcal{G} \Rightarrow \mathcal{F} \equiv_W^{stb} \mathcal{G} \Rightarrow \mathcal{F} \equiv^{st} \mathcal{G}$. Since stable semantics satisfy regularity, i.e. expansion equivalent AFs have to share the same arguments

Chapter 7. Enforcing and Minimal Change

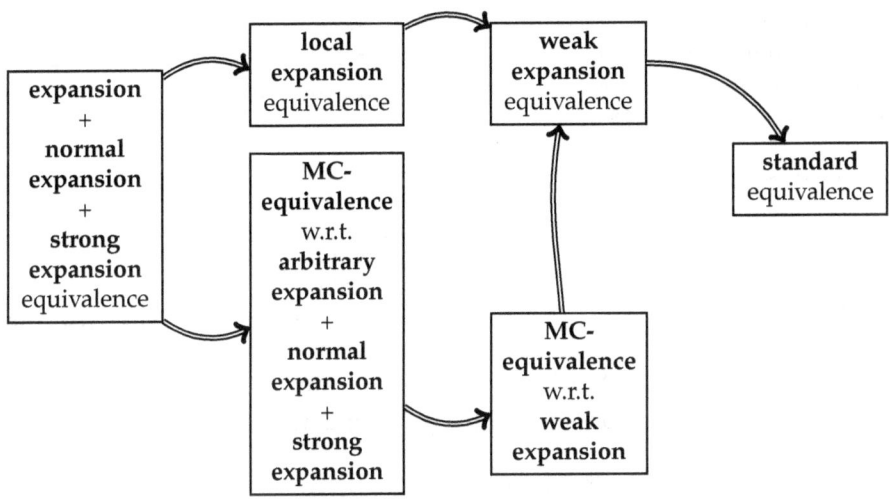

Figure 7.18: Stable Semantics (Arbitrary AFs)

(compare Proposition 5.48) we conclude that $\mathcal{F} \equiv_E^{stb} \mathcal{G} \Rightarrow \mathcal{F} \equiv_E^{stb,MC} \mathcal{G}$ (Theorem 7.54). Furthermore, by applying Theorems 7.51 and 7.52 we deduce $\mathcal{F} \equiv_E^{stb,MC} \mathcal{G} \Leftrightarrow \mathcal{F} \equiv_N^{stb,MC} \mathcal{G} \Leftrightarrow \mathcal{F} \equiv_S^{stb,MC} \mathcal{G}$.

We will show now that $\mathcal{F} \equiv_S^{stb,MC} \mathcal{G} \Rightarrow \mathcal{F} \equiv_W^{stb,MC} \mathcal{G}$. Assume $\mathcal{F} \equiv_S^{stb,MC} \mathcal{G}$ and $\mathcal{F} \not\equiv_W^{stb,MC} \mathcal{G}$. Note that the first assumption implies that $A(\mathcal{F}) = A(\mathcal{G})$. The second assumption means that there is a set E, such that $N_{stb,W}^{\mathcal{F}}(E) \neq N_{stb,W}^{\mathcal{G}}(E)$. Without loss of generality we assume $N_{stb,W}^{\mathcal{F}}(E) = \infty$ and $N_{stb,W}^{\mathcal{G}}(E) = 0$ (Theorem 7.21). Since the characteristic with respect to strong expansions does not exceed the characteristic with respect to weak expansions we have $N_{stb,S}^{\mathcal{G}}(E) = 0$ (Proposition 7.30). Consequently (first assumption), $N_{stb,S}^{\mathcal{F}}(E) = 0$ in contradiction to $N_{stb,W}^{\mathcal{F}}(E) = \infty$ which proves the claimed implication.

We show now that $\mathcal{F} \equiv_W^{stb,MC} \mathcal{G} \Rightarrow \mathcal{F} \equiv_W^{stb} \mathcal{G}$. Suppose, towards a contradiction, that $\mathcal{F} \equiv_W^{stb,MC} \mathcal{G}$ and $\mathcal{F} \not\equiv_W^{stb} \mathcal{G}$. First, minimal change equivalence implies $A(\mathcal{F}) = A(\mathcal{G})$. In Theorem 5.38 it was shown that two AFs are weak expansion equivalent with respect to stable semantics iff i) $A(\mathcal{F}) = A(\mathcal{G})$ and $\mathcal{E}_{stb}(\mathcal{F}) = \mathcal{E}_{stb}(\mathcal{G})$ or ii) $\mathcal{E}_{stb}(\mathcal{F}) = \mathcal{E}_{stb}(\mathcal{G}) = \emptyset$. Consequently, $\mathcal{E}_{stb}(\mathcal{F}) \neq \mathcal{E}_{stb}(\mathcal{G})$. Let $E \in \mathcal{E}_{stb}(\mathcal{F})$ and $E \notin \mathcal{E}_{stb}(\mathcal{G})$. Hence, $N_{stb,W}^{\mathcal{F}}(E) = 0$. Since minimal change equivalence is assumed, $N_{stb,W}^{\mathcal{G}}(E) = 0$. Since we assumed $E \notin \mathcal{E}_{stb}(\mathcal{G})$ there has to be a proper superset E' of E, such that $E' \in \mathcal{E}_{stb}(\mathcal{G})$. Consequently, $N_{stb,W}^{\mathcal{G}}(E') = 0$ and therefore $N_{stb,W}^{\mathcal{F}}(E') = 0$. This means there is a superset E'' of E', such that $E'' \in \mathcal{E}_{stb}(\mathcal{F})$. This means, there are two stable extensions E, E'' of $\mathcal{F}$, such that $E \subset E''$. This is impossible because stable semantics satisfies the I-maximality principle (compare Figure 2.4). Altogether, the claimed implications are shown.

7.4. Minimal Change Equivalence

Now we present some counter-examples showing that the converse directions do not hold. It suffices to consider the following four cases. The other non-relations can be easily obtained by using the already shown relations presented in Figure 7.18.

1. $\mathcal{F} \equiv^{stb} \mathcal{G} \not\Rightarrow \mathcal{F} \equiv^{stb}_W \mathcal{G}$.

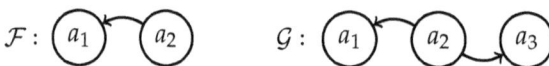

We have $\mathcal{E}_{stb}(\mathcal{F}) = \mathcal{E}_{stb}(\mathcal{G}) = \{\{a_2\}\} \neq \emptyset$ and obviously, $A(\mathcal{F}) \neq A(\mathcal{G})$. In Theorem 5.38 it was shown that two AFs are weak expansion equivalent with respect to stable semantics iff i) $A(\mathcal{F}) = A(\mathcal{G})$ and $\mathcal{E}_{stb}(\mathcal{F}) = \mathcal{E}_{stb}(\mathcal{G})$ or ii) $\mathcal{E}_{stb}(\mathcal{F}) = \mathcal{E}_{stb}(\mathcal{G}) = \emptyset$. Consequently, $\mathcal{F} \not\equiv^{stb}_W \mathcal{G}$ and at the same time $\mathcal{F} \equiv^{stb} \mathcal{G}$.

2. $\mathcal{F} \equiv^{stb,MC}_\Phi \mathcal{G} \not\Rightarrow \mathcal{F} \equiv^{stb}_L \mathcal{G}$ for each $\Phi \in \{E, N, S\}$.

Both AFs share the same arguments. Furthermore, $\mathcal{E}_{stb}(\mathcal{F}) = \mathcal{E}_{stb}(\mathcal{G}) = \{\{a_1, a_3\}\}$. Applying Theorem 7.29 we conclude: First, for any $E \subseteq \{a_1, a_3\}$, we have $N^{\mathcal{F}}_{stb,S}(E) = N^{\mathcal{G}}_{stb,S}(E) = 0$. Second, $N^{\mathcal{F}}_{stb,S}(\{a_1\}) = N^{\mathcal{G}}_{stb,S}(\{a_1\}) = 1$ and third, for all not mentioned subsets C of $A(\mathcal{F})$, $N^{\mathcal{F}}_{stb,S}(C) = N^{\mathcal{G}}_{stb,S}(C) = \infty$ because they contain at least one conflict. This verifies $\mathcal{F} \equiv^{stb,MC}_\Phi \mathcal{G}$ for each $\Phi \in \{E, N, S\}$ (Theorems 7.29, 7.51 and 7.52). Consider the AFs $\mathcal{H} = (\{a_2, a_3\}, \{(a_2, a_3)\})$. We observe that $\mathcal{E}_{stb}(\mathcal{F} \cup \mathcal{H}) = \{\{a_1, a_3\}, \{a_2\}\} \neq \{\{a_1, a_3\}\} = \mathcal{E}_{stb}(\mathcal{G}) = \mathcal{E}_{stb}(\mathcal{G} \cup \mathcal{H})$. Thus, $\mathcal{F} \not\equiv^{stb}_L \mathcal{G}$.

3. $\mathcal{F} \equiv^{stb,MC}_W \mathcal{G} \not\Rightarrow \mathcal{F} \equiv^{stb,MC}_\Phi \mathcal{G}$ for each $\Phi \in \{E, N, S\}$.

$\mathcal{F}: (a_1) \quad (a_2) \quad (a_3) \qquad \mathcal{G}: (a_1) \quad (a_2) \quad (a_3)$

Both AFs share the same arguments and $\mathcal{E}_{stb}(\mathcal{F}) = \mathcal{E}_{stb}(\mathcal{G}) = \{\{a_2\}\}$. Thus, $N^{\mathcal{F}}_{stb,W}(\emptyset) = N^{\mathcal{G}}_{stb,W}(\emptyset) = N^{\mathcal{F}}_{stb,W}(\{a_2\}) = N^{\mathcal{G}}_{stb,W}(\{a_2\}) = 0$. Furthermore, for any other subset C of $A(\mathcal{F})$, $N^{\mathcal{F}}_{stb,W}(C) = N^{\mathcal{G}}_{stb,W}(C) = \infty$ because they are not contained in an extension (Theorem 7.21, Definition 7.18). Consequently, $\mathcal{F} \equiv^{stb,MC}_W \mathcal{G}$. On the other hand, $N^{\mathcal{F}}_{stb,S}(\{a_1\}) = 1 \neq 2 = N^{\mathcal{G}}_{stb,S}(\{a_1\})$ (compare Theorem 7.29, Definition 7.24). This means, $\mathcal{F} \not\equiv^{stb,MC}_\Phi \mathcal{G}$ for each $\Phi \in \{E, N, S\}$.

4. $\mathcal{F} \equiv^{stb}_L \mathcal{G} \not\Rightarrow \mathcal{F} \equiv^{stb,MC}_W \mathcal{G}$.

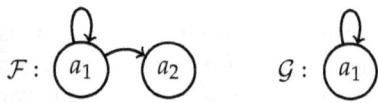

Since minimal change equivalence implies sharing the same arguments

Chapter 7. Enforcing and Minimal Change

we state $\mathcal{F} \not\equiv_W^{stb,MC} \mathcal{G}$. Furthermore, it can be easily checked that for any AF $\mathcal{H}$, such that $A(\mathcal{H}) \subseteq \{a_1, a_2\}$, we have $\mathcal{E}_{stb}(\mathcal{F} \cup \mathcal{H}) = \mathcal{E}_{stb}(\mathcal{G} \cup \mathcal{H})$. Hence, $\mathcal{F} \equiv_L^{stb} \mathcal{G}$. □

How does the situation change if we restrict our considerations to AFs possessing the same arguments? It turns out that the equivalence zoo collapses to only 3 distinct classes, and in contrast to the general case local expansion equivalence and the different forms of minimal change equivalence become comparable.

Proposition 7.58. *For any AFs $\mathcal{F}$ and $\mathcal{G}$ such that $A(\mathcal{F}) = A(\mathcal{G})$ we have:*

- $\mathcal{F} \equiv_E^{stb} \mathcal{G} \Leftrightarrow \mathcal{F} \equiv_N^{stb} \mathcal{G} \Leftrightarrow \mathcal{F} \equiv_S^{stb} \mathcal{G} \Leftrightarrow \mathcal{F} \equiv_L^{stb} \mathcal{G} \Rightarrow \mathcal{F} \equiv_E^{stb,MC} \mathcal{G}$,

- $\mathcal{F} \equiv_E^{stb,MC} \mathcal{G} \Leftrightarrow \mathcal{F} \equiv_N^{stb,MC} \mathcal{G} \Leftrightarrow \mathcal{F} \equiv_S^{stb,MC} \mathcal{G} \Rightarrow \mathcal{F} \equiv_W^{stb} \mathcal{G}$,

- $\mathcal{F} \equiv_W^{stb} \mathcal{G} \Leftrightarrow \mathcal{F} \equiv^{stb} \mathcal{G} \Leftrightarrow \mathcal{F} \equiv_W^{stb,MC} \mathcal{G}$.

Here is the graphical representation of the result:

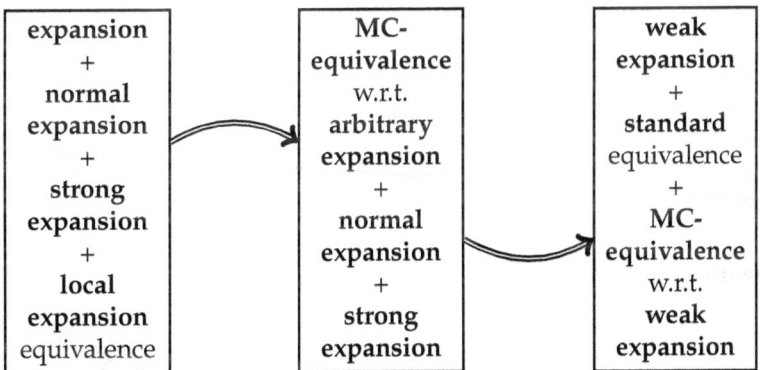

Figure 7.19: Stable Semantics (Same Arguments)

Proof. For this proof we consider AFs sharing the same arguments, i.e. $A(\mathcal{F}) = A(\mathcal{G})$. Using the results presented in Figure 7.18 it suffices to show the following three implications.

First, we will show that $\mathcal{F} \equiv_L^{stb} \mathcal{G} \Rightarrow \mathcal{F} \equiv_E^{stb} \mathcal{G}$. In [Oikarinen and Woltran, 2011, Theorem 9] it is proven that two AFs are local expansion equivalent iff i) $\mathcal{F} \equiv_E^{stb} \mathcal{G}$ or ii) $\mathcal{E}_{stb}(\mathcal{F}) = \mathcal{E}_{stb}(\mathcal{G})$ and there is an argument $a \in (A(\mathcal{F}) \setminus A(\mathcal{G})) \cup (A(\mathcal{F}) \setminus A(\mathcal{G}))$ satisfying certain properties. Since $A(\mathcal{F}) = A(\mathcal{G})$ is assumed, $\mathcal{F} \equiv_E^{stb} \mathcal{G}$ follows because there are no arguments in $(A(\mathcal{F}) \setminus A(\mathcal{G})) \cup (A(\mathcal{F}) \setminus A(\mathcal{G}))$.

7.4. Minimal Change Equivalence

We will show now that $\mathcal{F} \equiv^{stb} \mathcal{G} \Rightarrow \mathcal{F} \equiv^{stb}_W \mathcal{G}$. Theorem 5.38 shows that two AFs are weak expansion equivalent with respect to stable semantics iff i) $A(\mathcal{F}) = A(\mathcal{G})$ and $\mathcal{E}_{stb}(\mathcal{F}) = \mathcal{E}_{stb}(\mathcal{G})$ or ii) $\mathcal{E}_{stb}(\mathcal{F}) = \mathcal{E}_{stb}(\mathcal{G}) = \emptyset$. Consequently, standard equivalence, i.e. $\mathcal{E}_{stb}(\mathcal{F}) = \mathcal{E}_{stb}(\mathcal{G})$ together with the assumption $A(\mathcal{F}) = A(\mathcal{G})$ implies weak expansion equivalence, i.e. $\mathcal{F} \equiv^{stb}_W \mathcal{G}$ is shown.

Finally, we show that $\mathcal{F} \equiv^{stb}_W \mathcal{G} \Rightarrow \mathcal{F} \equiv^{stb,MC}_W \mathcal{G}$. Assume $\mathcal{F} \equiv^{stb}_W \mathcal{G}$ and $\mathcal{F} \not\equiv^{stb,MC}_W$. Using the characterization theorem 5.38 we deduce $\mathcal{E}_{stb}(\mathcal{F}) = \mathcal{E}_{stb}(\mathcal{G})$. Since we assumed that $\mathcal{F}$ and $\mathcal{G}$ are not minimal change equivalent we deduce $N^{\mathcal{F}}_{stb,W}(E) \neq N^{\mathcal{G}}_{stb,W}(E)$ for some $E \subseteq A(\mathcal{F})(= A(\mathcal{G}))$. Without loss of generality we assume $N^{\mathcal{F}}_{stb,W}(E) = 0$ and $N^{\mathcal{G}}_{stb,W}(E) = \infty$ (Theorem 7.21, Definition 7.18). This means there is a superset E' of E, such that $E' \in \mathcal{E}_{stb}(\mathcal{F})$. Consequently, $E' \in \mathcal{E}_{stb}(\mathcal{G})$ in contradiction to $N^{\mathcal{G}}_{stb,W}(E) = \infty$.

In consideration of the counter-examples 2 and 3 it follows that the converse directions do not hold because the considered AFs share the same arguments. □

The role of self-loops is somewhat controversial in the literature. It is sometimes argued such self-attacks are necessary as they model paradoxical statements. On the other hand, it was shown (see Theorem 4.13 in [Besnard and Hunter, 2001]) that self-attacking arguments do not occur if Dung-style AFs are considered as instantiations of classical logic-based frameworks. At least in such contexts investigating AFs without self-loops is of interest. For this reason we present the equivalence zoo restricted to self-loop-free AFs. In contrast to the general case, local expansion equivalence coincides with strong, normal expansion and strong expansion equivalence and thus, the equivalence zoo becomes totally ordered.

Proposition 7.59. *For any self-loop-free AFs $\mathcal{F}$ and $\mathcal{G}$,*

- $\mathcal{F} \equiv^{stb}_E \mathcal{G} \Leftrightarrow \mathcal{F} \equiv^{stb}_N \mathcal{G} \Leftrightarrow \mathcal{F} \equiv^{stb}_S \mathcal{G} \Leftrightarrow \mathcal{F} \equiv^{stb}_L \mathcal{G} \Rightarrow \mathcal{F} \equiv^{stb,MC}_E \mathcal{G}$,

- $\mathcal{F} \equiv^{stb,MC}_E \mathcal{G} \Leftrightarrow \mathcal{F} \equiv^{stb,MC}_N \mathcal{G} \Leftrightarrow \mathcal{F} \equiv^{stb,MC}_S \mathcal{G} \Rightarrow \mathcal{F} \equiv^{stb,MC}_W \mathcal{G}$,

- $\mathcal{F} \equiv^{stb,MC}_W \mathcal{G} \Rightarrow \mathcal{F} \equiv^{stb}_W \mathcal{G}$,

- $\mathcal{F} \equiv^{stb}_W \mathcal{G} \Rightarrow \mathcal{F} \equiv^{stb} \mathcal{G}$.

Here is again the graphical representation of the result:

Chapter 7. Enforcing and Minimal Change

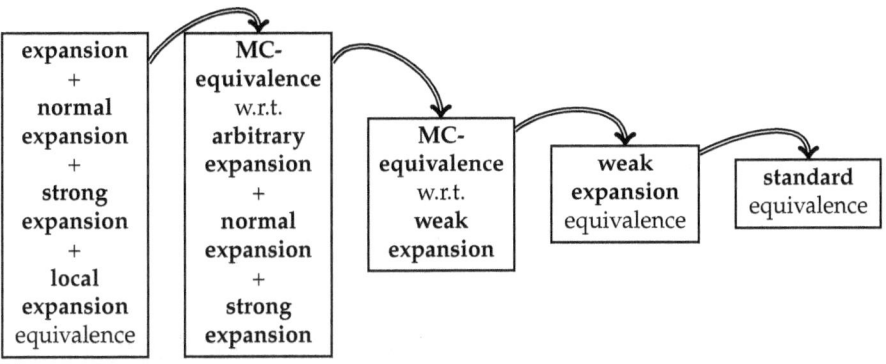

Figure 7.20: Stable Semantics (Self-loop-free AFs)

Proof. In consideration of the counter-examples given in the proof for stable semantics without restrictions we observe that only the fourth example showing that $\mathcal{F} \equiv_L^{stb} \mathcal{G} \not\Rightarrow \mathcal{F} \equiv_W^{stb,MC} \mathcal{G}$ contains self-loops. This is not a coincidence because local expansion equivalence coincides with strong equivalence in case of self-loop-free AFs. This follows immediately by [Oikarinen and Woltran, 2011, Theorem 13] and Proposition 5.47. Finally, we present a counter-example showing that $\mathcal{F} \equiv_W^{stb} \mathcal{G} \not\Rightarrow \mathcal{F} \equiv_W^{stb,MC} \mathcal{G}$.

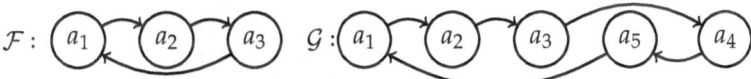

Two AFs are weak expansion equivalent with respect to stable semantics iff i) $A(\mathcal{F}) = A(\mathcal{G})$ and $\mathcal{E}_{stb}(\mathcal{F}) = \mathcal{E}_{stb}(\mathcal{G})$ or ii) $\mathcal{E}_{stb}(\mathcal{F}) = \mathcal{E}_{stb}(\mathcal{G}) = \emptyset$ (Theorem 5.38). The second condition holds for the considered AFs. Furthermore, they are not minimal change equivalent with respect to weak expansions since they do not share the same arguments. □

7.4.2.2 Preferred Semantics: The Full Picture

How does the equivalence zoo look if we turn to the more relaxed notion of preferred semantics? The following result presents the interrelations if we put no restriction on the considered AFs. We observe that as in the case of stable semantics there is no total ordering of the equivalence relations in the equivalence zoo. In particular, weak expansion equivalence is not comparable with strong expansion equivalence and minimal change equivalence with respect to arbitrary, normal and strong expansions. Furthermore, members of the family of minimal change equivalence relations *are shown to be intermediate forms between strong expansion and standard equivalence.* Interestingly, weak expansion equivalence and minimal change equivalence with respect to weak expansions change their position in comparison to stable semantics.

7.4. Minimal Change Equivalence

Proposition 7.60. *For any AFs $\mathcal{F}$ and $\mathcal{G}$,*

- $\mathcal{F} \equiv_E^{pr} \mathcal{G} \Leftrightarrow \mathcal{F} \equiv_N^{pr} \mathcal{G} \Leftrightarrow \mathcal{F} \equiv_L^{pr} \mathcal{G} \Rightarrow \mathcal{F} \equiv_S^{pr} \mathcal{G} \Rightarrow \mathcal{F} \equiv_E^{pr,MC} \mathcal{G}$,

- $\mathcal{F} \equiv_E^{pr} \mathcal{G} \Rightarrow \mathcal{F} \equiv_W^{pr} \mathcal{G} \Rightarrow \mathcal{F} \equiv_W^{pr,MC} \mathcal{G} \Rightarrow \mathcal{F} \equiv^{pr} \mathcal{G}$,

- $\mathcal{F} \equiv_E^{pr,MC} \mathcal{G} \Leftrightarrow \mathcal{F} \equiv_N^{pr,MC} \mathcal{G} \Leftrightarrow \mathcal{F} \equiv_S^{pr,MC} \mathcal{G} \Rightarrow \mathcal{F} \equiv_W^{pr,MC} \mathcal{G}$.

The graphical representation of the result is depicted in Figure 7.21.

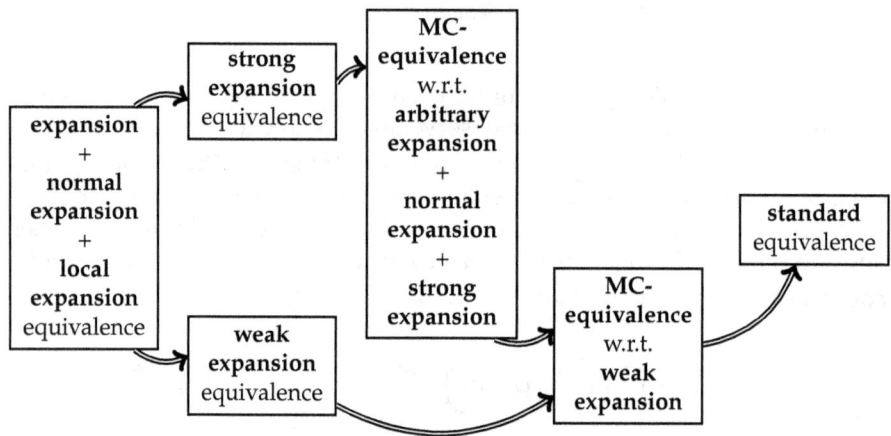

Figure 7.21: Preferred Semantics (Arbitrary AFs)

Proof. In Theorem 5.46 it was already shown that $\mathcal{F} \equiv_E^{pr} \mathcal{G} \Leftrightarrow \mathcal{F} \equiv_N^{pr} \mathcal{G} \Leftrightarrow \mathcal{F} \equiv_L^{pr} \mathcal{G} \Rightarrow \mathcal{F} \equiv_S^{pr} \mathcal{G}, \mathcal{F} \equiv_W^{pr} \mathcal{G} \Rightarrow \mathcal{F} \equiv^{pr} \mathcal{G}$. In the following we prove the remaining interrelations.

First, we will show that weak expansion equivalence implies minimal change equivalence with respect to weak expansions, i.e. $\mathcal{F} \equiv_W^{pr} \mathcal{G} \Rightarrow \mathcal{F} \equiv_W^{pr,MC} \mathcal{G}$. Applying Theorem 5.40 we deduce $A(F) = A(G)$ and $\mathcal{E}_{pr}(\mathcal{F}) = \mathcal{E}_{pr}(\mathcal{G})$. If $\mathcal{F} \not\equiv_W^{pr,MC} \mathcal{G}$, then there is a set $E \subseteq A(F)$, such that $N_{pr,W}^{\mathcal{F}}(E) \neq N_{pr,W}^{\mathcal{G}}(E)$. Without loss of generality we assume $N_{pr,W}^{\mathcal{F}}(E) = 0$ and $N_{pr,W}^{\mathcal{G}}(E) = \infty$ (compare Theorem 7.21, Definition 7.18). Hence, there is a superset E' of E, such that $E' \in \mathcal{E}_{pr}(\mathcal{F})$ and $E' \notin \mathcal{E}_{pr}(\mathcal{G})$ in contradiction to $\mathcal{E}_{pr}(\mathcal{F}) = \mathcal{E}_{pr}(\mathcal{G})$.

Since preferred semantics satisfies I-maximality (compare Figure 2.4), i.e. no extension can be a proper subset of another one, we conclude $\mathcal{F} \equiv_W^{pr,MC} \mathcal{G} \Rightarrow \mathcal{F} \equiv^{pr} \mathcal{G}$ (Theorem 7.55).

We show now that $\mathcal{F} \equiv_S^{pr} \mathcal{G}$ implies $\mathcal{F} \equiv_\Phi^{pr,MC} \mathcal{G}$ for each $\Phi \in \{E, N, S\}$. We have already shown that minimal change equivalence with respect to arbitrary, normal and strong expansions coincide (Theorems 7.51 and 7.52).

Chapter 7. Enforcing and Minimal Change

Hence, it suffices to prove that $\mathcal{F} \equiv_S^{pr} \mathcal{G}$ and $\mathcal{F} \not\equiv_S^{pr,MC} \mathcal{G}$ yields a contradiction. Since strong expansion equivalence implies sharing the same arguments (Proposition 5.48) it follows the existence of a subset $E \subseteq A(\mathcal{F}) = A(\mathcal{G})$, such that $\mathcal{F} \not\equiv_S^{pr,E} \mathcal{G}$. Consequently, $N_{pr,S}^{\mathcal{F}}(E) \neq N_{pr,S}^{\mathcal{G}}(E)$. Let $N_{pr,S}^{\mathcal{F}}(E) = k_1 < k_2 = N_{pr,S}^{\mathcal{G}}(E)$ where $k_1, k_2 \in \mathbb{N}_\infty$. Note that $k_1 = 0$ yields a contradiction because strong expansion equivalence implies standard equivalence, i.e. $\mathcal{E}_{pr}(\mathcal{F}) = \mathcal{E}_{pr}(\mathcal{G})$ (Proposition 2.18). Assume $k_1 \neq 0$. Consequently, there are an AF $\mathcal{H}$ and a set $E' \subseteq A(\mathcal{H})$, such that $\mathcal{F} \leq_S \mathcal{H}$, $d(\mathcal{F}, \mathcal{H}) = k_1$ and $E \subseteq E' \in \mathcal{E}_{pr}(\mathcal{H})$. Without loss of generality there exists an AF $\mathcal{H}'$, such that $R(\mathcal{F}) \cap R(\mathcal{H}') = \emptyset$ and any attack in $R(\mathcal{H}')$ contains at least one fresh argument and $\mathcal{H} = \mathcal{F} \cup \mathcal{H}'$ (compare Definition 2.13). Since $\mathcal{F} \equiv_S^{pr} \mathcal{G}$ is assumed and $A(\mathcal{F}) = A(\mathcal{G})$ is already shown we conclude $\mathcal{G} \leq_S \mathcal{G} \cup \mathcal{H}'$ and therefore $E' \in \mathcal{E}_{pr}(\mathcal{G} \cup \mathcal{H}')$. It can be easily seen that $d(\mathcal{G}, \mathcal{G} \cup \mathcal{H}') = k_1$. Thus, $k_2 = N_{pr,S}^{\mathcal{G}}(E) = k_1$ in contradiction to $k_1 < k_2$. Consequently, $\mathcal{F} \equiv_S^{pr} \mathcal{G} \Rightarrow \mathcal{F} \equiv_\Phi^{pr,MC} \mathcal{G}$ for each $\Phi \in \{E, N, S\}$ is shown.

Finally, we will show that $\mathcal{F} \equiv_\Phi^{pr,MC} \mathcal{G} \Rightarrow \mathcal{F} \equiv_W^{pr,MC} \mathcal{G}$ for each $\Phi \in \{E, N, S\}$. Again, it suffices to show that $\mathcal{F} \equiv_S^{pr,MC} \mathcal{G}$ and $\mathcal{F} \not\equiv_W^{pr,MC} \mathcal{G}$ yields a contradiction (Theorems 7.51 and 7.52). The first assumption implies $A(\mathcal{F}) = A(\mathcal{G})$. The second assumption means that there is a set E, such that $N_{pr,W}^{\mathcal{F}}(E) \neq N_{pr,W}^{\mathcal{G}}(E)$. Let $N_{pr,W}^{\mathcal{F}}(E) = \infty$ and $N_{pr,W}^{\mathcal{G}}(E) = 0$ (Theorem 7.21, Definition 7.50). Recalling that the characteristic with respect to strong expansions does not exceed the characteristic with respect to weak expansions (Proposition 7.30) we conclude $N_{pr,S}^{\mathcal{G}}(E) = 0$. Hence, applying minimal change equivalence with respect to strong expansions we deduce $N_{pr,S}^{\mathcal{F}}(E) = 0$. This means, there is a superset E' of E, such that $E' \in \mathcal{E}_{pr}(\mathcal{F})$. Consequently, $N_{pr,W}^{\mathcal{F}}(E) = \infty$ is impossible concluding the proof.

For the sake of completeness we will present some counterexamples showing that the converse directions do not hold. It suffices to check the following four cases. The other non-relations can be easily obtained by using the already shown relations depicted in Figure 7.21.

1. $\mathcal{F} \equiv^{pr} \mathcal{G} \not\Rightarrow \mathcal{F} \equiv_W^{pr,MC} \mathcal{G}$.

$\mathcal{F}:$ a_1 a_2 $\mathcal{G}:$ a_1 a_2 a_3

Obviously, $\mathcal{E}_{pr}(\mathcal{F}) = \mathcal{E}_{pr}(\mathcal{G}) = \{\{a_1\}\}$. Furthermore, $\mathcal{F} \not\equiv_W^{pr,MC} \mathcal{G}$ since minimal change equivalence guarantees sharing the same arguments (compare Definition 7.50) but $A(\mathcal{F}) \neq A(\mathcal{G})$.

2. $\mathcal{F} \equiv_\Phi^{pr,MC} \mathcal{G} \not\Rightarrow \mathcal{F} \equiv_S^{pr} \mathcal{G}$ for each $\Phi \in \{E, N, S\}$.

$\mathcal{F}:$ a_1 a_2 a_3 $\mathcal{G}:$ a_1 a_2 a_3

185

7.4. Minimal Change Equivalence

It can be checked (compare Theorem 7.29, Definition 7.24) that $N^{\mathcal{F}}_{pr,S}(\{a_1\})$ = $N^{\mathcal{G}}_{pr,S}(\{a_1\})$ = $N^{\mathcal{F}}_{pr,S}(\{a_3\})$ = $N^{\mathcal{G}}_{pr,S}(\{a_3\})$ = 1 and $N^{\mathcal{F}}_{pr,S}(\{a_2\})$ = $N^{\mathcal{G}}_{pr,S}(\{a_2\})$ = $N^{\mathcal{F}}_{pr,S}(\emptyset)$ = $N^{\mathcal{G}}_{pr,S}(\emptyset)$ = 0 and for any other $E \subseteq A(\mathcal{F}) = A(\mathcal{G})$ (thus, $E \notin cf(\mathcal{F}) = cf(\mathcal{G})$), we have $N^{\mathcal{F}}_{pr,S}(E) = N^{\mathcal{G}}_{pr,S}(E) = \infty$. Hence, $\mathcal{F} \equiv^{pr,MC}_{S} \mathcal{G}$. Furthermore, $\mathcal{F} \not\equiv^{pr}_{S} \mathcal{G}$ because $\mathcal{F}$ and $\mathcal{G}$ are self-loop-free but not syntactically identical (compare Proposition 5.47).

3. $\mathcal{F} \equiv^{pr}_{W} \mathcal{G} \not\Rightarrow \mathcal{F} \equiv^{pr,MC}_{\Phi} \mathcal{G}$ for each $\Phi \in \{E, N, S\}$.

$\mathcal{F}: (a_1) \quad (a_2) \quad (a_3) \quad \mathcal{G}: (a_1) \rightleftarrows (a_2) \rightleftarrows (a_3)$

Since $A(\mathcal{F}) = A(\mathcal{G})$, $\mathcal{E}_{pr}(\mathcal{F}) = \mathcal{E}_{pr}(\mathcal{G}) = \{\{a_2\}\}$ and $U^{\mathcal{F}}_{\{a_2\}} = \emptyset = U^{\mathcal{G}}_{\{a_2\}}$ we conclude $\mathcal{F} \equiv^{pr}_{W} \mathcal{G}$ (Theorem 5.40). Furthermore, $N^{\mathcal{F}}_{pr,S}(\{a_1\}) = 1 \neq 2 = N^{\mathcal{G}}_{pr,S}(\{a_1\})$ (compare Theorem 7.29, Definition 7.24). Consequently, $\mathcal{F} \not\equiv^{pr,MC}_{S} \mathcal{G}$.

4. $\mathcal{F} \equiv^{pr}_{S} \mathcal{G} \not\Rightarrow \mathcal{F} \equiv^{pr}_{W} \mathcal{G}$.

$\mathcal{F}: (a_1) \quad (a_2)\circlearrowleft \quad \mathcal{G}: (a_1) \rightleftarrows (a_2)\circlearrowleft$

Since $\mathcal{F}$ and $\mathcal{G}$ possess identical admissible-*-kernels, namely $\mathcal{F} = \mathcal{F}^{k^*(ad)} = \mathcal{G}^{k^*(ad)}$ we deduce $\mathcal{F} \equiv^{pr}_{S} \mathcal{G}$ (Theorem 5.16). Furthermore, $\mathcal{E}_{pr}(\mathcal{F}) = \mathcal{E}_{pr}(\mathcal{G}) = \{\{a_1\}\}$ but $U^{\mathcal{F}}_{\{a_1\}} = \{a_2\} \neq \emptyset = U^{\mathcal{G}}_{\{a_1\}}$. Hence, $\mathcal{F} \not\equiv^{pr}_{W} \mathcal{G}$ (Theorem 5.40). □

Restricting our considerations to AFs sharing the same arguments does not have a big effect in comparison to the general case. We state a slight difference only, namely standard equivalence of two AFs becomes sufficient for their minimal change equivalence with respect to weak expansions.

Proposition 7.61. *For any AFs $\mathcal{F}$ and $\mathcal{G}$ such that $A(\mathcal{F}) = A(\mathcal{G})$ we have:*

- $\mathcal{F} \equiv^{pr}_{E} \mathcal{G} \Leftrightarrow \mathcal{F} \equiv^{pr}_{N} \mathcal{G} \Leftrightarrow \mathcal{F} \equiv^{pr}_{L} \mathcal{G} \Rightarrow \mathcal{F} \equiv^{pr}_{S} \mathcal{G} \Rightarrow \mathcal{F} \equiv^{pr,MC}_{E} \mathcal{G}$,

- $\mathcal{F} \equiv^{pr}_{E} \mathcal{G} \Rightarrow \mathcal{F} \equiv^{pr}_{W} \mathcal{G} \Rightarrow \mathcal{F} \equiv^{pr,MC}_{W} \mathcal{G} \Rightarrow \mathcal{F} \equiv^{pr} \mathcal{G}$,

- $\mathcal{F} \equiv^{pr,MC}_{W} \mathcal{G} \Leftrightarrow \mathcal{F} \equiv^{pr} \mathcal{G}$,

- $\mathcal{F} \equiv^{pr,MC}_{E} \mathcal{G} \Leftrightarrow \mathcal{F} \equiv^{pr,MC}_{N} \mathcal{G} \Leftrightarrow \mathcal{F} \equiv^{pr,MC}_{S} \mathcal{G} \Rightarrow \mathcal{F} \equiv^{pr,MC}_{W} \mathcal{G}$.

Graphically:

Chapter 7. Enforcing and Minimal Change

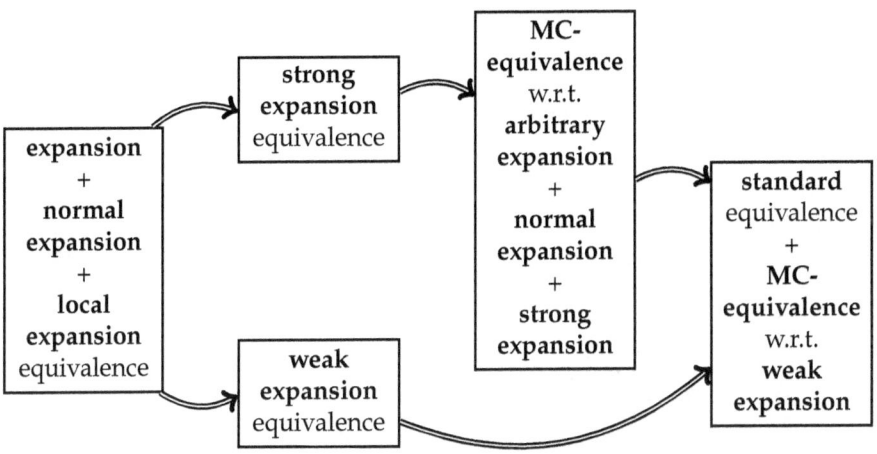

Figure 7.22: Preferred Semantics (Same Arguments)

Proof. Consider again the counter-examples given in the proof before showing that some relations do not hold. We observe that only the first counter-example (showing that $\mathcal{F} \equiv^{pr} \mathcal{G} \not\Rightarrow \mathcal{F} \equiv_W^{pr,MC} \mathcal{G}$) deals with AFs which do not share the same arguments. This is not a coincidence as the following proof shows.

We assume $A(\mathcal{F}) = A(\mathcal{G})$ and $\mathcal{F} \equiv^{pr} \mathcal{G}$. Standard equivalence of two AFs, i.e. $\mathcal{E}_{pr}(\mathcal{F}) = \mathcal{E}_{pr}(\mathcal{G})$ together with the assumption guarantees that for any set $E \subseteq A(\mathcal{F})$, either $N_{pr,W}^{\mathcal{F}}(E) = N_{pr,W}^{\mathcal{G}}(E) = 0$ or $N_{pr,W}^{\mathcal{F}}(E) = N_{pr,W}^{\mathcal{G}}(E) = \infty$ (compare Definition 7.18). Consequently, $\mathcal{F} \equiv_W^{pr,MC} \mathcal{G}$ is shown (Theorem 7.21). □

Finally, we consider the class of self-loop-free AFs. It turns out that in this case stable and preferred semantics behave in a very similar manner. The only difference is the role (or better, position) of weak expansion equivalence and minimal change equivalence with respect to weak expansions.

Proposition 7.62. *For any self-loop-free AFs $\mathcal{F}$ and $\mathcal{G}$,*

- $\mathcal{F} \equiv_E^{pr} \mathcal{G} \Leftrightarrow \mathcal{F} \equiv_N^{pr} \mathcal{G} \Leftrightarrow \mathcal{F} \equiv_S^{pr} \mathcal{G} \Leftrightarrow \mathcal{F} \equiv_L^{pr} \mathcal{G} \Rightarrow \mathcal{F} \equiv_E^{pr,MC} \mathcal{G}$,

- $\mathcal{F} \equiv_E^{pr,MC} \mathcal{G} \Leftrightarrow \mathcal{F} \equiv_N^{pr,MC} \mathcal{G} \Leftrightarrow \mathcal{F} \equiv_S^{pr,MC} \mathcal{G} \Rightarrow \mathcal{F} \equiv_W^{pr} \mathcal{G}$,

- $\mathcal{F} \equiv_W^{pr} \mathcal{G} \Rightarrow \mathcal{F} \equiv_W^{pr,MC} \mathcal{G}$,

- $\mathcal{F} \equiv_W^{pr,MC} \mathcal{G} \Rightarrow \mathcal{F} \equiv^{pr} \mathcal{G}$.

Graphically:

7.4. Minimal Change Equivalence

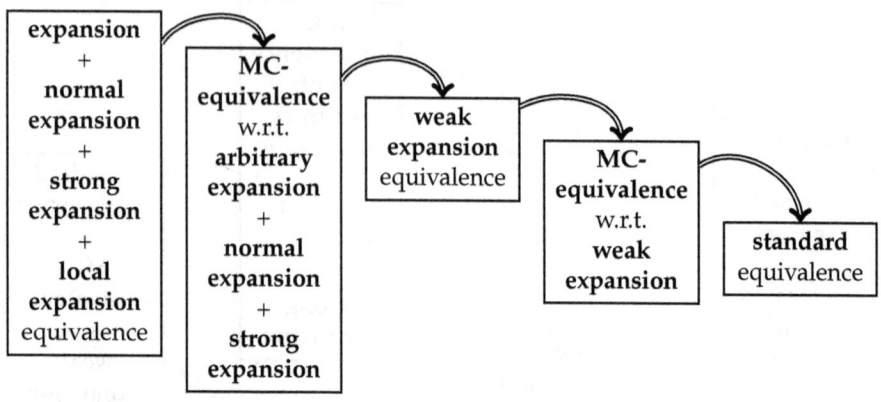

Figure 7.23: *Preferred Semantics (Self-loop-free AFs)*

Proof. In this proof we assume that the considered AFs do not possess self-loops. Consequently, in consideration of the results presented in Figure 7.21 it suffices to show the following two relations. First, $\mathcal{F} \equiv_S^{pr} \mathcal{G} \Rightarrow \mathcal{F} \equiv_N^{pr} \mathcal{G}$ (already shown in Proposition 5.47) and second, $\mathcal{F} \equiv_S^{pr,MC} \mathcal{G} \Rightarrow \mathcal{F} \equiv_W^{pr} \mathcal{G}$.

Given $\mathcal{F} \equiv_S^{pr,MC} \mathcal{G}$ we deduce $A(\mathcal{F}) = A(\mathcal{G})$ and $\mathcal{E}_{pr}(\mathcal{F}) = \mathcal{E}_{pr}(\mathcal{G})$. Consequently, if $\mathcal{F} \not\equiv_W^{pr} \mathcal{G}$, then there is an extension $E \in \mathcal{E}_{pr}(\mathcal{F})$, such that $U_E^\mathcal{F} \neq U_E^\mathcal{G}$ (Theorem 5.40). Without loss of generality let $a \in U_E^\mathcal{F} \setminus U_E^\mathcal{G}$. This means $a \notin E$, $(E,a) \notin R(\mathcal{F})$ and $(E,a) \in R(\mathcal{G})$. Consequently, $(a,E) \notin R(\mathcal{F})$ because E is assumed to be preferred in $\mathcal{F}$. Furthermore, we deduce $E \cup \{a\} \in cf(\mathcal{F})$ since we consider self-loop-free AFs. In Corollary 7.23 it was shown that whenever a set C is conflict-free we may enforce C in finitely many steps. This means, $N_{pr,S}^\mathcal{F}(E \cup \{a\}) < \infty$. On the other hand, we have $E \cup \{a\} \notin cf(\mathcal{G})$ because $(E,a) \in R(\mathcal{G})$ is already shown. Thus, $N_{pr,S}^\mathcal{F}(E \cup \{a\}) = \infty$ (Theorem 7.29, Definition 7.50). This means, $\mathcal{F} \not\equiv_S^{pr,MC} \mathcal{G}$ is implied (in contradiction to the assumption) concluding the proof.

Consider again the counter-examples given in the proof of the relations depicted in Figure 7.21. We observe that the counter-examples 1-3 do not possess self-loops. Hence, these (non)-relations do not hold here either. A counter-example remains to be given for $\mathcal{F} \equiv_W^{pr,MC} \mathcal{G} \not\Rightarrow \mathcal{F} \equiv_W^{pr} \mathcal{G}$.

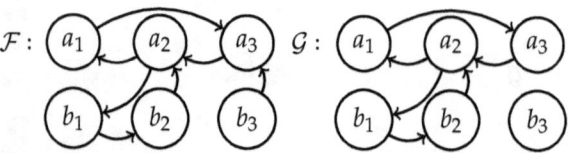

One may check that $\mathcal{E}_{pr}(\mathcal{F}) = \mathcal{E}_{pr}(\mathcal{G}) = \{\{b_3\}\}$. Furthermore, $A(\mathcal{F}) = A(\mathcal{G})$. Consequently, for any set $E \subseteq A(\mathcal{F})$, either $N_{pr,W}^{\mathcal{F}}(E) = N_{pr,W}^{\mathcal{G}}(E) = 0$ or $N_{pr,W}^{\mathcal{F}}(E) = N_{pr,W}^{\mathcal{G}}(E) = \infty$ (compare Definition 7.18). Hence, $\mathcal{F} \equiv_W^{pr,MC} \mathcal{G}$ is shown (Theorems 7.51 and 7.52). On the other hand, $U_{\{b_3\}}^{\mathcal{F}} = \{a_1, a_2, b_1, b_2\} \neq \{a_1, a_2, a_3, b_1, b_2\} = U_{\{b_3\}}^{\mathcal{G}}$. Thus, $\mathcal{F} \not\equiv_W^{pr} \mathcal{G}$ (Theorem 5.40). □

7.5 Conclusions and Related Work

In this last chapter we presented a study of three main problems regarding dynamical aspects of argumentation, namely the enforcing problem [Baumann and Brewka, 2010], the minimal change problem [Baumann, 2012b] as well as the spectrum problem [Baumann and Brewka, 2013b]. The first problem deals with the question whether it is possible to modify a given AF such that a certain set of arguments becomes a subset of an extension of the revised AF. This question is of high interest from a strategical point of view since the very nature of a dispute is putting forward arguments or counterarguments with the objective to convince the opponent. We therefore investigated conditions for the possibility as well as impossibility of enforcements.

In addition to clarifying the possibility of enforcing certain arguments, a natural further question we dealt with is concerned with the effort needed for the enforcements. This more general problem of minimal change can be precisely formulated as follows: what is the minimal number of modifications (additions or removals of attacks) needed to reach an enforcement of E? We defined the so-called characteristics which represent the problem formally. We then showed that for stable, preferred, complete and admissible semantics that this problem can be decided by local criteria encoded by the so-called value functions. The most remarkable result is that the characteristic does not change if we switch simultaneously between strong, normal or arbitrary expansions and preferred, complete or admissible semantics.

After the characterization of the minimal change problem we turned to an analysis of the space of possibilities of characteristics. We therefore introduced the so-called spectra which describe, for a collection of chosen semantics, the range of possible minimal efforts needed to enforce a set of arguments. Due to the coincidence of preferred, complete and admissible semantics with respect to minimal change we focused only on stable, semi-stable and preferred semantics. We were able to fully characterize the spectra for strong, normal and arbitrary expansions. This analysis revealed the surprising result that, although the three semantics are closely related, it may be arbitrarily more difficult to enforce arguments using stable rather than semi-stable semantics, and also using semi-stable rather than preferred semantics. The analysis of the spectrum for weak expansions turned out to be more difficult. We provided an almost complete characterization only including several impossibility results. A detailed analysis of the spectrum with respect to arbitrary modifications is part of future work.

There are a number of natural directions in which this research can be further pursued. For instance, a detailed classification of different kinds of changes with respect to the set of extensions if new information is added. A preliminary study (one argument, one interaction) is to be found in [Cayrol et al., 2008]. Since accepted arguments and therefore extensions possess a characteristic of zero a change of the outcome corresponds to a change of characteristics and thus can be analyzed by using the introduced value-functions. For the same reason, another important problem in dynamic argumentation is closely related to our study, namely the determination of the status of an argument. Since computing the justification state of an argument from scratch each time new information is added is very inefficient the possibility to *reuse* some previous computations would be a great advantage. Some results in this line of research (compare [Liao et al., 2011] as well as Sections 3.2 and 4.1.4) show that reusing is possible if the considered semantics satisfy the directionality criterion.

Further future work may include the consideration of a stronger form of enforcement where the enforced set of arguments has to be contained in *all* extensions rather than in *some* extension. We also might want to enforce a set of arguments C and at the same time exclude another, disjoint set D, that is, we might be interested in modifications leading to an AF possessing an extension E such that $C \subseteq E$ and $E \cap D = \emptyset$.

Booth and colleagues investigated several quantitative distance measures for argumentation [Booth et al., 2012]. In contrast to our work where the focus is on distances among different argumentation frameworks, the distance in that paper measures how far apart two labellings representing two complete extensions of the same argumentation framework are. This has applications in argument-based belief revision (e.g. if an agent is forced to switch to another extension and tries to identify the one closest to his original extension) and in judgement aggregation. Although the goals of this work are different from ours, it remains to be seen whether results from that work can be reused for our purposes.

The fourth main topic we studied in this chapter was replacements without loss of informations with respect to minimal change. In particular, we introduced and characterized minimal-D-equivalence and the more general minimal change equivalence. We present characterization theorems for several Dung semantics and furthermore, we showed the relations to standard and expansion equivalence for a whole range of semantics satisfying abstract principles like regularity and/or I-maximality. In case of stable and preferred semantics we fully clarified the relationship among all equivalence notions for AFs presented in this book. We provided an analysis for the whole class of AFs as well as for two important subclasses, namely AFs sharing the same arguments and self-loop-free AFs. The most relevant "take home" message following from our results is that the different notions of minimal change equivalence fit nicely into the global picture of other equivalence notions in the sense that they constitute alternative notions in between expansion and standard equivalence.

Chapter 7. Enforcing and Minimal Change

In [Baroni et al., 2012] so-called input/output argumentation frameworks were introduced, an approach to characterize the behavior of an argumentation framework as sort of a black box with a well-defined external interface. The paper defines the notion of semantics decomposability and analyzes complete, stable, grounded and preferred semantics in this regard. It turns out that, under grounded, complete, stable and credulous preferred semantics, input/output argumentation frameworks with the same behaviour can be exchanged without affecting the results of the evaluation of other interacting arguments. Since replaceability is one of the main motivations for studying equivalence notions, we plan to explore connections between equivalence and decomposability in the near future.

7.5. Conclusions and Related Work

Chapter 8

Conclusion

Finally, we briefly summarize the main results of this book, sketch open problems and give pointers to future research directions.

8.1 Summary

We have presented a comprehensive analysis of meta-logical properties in abstract argumentation. One can conclude that although AFs are among the simplest nonmonotonic systems one can think of, they give rise to deep and interesting mathematical questions. Indeed, in terms of the complexity and subtleness of the established results AFs are certainly placed alongside the well-established nonmonotonic formalisms like auto-epistemic logic, logic programs and default logic (cf. [Brewka, 1991] for an excellent overview). In the following we review our main contributions and refer to the sections "Conclusions and Related Work" (3.4, 4.4, 5.6, 6.4 and 7.5) for more details.

In Chapter 3 we studied the behavior of the extensions of an argumentation framework if finitely or infinitely many new arguments are added. In general, adding new arguments and their associated interactions obviously may change the outcome of an AF in a nonmonotonic way. This means, arguments accepted earlier may become unaccepted, others become accepted and the number of extensions may shrink or increase, depending on the new arguments. In contrast to this general observation we identified sufficient conditions for monotonic evolutions (Theorem 3.2). Roughly speaking, the class of weak expansions and semantics satisfying directionality play the decisive role. We showed how to simplify the computation of justification states (Proposition 3.6). In particular, the results allow to consider finite subframeworks even if the underlying argumentation framework is infinite. In the last part of that chapter we showed how to use abstract argumentation to reason about actions. The theoretical results proven before played a key role in showing that our approach can be efficiently implemented (Proposition 3.29).

In propositional logic we have: $Mod(S \cup T) = Mod(S) \cap Mod(T)$ for any theory S and T. This means, it is possible to divide a theory in subtheor-

8.1. Summary

ies such that the formal semantics of the entire theory can be obtained by constructing the semantics of the subtheories. In Chapter 4 we studied this possibility in case of abstract argumentation. We have shown that the extensions of an AF can be alternatively obtained if one splits the considered AF into two parts such that the remaining attacks are restricted to a single direction. The procedure is as follows: one computes an extension E_1 of the first part, uses E_1 to reduce and modify the second part, computes an extension E_2 of the modification and then simply combines E_1 and E_2 (Theorems 4.10). In case of stable semantics we even prove a generalized splitting result allowing for arbitrary splits (Theorem 4.28). One exciting result is the establishment of a general relation between universal definedness, the splitting property and the directionality principle (Theorem 4.13).

Argumentation is inherently dynamic. Thus, the question whether two frameworks represent the same knowledge even if this knowledge is not captured by the actual extensions, but can be made explicit by augmenting the framework under consideration is of great importance, for instance in the light of replaceability and therefore simplification issues. Chapter 5 was dedicated to this problem. In particular, we provided characterization theorems for normal and strong expansion equivalence arguably the most important notions with respect to dynamic evolvements. It turned out that syntactical criteria, so-called kernels, play a crucial role (see Figure 5.18). Quite surprisingly, we showed that for any semantics considered in this book expansion equivalence and normal expansion equivalence coincide. This means, whenever two AFs are semantically distinguishable via an arbitrary expansion, then this can be made explicit by a normal expansion too. A further main result is that any equivalence notion on the set of AFs characterizable through kernels presented in this book collapses to identity if we restrict ourselves to self-loop-free AFs (Theorem 5.47). This underlines the thesis that abstract argumentation frameworks provide a very "compact" KR formalism.

In Chapter 6 we conducted a detailed analytical and empirical study on the maximal and average numbers of stable extensions in abstract argumentation frameworks. We proved a non-trivial tight upper bound on the maximal number of stable extensions. In particular, given an AF possessing n arguments, then the number of its stable extensions does not exceed $3^{\frac{n}{3}}$ (Theorem 6.5). For specific numbers of attacks, we have also given the precise average number of stable extensions in terms of closed-form expressions (Proposition 6.8). Our empirical results show that attacks cannot simply be thought of as constraining: adding an attack may sometimes increase and sometimes decrease the number of stable extensions. Although this might be obvious in general to argumentation researchers, for the first time we were able to present some precise numerical figures around this phenomenon (cf. Figures 6.4 and 6.5). Furthermore, the results indicate that the distribution of stable extensions as a function of the number of attacks in the framework seems to follow a universal pattern that is independent of the number of arguments.

In Chapter 7 we presented a study of several problems regarding dynamical aspects of argumentation. The enforcing problem deals with the question

Chapter 8. Conclusion

whether it is possible to modify a given AF such that a certain set of arguments becomes a subset of an extension of the revised AF. This question is of high interest from a strategical point of view since the very nature of a dispute is putting forward arguments or counterarguments with the objective to convince the opponent. We therefore investigated conditions for the possibility as well as impossibility of enforcements (Theorems 7.5, 7.6 and 7.7). In particular, enforcing is always possible given that the desired set of arguments is conflict-free. Then, in addition to clarify the possibility of enforcing we studied the minimal number of modifications (additions or removals of attacks) needed to reach an enforcement, the so-called minimal change problem. We presented characterization theorems for weak, strong, normal, arbitrary expansions as well as arbitrary modifications for the stable, preferred, complete and admissible semantics. The most remarkable result is that the characteristic (representing the minimal change problem formally) does not change if we switch simultaneously between strong, normal or arbitrary expansions and preferred, complete or admissible semantics (Theorem 7.29). In Section 7.3 we provided a precise mathematical analysis of characteristics. The spectrum problem which we studied there can be summarized as follows: given a collection of semantics and a modification type, what are the corresponding tuples of characteristics one may obtain for an arbitrary argumentation framework and set of arguments. In other words, determine the set of all tuples of natural numbers which may occur as characteristics simultaneously, the so-called spectrum. In case of stable, semi-stable and preferred semantics we provided a complete characterization with respect to strong, normal and arbitrary expansions (Theorem 7.43). The study of spectra is not an academic exercise only. To the contrary, the investigation of spectra reveals interesting and sometimes surprising insights into the relationship among several semantics. For instance, we have shown that it may be arbitrarily more difficult to enforce arguments using stable rather than semi-stable semantics, and also using semi-stable rather than preferred semantics. Additionally, we presented some first results for the spectra with respect to weak expansions as well as arbitrary modifications. In Section 7.4 we studied notions of equivalence which guarantee equal minimal efforts, namely minimal-D-equivalence and the more general minimal change equivalence. We presented characterization theorems (Theorems 7.51, 7.52) and provided the relation between minimal change equivalence and expansion or standard equivalence in general via using abstract principles (Theorems 7.54 and 7.55). Furthermore, in case of stable and preferred semantics we gave the complete picture for all equivalence relations considered in this book. In particular, in case of stable semantics the different forms of minimal change equivalence are shown to be intermediate forms between strong expansion and weak expansion equivalence (cf. Figure 7.18).

8.2 Open Problems and Future Research

Some obvious further work remains to be done. In particular, it would be worthwhile to extend our analysis to further (more exotic) argumentation semantics like $cf2$ semantics, and the recently introduced variant, $stage2$ semantics [Baroni et al., 2005; Dvořák and Gaggl, 2012]. In contrast to the semantics studied in this book both are representatives of *redundancy-free* semantics. This means, any attack may play a crucial role with respect to further evaluation or more precisely, it is shown that for both expansion equivalence and syntactical identity coincide [Dvořák and Gaggl, 2012; Gaggl and Woltran, 2012]. It would be interesting to check whether this behaviour carries over to normal and/or strong expansion equivalence. Further semantics which have not been considered in this book at all are *robust* [Jakobovits and Vermeir, 1999] as well as *sustainable* and *tolerant* [Bodanza and Tohmé, 2009] semantics. The study of these semantics, especially in the context of equivalence might be interesting since these approaches actually are designed to treat self-attacks different to standard semantics.

A more promising but substantially more difficult task is instead of considering particular semantics to alternatively look at general criteria sufficient and/or necessary for being in a particular interrelation or satisfying a certain desired property. Note that we already achieved some results in this direction, for instance concerning the splitting property (Theorem 4.13) as well as general relations between minimal change, expansion and standard equivalence (Theorems 7.54 and 7.55).

Another possible future direction would be the study of meta-logical properties like monotonicity and replacement for certain generalizations of Dung's abstract argumentation frameworks. We mention here the recently proposed extended argumentation frameworks (EAFs) [Modgil, 2009], abstract dialectical frameworks (ADFs) [Brewka and Woltran, 2010] and the promising argumentation frameworks with recursive attacks (AFRAs) [Baroni et al., 2011b].

In Section 5.3 we have proven that normal expansion equivalence equals expansion equivalence for any semantics considered in this book. This result was more than unexpected since the class of normal expansions is a strict subset of arbitrary expansion. Although we have shown the coincidence in a rigorous manner there still remains the question *why?*. What do these semantics all have in common? Is there a certain abstract property guaranteeing this equality? Does a reasonable semantics exists such that normal expansion equivalence is strictly coarser than expansion equivalence?

As a final remark concerning future work, we pose the question whether there is an analogue to *Arrow's impossibility theorem* in social choice theory [Arrow, 1950]. More precisely, is there a set of desirable properties for argumentation semantics which cannot be satisfied simultaneously. Without doubt, such a result would be of enormous interest to the argumentation community.

Bibliography

[Alchourrón et al., 1985] Alchourrón, C. E., Gärdenfors, P., and Makinson, D. (1985). On the logic of theory change: Partial meet contraction and revision functions. *Journal of Symbolic Logic*, pages 510–530.

[Amgoud and Besnard, 2010] Amgoud, L. and Besnard, P. (2010). A formal analysis of logic-based argumentation systems. In *Conference on Scalable Uncertainty Management*, pages 42–55.

[Amgoud and Cayrol, 2002] Amgoud, L. and Cayrol, C. (2002). A reasoning model based on the production of acceptable arguments. *Annals of Mathematics and Artificial Intelligence*, pages 197–215.

[Amgoud and Vesic, 2011] Amgoud, L. and Vesic, S. (2011). On the equivalence of logic-based argumentation systems. In Benferhat, S. and Grant, J., editors, *Scalable Uncertainty Management*, volume 6929 of *Lecture Notes in Computer Science*, pages 123–136. Springer.

[Arrow, 1950] Arrow, K. J. (1950). A difficulty in the concept of social welfare. *Journal of Political Economy*, page 328.

[Baroni et al., 2012] Baroni, P., Boella, G., Cerutti, F., Giacomin, M., van der Torre, L. W. N., and Villata, S. (2012). On input/output argumentation frameworks. In *Conference on Computational Models of Argument*, pages 358–365.

[Baroni et al., 2011a] Baroni, P., Caminada, M., and Giacomin, M. (2011a). An introduction to argumentation semantics. *The Knowledge Engineering Review*, pages 365–410.

[Baroni et al., 2011b] Baroni, P., Cerutti, F., Dunne, P. E., and Giacomin, M. (2011b). Computing with infinite argumentation frameworks: The case of afras. In *Workshop on Theory and Applications of Formal Argumentation*, pages 197–214.

[Baroni et al., 2009] Baroni, P., Cerutti, F., Giacomin, M., and Guida, G. (2009). Encompassing attacks to attacks in abstract argumentation frameworks. In *European Conference on Symbolic and Quantitative Approaches to Reasoning with Uncertainty*, pages 83–94.

[Baroni et al., 2010] Baroni, P., Dunne, P. E., and Giacomin, M. (2010). On extension counting problems in argumentation frameworks. In *Conference on Computational Models of Argument*, pages 63–74.

[Baroni et al., 2011c] Baroni, P., Dunne, P. E., and Giacomin, M. (2011c). On the resolution-based family of abstract argumentation semantics and its grounded instance. *Artificial Intelligence*, pages 791–813.

[Baroni and Giacomin, 2004] Baroni, P. and Giacomin, M. (2004). A general recursive schema for argumentation semantics. In *European Conference on Artificial Intelligence*, pages 783–787.

[Baroni and Giacomin, 2007] Baroni, P. and Giacomin, M. (2007). On principle-based evaluation of extension-based argumentation semantics. *Artificial Intelligence*, pages 675–700.

[Baroni and Giacomin, 2009] Baroni, P. and Giacomin, M. (2009). Semantics of abstract argument systems. In Simari, G. and Rahwan, I., editors, *Argumentation in Artificial Intelligence*, pages 25–44. Springer US.

[Baroni et al., 2005] Baroni, P., Giacomin, M., and Guida, G. (2005). Scc-recursiveness: a general schema for argumentation semantics. *Artificial Intelligence*, pages 162–210.

[Baumann, 2011] Baumann, R. (2011). Splitting an argumentation framework. In Delgrande, J. P. and Faber, W., editors, *Logic Programming and Non-Monotonic Reasoning*, volume 6645 of *Lecture Notes in Computer Science*, pages 40–53. Springer.

[Baumann, 2012a] Baumann, R. (2012a). Normal and strong expansion equivalence for argumentation frameworks. *Artificial Intelligence*, pages 18–44.

[Baumann, 2012b] Baumann, R. (2012b). What does it take to enforce an argument? minimal change in abstract argumentation. In *European Conference on Artificial Intelligence*, pages 127–132.

[Baumann and Brewka, 2010] Baumann, R. and Brewka, G. (2010). Expanding argumentation frameworks: Enforcing and monotonicity results. In Baroni, P., Cerutti, F., Giacomin, M., and Simari, G. R., editors, *Conference on Computational Models of Argument*, volume 216 of *Frontiers in Artificial Intelligence and Applications*, pages 75–86. IOS Press.

[Baumann and Brewka, 2013a] Baumann, R. and Brewka, G. (2013a). Analyzing the equivalence zoo in abstract argumentation. In *International Workshop on Computational Logic in Multi-Agent Systems*, pages 18–33.

[Baumann and Brewka, 2013b] Baumann, R. and Brewka, G. (2013b). Spectra in abstract argumentation: An analysis of minimal change. In *Logic Programming and Non-Monotonic Reasoning*. Springer.

BIBLIOGRAPHY

[Baumann et al., 2012] Baumann, R., Brewka, G., Dvorak, W., and Woltran, S. (2012). Parameterized splitting: A simple modification-based approach. In Erdem, E., Lee, J., Lierler, Y., and Pearce, D., editors, *Correct Reasoning - Essays on Logic-Based AI in Honour of Vladimir Lifschitz*, volume 7625 of *Lecture Notes in Computer Science*, pages 57–71. Springer.

[Baumann et al., 2010] Baumann, R., Brewka, G., Strass, H., Thielscher, M., and Zaslawski, V. (2010). State defaults and ramifications in the unifying action calculus. In *International Conference on Principles of Knowledge Representation and Reasoning*, pages 435–444.

[Baumann et al., 2011] Baumann, R., Brewka, G., and Wong, R. (2011). Splitting argumentation frameworks: An empirical evaluation. In Modgil, S., Oren, N., and Toni, F., editors, *Workshop on Theorie and Applications of Formal Argumentation*, volume 7132 of *Lecture Notes in Computer Science*, pages 17–31. Springer.

[Baumann and Strass, 2012] Baumann, R. and Strass, H. (2012). Default reasoning about actions via abstract argumentation. In *Conference on Computational Models of an Argument*, pages 297–309.

[Baumann and Strass, 2013] Baumann, R. and Strass, H. (2013). On the maximal and average numbers of stable extensions. In *Workshop Theory and Applications of Formal Argumentation*.

[Baumann and Woltran, 2014] Baumann, R. and Woltran, S. (2014). The role of self-attacking arguments in characterizations of equivalence notions (accepted). *Journal of Logic and Computation: Special Issue on Loops in Argumentation*.

[Bench-Capon, 2003] Bench-Capon, T. J. M. (2003). Persuasion in practical argument using value-based argumentation frameworks. *Journal of Logic and Computation*, pages 429–448.

[Besnard and Hunter, 2001] Besnard, P. and Hunter, A. (2001). A logic-based theory of deductive arguments. *Artificial Intelligence*, pages 203–235.

[Besnard and Hunter, 2008] Besnard, P. and Hunter, A. (2008). *Elements of Argumentation*. MIT Press.

[Besnard and Hunter, 2009] Besnard, P. and Hunter, A. (2009). *Argumentation Based on Classical Logic*, chapter 7 of Argumetation of Artificial Intelligence, pages 133–152. Springer.

[Bisquert et al., 2011] Bisquert, P., Cayrol, C., Dupin de Saint-Cyr, F., and Lagasquie-Schiex, M.-C. (2011). Change in argumentation systems: exploring the interest of removing an argument. In Benferhat, S. and Grant, J., editors, *International Conference on Scalable Uncertainty Management*, volume LNCS 6929, pages 275–288. Springer.

BIBLIOGRAPHY

[Bodanza and Tohmé, 2009] Bodanza, G. A. and Tohmé, F. A. (2009). Two approaches to the problems of self-attacking arguments and general odd-length cycles of attack. *Journal of Applied Logic*, pages 403–420.

[Boella et al., 2009] Boella, G., Kaci, S., and van der Torre, L. (2009). Dynamics in argumentation with single extensions: Abstraction principles and the grounded extension. In *European Conferences on Symbolic and Quantitative Approaches to Reasoning with Uncertainty*, pages 107–118.

[Booth et al., 2012] Booth, R., Caminada, M., Podlaszewski, M., and Rahwan, I. (2012). Quantifying disagreement in argument-based reasoning. In *International Conference on Autonomous Agents and Multiagent Systems*, pages 493–500.

[Booth et al., 2013] Booth, R., Kaci, S., Rienstra, T., and van der Torre, L. (2013). A logical theory about dynamics in abstract argumentation. In *Theorie and Applications of Formal Argumentation*.

[Brewka, 1991] Brewka, G. (1991). *Nonmonotonic Reasoning: Logical Foundations of Commonsense*. Cambridge University Press, Cambridge.

[Brewka et al., 2011] Brewka, G., Eiter, T., and Truszczynski, M. (2011). Answer set programming at a glance. *Communications of the ACM*, pages 92–103.

[Brewka and Woltran, 2010] Brewka, G. and Woltran, S. (2010). Abstract dialectical frameworks. In *Principles of Knowledge Representation and Reasoning*, pages 102–111.

[Caminada, 2005] Caminada, M. (2005). Contamination in formal argumentation systems. In Verbeeck, K., Tuyls, K., Nowé, A., Manderick, B., and Kuijpers, B., editors, *Benelux Conference on Artificial Intelligence*, pages 59–65. Koninklijke Vlaamse Academie van Belie voor Wetenschappen en Kunsten.

[Caminada, 2006] Caminada, M. (2006). On the issue of reinstatement in argumentation. In *European Conference on Logics in Artificial Intelligence*, pages 111–123.

[Caminada, 2007] Caminada, M. (2007). Comparing two unique extension semantics for formal argumentation: ideal and eager. In *Belgian-Dutch Conference on Artificial Intelligence*, pages 81–87.

[Caminada and Amgoud, 2007] Caminada, M. and Amgoud, L. (2007). On the evaluation of argumentation formalisms. *Artificial Intelligence*, pages 286–310.

[Caminada et al., 2012] Caminada, M. W., Carnielli, W. A., and Dunne, P. E. (2012). Semi-stable semantics. *Journal of Logic and Computation*, pages 1207–1254.

BIBLIOGRAPHY

[Cayrol et al., 2008] Cayrol, C., de Saint-Cyr, F. D., and Lagasquie-Schiex, M.-C. (2008). Revision of an argumentation system. In *International Conference on Principles of Knowledge Representation and Reasoning*, pages 124–134.

[Cayrol et al., 2010] Cayrol, C., de Saint-Cyr, F. D., and Lagasquie-Schiex, M.-C. (2010). Change in abstract argumentation frameworks: Adding an argument. *Journal of Artificial Intelligence Research*, pages 49–84.

[Dung et al., 2006] Dung, P., Kowalski, R., and Toni, F. (2006). Dialectic proof procedures for assumption-based, admissible argumentation. *Artificial Intelligence*, pages 114 – 159.

[Dung, 1995] Dung, P. M. (1995). On the acceptability of arguments and its fundamental role in nonmonotonic reasoning, logic programming and n-person games. *Artificial Intelligence*, pages 321 – 357.

[Dunne et al., 2013] Dunne, P. E., Dvořák, W., Linsbichler, T., and Woltran, S. (2013). Characteristics of multiple viewpoints in abstract argumentation. In *Workshop on Dynamics of Knowledge and Belief*.

[Dunne and Wooldridge, 2009] Dunne, P. E. and Wooldridge, M. (2009). Complexity of abstract argumentation. In Simari, G. and Rahwan, I., editors, *Argumentation in Artificial Intelligence*, pages 85–104. Springer US.

[Dvořák, 2012] Dvořák, W. (2012). *Computational Aspects of Abstract Argumentation*. PhD thesis, Technische Universität Wien.

[Dvořák and Gaggl, 2012] Dvořák, W. and Gaggl, S. A. (2012). Incorporating stage semantics in the scc-recursive schema for argumentation semantics. In *International Workshop on Non-Monotonic Reasoning*.

[Eiter and Fink, 2003] Eiter, T. and Fink, M. (2003). Uniform equivalence of logic programs under the stable model semantics. In Palamidessi, C., editor, *International Conference on Logic Programming*, volume 2916 of *Lecture Notes in Computer Science*, pages 224–238. Springer.

[Falappa et al., 2009] Falappa, M. A., Kern-Isberner, G., and Simari, G. R. (2009). Belief revision and argumentation theory. In Rahwan, I. and Simari, G. R., editors, *Argumentation in Artificial Intelligence*, pages 341–360. Springer.

[Gabbay, 2013] Gabbay, D. (2013). *Meta-logical Investigations in Argumentation Networks*. College Publications.

[Gaggl, 2013] Gaggl, S. A. (2013). *A Comprehensive Analysis of the cf2 Argumentation Semantics*. PhD thesis, Technische Universität Wien.

[Gaggl and Woltran, 2011] Gaggl, S. A. and Woltran, S. (2011). Strong equivalence for argumentation semantics based on conflict-free sets. In Liu, W., editor, *European Conference on Symbolic and Quantitative Approaches to Reasoning with Uncertainty*, volume 6716 of *Lecture Notes in Computer Science*, pages 38–49. Springer.

[Gaggl and Woltran, 2012] Gaggl, S. A. and Woltran, S. (2012). The cf2 argumentation semantics revisited. *Journal of Logic and Computation*.

[Gebser et al., 2011] Gebser, M., Kaminski, R., Kaufmann, B., Ostrowski, M., Schaub, T., and Schneider, M. (2011). Potassco: The potsdam answer set solving collection. *AI Communications*, pages 105–124. Available at http://potassco.sourceforge.net.

[Gelder et al., 1988] Gelder, A. V., Ross, K. A., and Schlipf, J. S. (1988). Unfounded sets and well-founded semantics for general logic programs. In *Symposium on Principles of Database Systems*, pages 221–230.

[Gelfond and Lifschitz, 1988] Gelfond, M. and Lifschitz, V. (1988). The stable model semantics for logic programming. In *Logic Programming: 5th Conference and Symposium*, pages 1070–1080.

[Gelfond and Lifschitz, 1991] Gelfond, M. and Lifschitz, V. (1991). Classical negation in logic programs and disjunctive databases. *New Generation Computing*, pages 365–386.

[Gelfond and Lifschitz, 1998] Gelfond, M. and Lifschitz, V. (1998). Action languages. *Electronic Transactions on Artificial Intelligence*, pages 193–210.

[Gelfond and Przymusinska, 1992] Gelfond, M. and Przymusinska, H. (1992). On consistency and completeness of autoepistemic theories. *Journal Fundamenta Informaticae*, pages 59–92.

[Gorogiannis and Hunter, 2011] Gorogiannis, N. and Hunter, A. (2011). Instantiating abstract argumentation with classical logic arguments: Postulates and properties. *Artificial Intelligence*, pages 1479–1497.

[Governatori et al., 2000] Governatori, G., Maher, M. J., Antoniou, G., and Billington, D. (2000). Argumentation semantics for defeasible logics. *Journal of Logic and Computation*, page 2004.

[Grossi and Gabbay, 2013] Grossi, D. and Gabbay, D. M. (2013). When are two arguments the same? equivalence in abstract argumentation. Technical report, University of Liverpool.

[Jakobovits and Vermeir, 1999] Jakobovits, H. and Vermeir, D. (1999). Robust semantics for argumentation frameworks. *Journal of Logic and Computation*, pages 215–261.

[Janhunen et al., 2009] Janhunen, T., Oikarinen, E., Tompits, H., and Woltran, S. (2009). Modularity aspects of disjunctive stable models. *Journal of Artificial Intelligence Research*, pages 813–857.

[Kakas et al., 2008] Kakas, A. C., Michael, L., and 0002, R. M. (2008). Fred meets tweety. In Ghallab, M., Spyropoulos, C. D., Fakotakis, N., and Avouris, N. M., editors, *European Conference on Artificial Intelligence*, volume 178 of *Frontiers in Artificial Intelligence and Applications*, pages 747–748. IOS Press.

BIBLIOGRAPHY

[Kakas et al., 1999] Kakas, A. C., Miller, R., and Toni, F. (1999). An argumentation framework of reasoning about actions and change. In *International Conference on Logic Programming and Nonmonotonic Reasoning*, pages 78–91.

[Lakemeyer and Levesque, 2009] Lakemeyer, G. and Levesque, H. J. (2009). A semantical account of progression in the presence of defaults. In Boutilier, C., editor, *International Joint Conference on Artificial Intelligence*, pages 842–847.

[Liao et al., 2011] Liao, B., Jin, L., and Koons, R. (2011). Dynamics of argumentation systems: A division-based method. *Artificial Intelligence*, pages 1790–1814.

[Lifschitz et al., 2001] Lifschitz, V., Pearce, D., and Valverde, A. (2001). Strongly equivalent logic programs. *ACM Transactions on Computational Logic*, pages 526–541.

[Lifschitz and Turner, 1994] Lifschitz, V. and Turner, H. (1994). Splitting a logic program. In Hentenryck, P. V., editor, *International Conference on Logic Programming*, pages 23–37. MIT Press.

[Lin and Shoham, 1989] Lin, F. and Shoham, Y. (1989). Argument systems: A uniform basis for nonmonotonic reasoning. In Brachman, R. J., Levesque, H. J., and Reiter, R., editors, *International Conference on Principles of Knowledge Representation and Reasoning*, pages 245–255. Morgan Kaufmann.

[Lonc and Truszczyński, 2011] Lonc, Z. and Truszczyński, M. (2011). On graph equivalences preserved under extensions. *Discrete Mathematics*, pages 966–977.

[Loui, 1987] Loui, R. P. (1987). Defeat among arguments: a system of defeasible inference. *Computational Intelligence*, pages 100–106.

[Maher, 1986] Maher, M. J. (1986). Eqivalences of logic programs. In *International Conference on Logic Programming*, pages 410–424.

[McCarthy, 1959] McCarthy, J. (1959). Programs with common sense. In *Teddington Conference on the Mechanization of Thought Processes*, pages 75–91.

[McCarthy, 1990] McCarthy, J. (1990). *Formalization of common sense*. Ablex, Norwood, NJ.

[McCarthy and Hayes, 1969] McCarthy, J. and Hayes, P. J. (1969). Some philosophical problems from the standpoint of artificial intelligence. In Meltzer, B., Michie, D., and Swann, M., editors, *Machine Intelligence 4*, pages 463–502. Edinburgh University Press, Edinburgh, Scotland.

[McDermott and Doyle, 1980] McDermott, D. V. and Doyle, J. (1980). Nonmonotonic logic. *Artificial Intelligence*, pages 41–72.

[Michael and Kakas, 2011] Michael, L. and Kakas, A. (2011). A unified argumentation-based framework for knowledge qualification. In Davis, E., Doherty, P., and Erdem, E., editors, *International Symposium on Logical Formalizations of Commonsense Reasoning*, Stanford, CA.

[Michael and Kakas, 2009] Michael, L. and Kakas, A. C. (2009). Knowledge qualification through argumentation. In Erdem, E., Lin, F., and Schaub, T., editors, *International Conference on Logic Programming and Nonmonotonic Reasoning*, volume 5753 of *Lecture Notes in Computer Science*, pages 209–222. Springer.

[Modgil, 2009] Modgil, S. (2009). Reasoning about preferences in argumentation frameworks. *Artificial Intelligence*, pages 901–934.

[Modgil and Caminada, 2009] Modgil, S. and Caminada, M. (2009). Proof theories and algorithms for abstract argumentation frameworks. In Rahwan, I. and Simari, G. R., editors, *Argumentation in Artificial Intelligence*, pages 105–132. Springer.

[Modgil et al., 2013] Modgil, S., Toni, F., Bex, F., Bratko, I., Chesnevar, C. I., Dvořák, W., Falappa, M. A., Fan, X., Gaggl, S. A., Garcia, A. J., Gonzalez, M. P., Gordon, T. F., Leite, J., Mozina, M., Reed, C., Simari, G. R., Szeider, S., Torroni, P., and Woltran, S. (2013). The added value of argumentation. In Ossowski, S., editor, *Agreement Technologies*, Law, Governance and Technology Series, pages 357–403. Springer Netherlands.

[Moon and Moser, 1965] Moon, J. and Moser, L. (1965). On cliques in graphs. *Israel Journal of Mathematics*, pages 23–28.

[Mueller, 2006] Mueller, E. (2006). *Commonsense Reasoning*. Morgan Kaufmann.

[Nielsen and Parsons, 2006] Nielsen, S. H. and Parsons, S. (2006). A generalization of dung's abstract framework for argumentation: Arguing with sets of attacking arguments. In *Workshop on Argumentation in Multi-Agent Systems*, pages 54–73.

[Oikarinen and Woltran, 2011] Oikarinen, E. and Woltran, S. (2011). Characterizing strong equivalence for argumentation frameworks. *Artificial Intelligence*, pages 1985–2009.

[Pollock, 1987] Pollock, J. L. (1987). Defeasible reasoning. *Cognitive Science*, pages 481–518.

[Pollock, 1994] Pollock, J. L. (1994). Justification and defeat. *Artificial Intelligence*, pages 377–407.

[Pollock, 1995] Pollock, J. L. (1995). *Cognitive Carpentry: A Blueprint for How to Build a Person*. The MIT Press, Cambridge, Massachusetts.

BIBLIOGRAPHY

[Prakken, 2009] Prakken, H. (2009). An abstract framework for argumentation with structured arguments. Technical Report UU-CS-2009-019, Department of Information and Computing Sciences, Utrecht University.

[Prakken and Vreeswijk, 2002] Prakken, H. and Vreeswijk, G. (2002). Logics for defeasible argumentation. In Gabbay, D. and Guenthner, F., editors, *Handbook of Philosophical Logic, second edition, vol. 4*, pages 219–318. Dordrecht.

[Rautenberg, 1996] Rautenberg, W. (1996). *Einführung in die mathematische Logik - ein Lehrbuch mit Berücksichtigung der Logikprogrammierung*. Vieweg.

[Reiter, 1980] Reiter, R. (1980). A logic for default reasoning. *Artificial Intelligence*, pages 81–132.

[Reiter, 2001] Reiter, R. (2001). *Knowledge in Action: Logical Foundations for Specifying and Implementing Dynamical Systems*. The MIT Press.

[Scholz, 1952] Scholz, H. (1952). Ein ungelöstes Problem in der symbolischen Logik. *Journal of Symbolic Logic*, page 160.

[Shmueli, 1987] Shmueli, O. (1987). Decidability and expressiveness aspects of logic queries. In *Principles of Database Systems*, PODS '87, pages 237–249. ACM.

[Sperner, 1928] Sperner, E. (1928). Ein satz ueber untermengen einer endlichen menge. *Mathematische Zeitschrift*, pages 544–548.

[Tarjan, 1972] Tarjan, R. E. (1972). Depth-first search and linear graph algorithms. *SIAM Journal on Computing*, pages 146–160.

[Tarski, 1955] Tarski, A. (1955). A lattice-theoretical fixpoint theorem and its applications. *Pacific Journal of Mathematics*, pages 285–309.

[Thielscher, 2011] Thielscher, M. (2011). A unifying action calculus. *Artificial Intelligence*, pages 120–141.

[Truszczynski, 2006] Truszczynski, M. (2006). Strong and uniform equivalence of nonmonotonic theories - an algebraic approach. *Annals of Mathematics and Artificial Intelligence*, pages 245–265.

[Turner, 1996] Turner, H. (1996). Splitting a default theory. In Clancey, W. J. and Weld, D. S., editors, *Conference on Artificial Intelligence and Innovative Applications of Artificial Intelligence Conference*, pages 645–651. AAAI Press / The MIT Press.

[Turner, 2004] Turner, H. (2004). Strong equivalence for causal theories. In *International Conference on Logic Programming and Nonmonotonic Reasoning*, pages 289–301.

[Verheij, 1996] Verheij, B. (1996). Two approaches to dialectical argumentation: Admissible sets and argumentation stages. In Meyer, J.-J. and van der Gaag, L., editors, *International Conference on Formal and Applied Practical Reasoning workshop*, pages 357–368. Utrecht University.

[Weydert, 2011] Weydert, E. (2011). Semi-stable extensions for infinite frameworks. In *Benelux Conference on Artificial Intelligence*, pages 336–343.

[Woltran, 2010] Woltran, S. (2010). Equivalence between extended datalog programs - a brief survey. In de Moor, O., Gottlob, G., Furche, T., and Sellers, A., editors, *Datalog Reloaded*, volume 6702 of *Lecture Notes in Computer Science*, pages 106–119. Springer. invited talk.

[Wong, 2013] Wong, R. (2013). Parameterized splitting for argumentation frameworks under stable semantics: Implementation and evaluation. Master's thesis, Institute of Computer Science, Leipzig University, Germany.

[Wood, 2011] Wood, D. (2011). On the number of maximal independent sets in a graph. *Discrete Mathematics & Theoretical Computer Science*, pages 17–20.

[Wooldridge et al., 2006] Wooldridge, M., Dunne, P. E., and Parsons, S. (2006). On the complexity of linking deductive and abstract argument systems. In *National Conference on Artificial intelligence*, AAAI'06, pages 299–304. AAAI Press.

[Wu et al., 2009] Wu, Y., Caminada, M., and Gabbay, D. M. (2009). Complete extensions in argumentation coincide with 3-valued stable models in logic programming. *Studia Logica*, pages 383–403.

www.ingramcontent.com/pod-product-compliance
Lightning Source LLC
Chambersburg PA
CBHW050145170426
43197CB00011B/1974